Online Territories

Steve Jones
General Editor

Vol. 61

PETER LANG
New York • Washington, D.C./Baltimore • Bern
Frankfurt • Berlin • Brussels • Vienna • Oxford

Online Territories

Globalization, Mediated Practice and Social Space

EDITED BY
Miyase Christensen, André Jansson
AND Christian Christensen

PETER LANG
New York • Washington, D.C./Baltimore • Bern
Frankfurt • Berlin • Brussels • Vienna • Oxford

Library of Congress Cataloging-in-Publication Data

Online territories: globalization, mediated practice, and social space /
edited by Miyase Christensen, André Jansson, Christian Christensen.
p. cm. — (Digital formations; v. 61)
Includes bibliographical references and index.
1. Online social networks. 2. Transnationalism.
3. Technological innovations—Social aspects. I. Christensen, Miyase.
II. Jansson, André. III. Christensen, Christian.
HM742.O55 303.48'3301—dc22 2011003401
ISBN 978-1-4331-0798-6 (hardcover)
ISBN 978-1-4331-0797-9 (paperback)
ISSN 1526-3169

Bibliographic information published by **Die Deutsche Nationalbibliothek.**
Die Deutsche Nationalbibliothek lists this publication in the "Deutsche Nationalbibliografie"; detailed bibliographic data is available on the Internet at http://dnb.d-nb.de/.

The paper in this book meets the guidelines for permanence and durability of the Committee on Production Guidelines for Book Longevity of the Council of Library Resources.

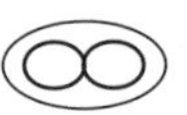

Printed in the United States of America

To our children Arman and Lara, Nora and Ilse

Contents

Preface

> "Doubt. What is doubt? You don't believe in doubt. Computer power eliminates doubt. All doubt rises from past experience. But the past is disappearing. We used to know the past but not the future. This is changing."
>
> DON DELILLO, *COSMOPOLIS*, 2003:86

The title of this volume, *Online Territories*, implicitly suggests an inclination towards reifying the uniqueness of "the Internet" as an empowering and power-infested domain, as a territory in its own right and autonomy. The stance we take here complicates such a vision considerably. Over the last two decades, a great deal of research and scholarship has been produced based on the premise, to put it simply, that there is something fundamentally life-changing, earth-shattering about the Internet and something truly unworldly about virtuality. Research produced particularly in the field of so-called Internet Studies often made reference to the *juxtapositions* of the online and the offline and the multivalent flexibilities afforded by virtual engagement from gender play to alternative politics. We do not disregard the intellectual value inherent in this body of scholarship. Our own thinking matured immensely both alongside and in opposition to the vision it provided. Yet, we believe firmly that it is crucial to continue to challenge the romanticizing attributes associated with the online, and accounts that celebrate or condemn its presumed distinctness, precisely because it *is* alien yet familiar, empowering yet easily co-optable and a territory both robust and ephemeral.

In this anthology, which brings together an interdisciplinary body of recent work, we argue with conviction and passion that formations of social constellations and the space and mediations that produce and produced by such togetherness are to be looked at in the light of increased mobility, interactivity and digitality on the one hand and materiality, power and fixity on the other. The chapters that constitute the volume address and exemplify the tension fields that underlie these two intertwined domains at a time when it is impossible to think about social space without mobile and online media, and the media without participatory engagement and situated practice. This is also a transitory time when the high-sublime of the online has long-since entered the realm of the banal and the virtual is every bit as material and socially stratifying as reality-and-order-unplugged, making the study of technology and the categories of online and offline spatiality ever more complex. It is our hope that this volume will contribute to the debates encapsulating globalization, mediation and space in a meaningful manner, not least by locating the social question back in *the social*.

In the process of conceptualizing and putting together this project, a number of institutions and individuals made highly valuable contributions. First, our gratitude to the Swedish Central Bank (Riksbankens Jubileumsfond) for awarding a generous grant to our Media and Community Research Group between 2008 and 2010. It was under this platform that the *Online Territories* idea and the thinking around it found the means to further flourish. It was the Riksbanken grant that funded a first thematic conference panel at the 2008 ECREA conference in Barcelona 2008, as well the two *Online Territories* international colloquia organized in May 2009 and May 2010 to address and discuss the multiplicity of questions that arose from the volume. We are most grateful to our colleagues and friends Des Freedman, Cornel Sandvoss, Olga Guedes Bailey, Kristina Riegert, Patrik Wikström, Johan Lidberg, Johan Fornäs, Shaun Moores, Maren Hartmann, Charlotte Kroløkke, Myria Georgiou, and Thomas Tufte for accepting to deliver keynote speeches and act as discussants during the two colloquia, from which we, as the editors, benefited immensely. Our thanks also to Mia Lindgren and Mia Ohlsson for their valuable contributions during the seminars that followed the first event.

We are sincerely indebted to David Morley for his support and encouragement since the launch of this book project and for the insightful afterword he provided to the volume, enriching, in many ways, the discussions put forth throughout the anthology.

Our thanks are due to all our colleagues, too many to name here, whom, over the years, we worked with and exchanged ideas and experiences with, continuously broadening the scope of our thinking. And, to our institutions Karlstad University and Uppsala University for supporting us, for providing work space and material resources, and for hosting the two colloquia.

Finally, our heartfelt thanks to our families for being there for us, and to our children Arman and Lara, and Nora and Ilse, to whom this book is dedicated, for being endless sources of inspiration, joy and wisdom.

Introduction: Globalization, Mediated Practice and Social Space

Assessing the Means and Metaphysics of Online Territories

MIYASE CHRISTENSEN, ANDRÉ JANSSON &
CHRISTIAN CHRISTENSEN

The purpose of this book is to bring together key research and writings in the interdisciplinary study of new media and society in order to address a number of questions arising from the ways in which online technologies are currently being envisioned, used and experienced. In doing so, our specific aim is to offer an up-to-date contextualization of online practices and to explore, from a variety of perspectives, the emergence of new experiences/routines in relation to, and new conceptions of, social space. A central rationale for this book is the need for further, research-based contextualization of preexisting theories related with, for example, globalization, mobility, citizenship and civic participation, socio-spatial dynamics and network society. This need is particularly acute considering the ever-growing popularity of online communications (due to new applications such as social network sites) and the diverse and complex shapes such practice takes.

Both the practical and scholarly conceptions *and* the very experience of media use and social space are shifting. However, in the totalizing rhetorics of globalization, late modernity and networked capitalism, too much has been made of the "disappearance of place" and "deterritorialization." As many scholars have recently argued, territories—understood as socially produced spaces with certain rules for inclusion and exclusion—do not vanish or become less significant through the expansion of networked media and increasingly ephemeral flows of capital and information. Still, there is a clear need for explorations of what constitutes social territories today, and to what extent they reside within the "placelessness" of online interaction. Our purpose here is situated within a wider concern for placing two specific, interlinked departure points at the heart of our exploration of online territories. The first is an understanding of online practices and spaces not as distinct and isolated pursuits, but as closely linked with the everyday and offline milieu. Thus, we construe online territories broadly to take into account not only relational spaces that are purely online, but also a variety of intersections between the online and the offline. This includes virtual spaces that are anchored offline, or extensions of offline entities, as well as those social territories that may emerge when online activities are "lived out" or re-enacted in other parts of everyday life. While the "social uses" of media are as old as the media themselves, and extensively explored in studies of media rituals, one of our ambitions is to capture the socio-spatial formations of community and practice under conditions of increased interactivity, mobility, and media convergence.

The second angle in our approach implies a consideration of online territories on multiple analytical levels. Recently, with the increased use of Web 2.0 applications, the-not-so-new shift from representation of the subject by the media to the self-presentation of and production by the subject her/himself (hence, the shift from *response* to *mediated practice*) has assumed a more accentuated and complex form. Thus, a complex understanding of online territories must account for the interplay between situated individual and social practice, and global processes. In this volume, we trace online territories in relation to three distinct *and* interrelated pathways—the *everyday*; the *civic* and the *public*; and the *transnational/translocal*—and we do so by taking mediation, communicative practice and social space as departure points.

The more specific implications of these three pathways will be delineated before each of the three parts of the book. Before reaching this far, however, we will engage more closely with the fundamental conceptualizations and theories that keep this project together. Our prime objective here is to anchor the key argument of (re)mediated extensions of more durable structures of social practice in a broader field of social, cultural and communication theory.

The New Means of Territorialization

As the title suggests, this book is about space, and more specifically about territory. Such a thrust may seem somehow one-sided, even conservative, in an era of ephemeral and boundless information flows. However, placing territory center-stage is not to deny the significance of deterritorialization, whether media-enhanced or not, but to bring the social logics of boundary making, maintenance and negotiation clearly into the vision of contemporary (online) media studies. As geographer Doreen Massey (2005, p. 91) contends "the really serious question which is raised by speed-up, by 'the communications revolution' and by cyberspace, is not whether space will be annihilated but what kinds of multiplicities (patternings of uniqueness) and relations will be co-constructed with these new kinds of spatial configurations." It is precisely these processes of co-construction, the interplay between structural forces and the social and cultural affordances of online media, that call for a critical re-examination of how territories are (re)produced and legitimized. Understanding the enduring role of such factors as the distribution of cultural and economic capital in society can open up for a more complex view of how global processes that in a certain light could be seen as deterritorialization may also entail, even depend upon, mutual processes of territorial struggle.

A decade ago David Morley (2000) in his book *Home Territories* highlighted this tension field, considering above all the ambivalent role of television. While transnational flows of programming in certain respects had displaced audiences' cultural frames of reference and produced deterritorialized modes of sociability, they also implicated boundaries: "Sociability, by definition, can only ever be produced in some particular cultural (and linguistic) form—and only those with access to the relevant forms of cultural capital will feel interpellated by and at home within the particular form of sociability offered by a given programme" (ibid., p. 111). In this way, television and other media have for a long time exercised a soft but pervasive form of symbolic violence through their very conception and targeting of 'audiences,' which in turn integrate advanced forms of market research and monitoring. This is indeed a coercive form of territorialization. To be 'at home' with the media is not an 'innocent' or 'natural' sense of belonging, but embedded in symbolic struggles, in which common sense understandings of territorial borders (typically of a geopolitical kind) interweave with more abstract or imaginary territorial constructs, such as taste cultures, genres, consumer segments, fan and supporter communities, and so on.

This is an illustration of how the media operate as machineries of spatial production, or, to be more specific, as *means of territorialization*. This particular function of the media is far from unitary, however. Different media hold different affordances when it comes to sustaining and negotiating boundaries. Nor is it a one-sided process. As already pointed out, territorial arrangements and understandings emerge through the interplay between media circulation, material as well as cultural, and social agency (what is sometimes referred to as 'audience activity'). If we turn to Morley's *Media, Modernity and Technology* from 2007 we can gather a somewhat different view of the territorializing role of media technologies—in the midst of mobilized telecommunication patterns. Mobile telephone networks, as opposed to land lines, accentuate the role of the geographical imagination, precisely because communicators can not be sure of where the other person is without asking. Thus,

Morley argues, the mobile becomes "a device for dealing with our anxieties about the problems of distance created by our newly mobile lifestyles" (ibid., p. 223). At the very same time, the very same medium may also function as an enclosed space in itself, a protective cocoon or capsule, which on the one hand lifts the individual out of a particular local context, and on the other hand brings him or her into a new kind of representational territory where a sense of familiarity and ease can prevail (ibid., p. 221). If we in this way consider the many overlapping functions and meanings of a single medium, generalized understandings of deterritorialization fall short unless they take into account the parallel social gravitation towards closure, distinction and the hearth, which is integral to most life forms, as well as an operationalized category among media and culture industries.

The point of departure for reaching a complex understanding of territorial reconfiguration must thus be to regard space as a *multidimensional* and *processual* category. This is why we in this book consistently envision space as a *social space*, following the legacies of Henri Lefebvre (1974/1991) and Pierre Bourdieu (1979/1984). While it is indeed a complicated task to elaborate a finite analytical position that brings together Lefebvre's and Bourdieu's views of social space, the important point here is that their perspectives entail a possibility to critically *think space and communication together*. In Lefebvre, this possibility stems from his notion of a *triadic social space*—constituted by perceived space (spatial practice, material formations), conceived space (representations of space, e.g., media texts) and lived space (spaces of representation, e.g., myths and imageries)—where none of the three realms can be separated from the others. In Bourdieu, it is especially his view of *habitus* as the structuring mechanisms between the space of social positions and the space of classifying and classified tastes and lifestyles that is helpful. Altogether, Lefebvre and Bourdieu share a view of space as a processual realm of social and cultural struggle, and a view of communication as a structured practice that (re)produces social space as a cultural-material formation.

Accordingly, communicative action must be thoroughly contextualized; approached as a social practice among others. While a particular kind of activity is classified according to the dominant structure of classification, corresponding to Bourdieu's hierarchies of social space, and thus associated with certain groups and their tastes, there is also a potential for slight variation and negotiation, which may gradually, albeit just very slowly, alter the structures themselves. As Bourdieu (ibid., pp. 209–11) points out in a discussion of sports, "[b]ecause agents apprehend objects through the schemes of perception and appreciation of their habitus, it would be naïve to suppose that all practioners of the same sport (or any other practice) confer the same meaning on their practice or even, strictly speaking, that they are practising the same practice." This observation leads us back to Doreen Massey's opening statement regarding spatial multiplicity, as well as to Morley's examples of mediated territories. Even though people in diverse spaces use similar media texts or genres, or interact through space-binding communication networks they do not merge into one entity. Rather, such globalizing processes generate new frontiers of social and symbolic struggle, in which issues of territorial belonging and control are inescapable.

Online territories are in this regard no different from other territories. If we by territory mean a bounded social space of individual or collective mastery, it follows that no territory can be understood as only material, only symbolic, or only imagined, but evolves through the triadic interplay suggested by Lefebvre. The mechanisms for inclusion and exclusion are anchored in material-economic realities, which in particular have to do with the very access to media and networking resources, as well as in cultural code systems and pre-understandings, which sometimes exclude entire social groups from making use of the media in a meaningful way. Typically these structures are intertwined with one another. The concept of online territory, then, does not refer to an exclusive realm of 'online practices' (especially since the latter notion is gravely problematic in itself). On the contrary, the concept highlights the extensions and reconfigurations of pre-existing means of territorialization, be they cultural, economic, or geopolitical, as well as the potential for new types of social territories to take shape, enabled by online connectivity and sociability.

If we, for instance, consider the form and content of a photo sharing site such as Flickr it may at first glance appear like an open-ended, deterritorialized realm of online interaction and appearance, where the boundaries between private and public are blurred, and where geographical distances are overcome. However, one must not underestimate the fact that pre-established territorial formations such as 'the home,' whether we regard such a category from the viewpoint of the nation-state or the domestic setting, still exist, and even resonate with the conceived space of online representations. It is an established conception that even mobile, networked media in most social groups function not as extensive 'technologies of cosmos,' but primarily as 'technologies of the hearth'; as "imperfect instruments by which people try, in conditions of mundane deterritorialization, to maintain something of the security of cultural location" (Tomlinson, 2008, p. 68). Here, we may envision not only the maintenance of social and cultural bonds among people on the move, whether commuters, tourists or migrant groups, but also a more general praxis of social reproduction.

What must be kept in mind too, is that territories are not sealed entities in the first place, but defined through the rules and resources through which socio-spatial control is exercised. Letting guests into our homes, or showing selected others pictures from our private lives, are integral parts of the production of 'home-territories.' But, again, such practices may be carried out in different ways, and mean different things to different groups. Online photo sharing is hence embraced by certain groups; detested and rejected by others, and to even others an inaccessible or alien phenomenon—which is why online spaces must be understood foremost as extensions of the symbolic struggles of social space, rather than as an exclusive realm of placeless interaction.

At the same time online media attain qualities that in their capacity of means of territorialization set them apart from for instance print media and broadcasting. This has to do with their socio-spatial quality as arenas for mutual cultural expression and networking, which at a general level means that people can engage, or indeed experiment, in a more unrestrained manner with a multiplicity of socio-cultural belongings. In this view, online spaces allow for the exploration and construction of liminal 'elsewheres,' more or less disembedded from the constraints of social space (see Hetherington, 1998, Ch 5). Such 'elsewheres' may for instance represent alternative tastes, lifestyles and ethical holdings, and are important sites of social imagination and reflexivity. For particular groups the availability of online social platforms may even mean that for the first time there is an actual opportunity to create a more enduring sense of identity, such as in the case of sexually alternative lifestyles, which have long been denied public access and appearance (see e.g. McGrath, 2004). The interaction and expressivity channeled through online media spaces may hence ignite the formation of new emancipatory territories—a potential that challenges the institutionalized exclusivity of most other parts of the public sphere.

We are here reminded of Roger Silverstone's (2007) vision of *mediapolis*, a descriptive and normative concept for investigating the wholeness of media culture: "The mediapolis is, I intend, the mediated space of appearance in which the world appears and in which the world is constituted in its worldliness, and through which we learn about those who are and who are not like us" (ibid., p. 31). Although Silverstone is careful to point out that mediapolis is indeed, like the original Greek *polis*, an often elitist and exclusive space, he also asserts that mediapolis carries the promise of an extended *cosmopolitan realism*. This means that the people of mediapolis (which then includes the totality of media forms) are potentially aware that there are no entirely separate worlds, but that 'the other' is both different and the same. Our suggestion that online media operate as new means of territorialization does not contradict Silverstone's view. On the contrary, it is precisely through the remediation of prevailing territories, and through the constitution of alternative ones, that the diversities and similarities of the world can be made to appear. In spite of the symbolic battles and social divisions that set different groups, communities and populations apart from one another, territorial representations (as an essential component of territorialization as such) can indeed contribute to the kind of ethic that Silverstone proclaims: "it is in the experienced dialectic of sameness and difference that the possibility of a personal or communal ethics emerges, and it is in the mediated *representation* of that dialectic that the equivalent possibility of a media ethics emerges" (ibid., p. 16, italics original). In this regard *online territory* is not merely a socio-critical concept, but a concept that entails a utopian impetus as well. Of course, a recognition of such an impetus should in no way preclude a critical analysis of *dys*topian elements ever-present in materialities that shape social territories. But the fact that such territories are always marked by flux and the potentiality such instability brings is what maintains the possibility of social change.

Power and Control in See-through Spaces

While power inequalities are far from leveled merely by virtue of the penetration of technology into almost every domain of life, the shape technology takes, in conjunction with other social dynamics, has impacted power geometries and social relations in significant ways, making the distinction between the "real" and the "virtual" even more obsolete. For one, the very inequalities and socio-political gridlocks (such as the environmental crisis) that mark social space and territorial politics are mirrored in the materialities that govern the politics of the online. The fact that computer waste is transported to "lesser developed" national territories to be dumped, creating health hazards for their human communities; and, that two Google searches from a desktop computer allegedly leads to the release of the same amount of carbon dioxide as boiling a kettle (and the irony that such information can only be retrieved through an online search) are but only two examples of the impossibility of conceiving of power and politics as simply *online vs. offline*. Added to this should be the fact that many of us are using/abusing technology from the safety of our social space in the prosperous West, yet, as a direct consequence of centuries-old spatial politics, there are violent territorial struggles "elsewhere" fueled by corporate giants that manufacture that very same technology—those elsewheres being mostly bypassed by the "new media revolution." As such, online and offline are spatial extensions of each other in reproducing power and social (dis)order.

A significant transformation that further makes such a distinction redundant has been the growing permeation of technologies that allow for both the pervasive monitoring of everyday life and the embedding of surveillance in networked sociality. Just as surveillance and administrative power through the collation of information and direct monitoring were, as Giddens (1985) famously argued, defining characteristics of modernity and the nation-state form, so a diffuse and complex entanglement of surveillant practice, resembling a rhizomatic whole (see Haggerty and Ericson, 2000 and Deleuze and Guattari, 1986) lies at the core of power and social relations in the late-modern era. In the face of digitization as Lyon discerns

(2007, p. 54), Giddens' distinction between supervision and coded information becomes obsolete (i.e. information collation becomes surveillance itself) and surveillance is no longer bound within the territory of the nation-state, but engaged in by an amalgamation of commercial, state, non-state and military entities, and used for a variety of governmental or non-governmental purposes.

The use of new applications such as online social networking and mobile communications is a case in point. Networked sociality leads to new experiences of community and community-building, new senses of security, control and freedom, and more vulnerability to the monitoring of communication and consumption habits, lifestyle choices and private lives. There is a dual dynamic at work: the global consumer culture feeds into and from such motives and their extensions in the form of new technological applications. The transformation of personal territory into public domain (and the accompanying commodification) is well accommodated by the architecture of online media. Consequently, places online continuously spatialize and publicize what formerly lay in the domain of the human and institutional selves. As such, online territories are governed by a transcending logic of social control and a multitude of power geometries of varying scale and form.

Furthermore, new modes of mediated sociality blur the boundaries between freedom/dependence and labor/leisure. There exists a social ambivalence between, on the one side, freedom and flexibility, and dependency and social control on the other. The increasing potential for freedom and flexibility can be found in, for example, the possibility of distance working, working while traveling, establishing professional contacts, marketing both products and oneself globally, and as taking care of "private" affairs while at work. These types of advantages are normally highlighted when new technologies are "sublime" (Mosco, 2004), and elevated to a dominant ideology.

Metaphysics of the Online and the Offline

Because what separates online media in their latest manifestation from earlier forms of new media is a marked switch from representation to presentation and produsage, to treat *online territory* as a conceptual framework also warrants discussion on the metaphysics of 'the online' and 'the offline.' Much of the enchantment related to "the online" (emancipation, connectivity, growth, etc.), can be translated into a form of ideologically fuelled metaphysic, as well as the opposite: the rejection of "the online" as a threat to stability, family values and even humanity as based on a metaphysic of the offline. In this respect, a particularly valuable and inspiring vein of thinking can be found in the writings of Cresswell (2002). Rooted in the work of scholars such as Relph (1976), Malkki (1992), Virilio (1986), Deleuze & Guattari (1986), and Clifford (1997), Cresswell outlines how the concepts of "sedentarist" and "nomadic" metaphysics can be useful tools for the investigation and understanding of "intimately connected concepts" of "place, mobility, representation and practice" (Creswell, 2002, p. 11).

For Cresswell, the *sedentarist metaphysic*—built upon the work of Heidegger—emphasizes the position that to be human is to both understand and know your place in the world; in short, to know one's home. To have and understand one's place and to know where one belongs are essential human needs, and "place as home is described as perhaps the most important significance-giving factor in human life." Place and home, in this respect, are moral concepts that are closely linked to notions of authenticity. From the perspective of sedentary metaphysics, mobility, on the other hand, is a problem to be solved "with recourse to place and roots," and marked by the absence of commitment, attachment and involvement, thus requiring us "to think of mobile people in wholly negative ways." With the evolution of feminist, poststructuralist and postmodern theory, however, sedentarist metaphysics was replaced by what Cresswell calls a *nomadic metaphysics* in which place is marginalized, and travel, flow, flexibility and transgressions are celebrated.

The concepts of sedentarist and nomadic metaphysics, and the concomitant epistemological arguments associated with these positions, are reflected in a great deal of academic and popular discourse in relation to online and offline activities (more often than not reflecting the aforementioned "enchantment" with the online). A particularly salient example of this is the popular discourse on the use of social media that emerged before, during and after the Iranian presidential elections of June 2009 (Christensen, 2009). The (supposed) use of social networking media such as Twitter, Flickr and YouTube by anti-government, pro-Mousavi protesters in Tehran in the weeks following the disputed elections was held up as an example of the cosmopolitan nature of young, educated Iranians frustrated with decades of repression. The appropriation of "modern" communication technologies for the purposes of organizing protest, informing the global news media of events and interacting with users outside of Iran (in short, the nomadic metaphysical represented by the young Iranian protesters) was seen to be in stark contrast with the repressive, backward, sedentarist metaphysical represented by the guardians of the 1979 Iranian Revolution: individuals trapped—according to popular discourse—in outdated notions of home, tradition, place and space. This constructed dichotomy of *traditional-religious* ("offline") versus *modern-secular* ("online"), in which the latter group is marked by technologically savvy, cosmopolitan sensibility and the production and reception of rapid flows of information, falls into line with what Sheller and Urry (2006) call the "grand narrative' of mobility, fluidity or liquidity" (p. 210). What the discourses surrounding the Iranian case—and other examples of popular and academic enchantment with the online—ignore, however, is the highly complex interplay between the online and the offline, and the ways in which even groups defined as sedentary and traditional (such as the Iranian power elite) extend their offline power into the online realm.

As we have argued throughout, as a concept, we construe *online territories* as not exclusively what lies inside the domain of the "virtual," as the shape that the online takes is also very much contingent upon what is excluded, voluntarily withdrawn, not there or simply not possible, and on territorial struggles (material or otherwise) that give way to new or altered forms of communicative practice. Indeed, it is more fruitful to consider the new and evolving "set of questions, theories and methodologies" raised by the new mobilities paradigm, rather than falling back upon "totalising or reductive descriptions" of the modern world. (Sheller & Urry, 2006, p. 210). What marks the realm(s) of online territories is the tension between, on the one hand, the affordances—such as complex forms of sociality and leisure, the restructuring of the public sphere and social space or the partial/perceived elimination of spatio-temporal borders—inherent in the architecture of new technologies, and the materialities—such as power relations, global human flows, access to economic resources or corporeal manifestations of urban space—of the offline. In essence, we would like to diverge from the vein of media and communication studies that subordinated sociologically informed questions related with technology to mythical accounts of the "Age of the Computer," to borrow Mosco's (2004) phrase. We further agree with Mosco that when technologies cease to be sublime and enter the realm of the banal, they become significant sources for social and economic change. As we write in 2010, the online has long-since joined the ranks of the banal, and virtuality is further and further distanced from fantasy and hyper-reality and is very much in the domain of mundane, everyday corporeality.

Ultimately, the emergence and evolution of online territories is an extremely complex, dialectical process, intertwined with the macro-dynamics that govern mode of production and economic relations, on the one hand, and mediation, negotiation and imagination on the other. Some in the form of case studies, and some more theoretically oriented, the chapters in this volume address a variety of key issues and questions within a conceptual scope accommodated by *online territories*.

The Structure of the Book: Three Pathways of Exploration

With the aim of situating each essay in a particular context, this book is organized around three conceptual constellations: *Everyday Intersections; Citizenship, Public Space and Communication Online*; and *Transnational/Translocal Nexuses*. Each constellation opens with a brief introduction and addresses a particular realm of social interaction, containing certain codes and conventions that are elaborated both online and offline—thus creating particular territories and identities. While the *Everyday* can be seen as the common ground for all social practice, the *Civic/Public* and the *Transnational/Translocal* consider the superstructural implications of interactive networks vis-à-vis political processes and global (trans)migration. The individual chapters then explore online territories in closer detail in order to provide a situated understanding of how social spaces and communities are (re)produced through media practices within the realms of the *Everyday*, the *Civic/Public*, and the *Transnational/Translocal*. The book has an afterword/response by David Morley.

References

Bourdieu, Pierre (1979/1984) *Distinction: A Social Critique of the Judgement of Taste*. London: Routledge.

Christensen, Christian (2009) Iran: Networked Dissent. *Le Monde Diplomatique*, July, 2009. URL: http://mondediplo.com/blogs/iran-networked-dissent

Clifford, James (1997). *Routes: Travel and Translation in the Later Twentieth Century*. Cambridge: Harvard University Press.

Cresswell, Timothy (2002) Theorizing Place. *Thamyris/Intersecting*, 9: 11–32.

Deleuze, Gilles and Guattari, Felix (1986) *Nomadology: The War Machine*. New York: Semiotext(e).

Giddens, Anthony (1985) *A Contemporary Critique of Historical Materialism. Vol. 2. The Nation State and Violence*. Cambridge: Polity.

Haggerty, Kevin D. & Ericson, Richard (2000) The Surveillant Assemblage. *British Journal of Sociology*, 51(4), 605–622.

Hetherington, Kevin (1998). *Expressions of Identity: Space, Performance, Politics*. London: Sage.

Lefebvre, Henri (1974/1991) *The Production of Space*. Oxford: Blackwell.

Lyon, D. (2007) *Surveillance Studies: An Overview*. Cambridge: Polity.

Malkki, Liisa (1992) "National Geographic: The Rooting of Peoples and the Territorialization of National Identity Among Scholars and Refugees. *Cultural Anthropology*, 7(1), 24–44.

Massey, Doreen (2005) *For Space*. London: Sage.

McGrath, John E. (2004) *Loving Big Brother: Performance, Privacy and Surveillance Space*. London: Routledge.

Morley, David (2000) *Home Territories: Media, Mobility and Identity*. London: Routledge.

Morley, David (2007) *Media, Modernity and Technology: The Geography of the New*. London: Routledge.

Mosco, Vincent (2004) *The Digital Sublime: Myth, Power and Cyberspace*. Cambridge: MIT Press.

Relph, Edward (1976) *Place and Placelessness*. London: Pion.

Sheller, Mimi, & Urry, John. (2006). "Introduction: Mobile Cities, Urban Mobilities." In M. Sheller & J. Urry (eds.), *Mobile Technologies of the City*. London: Routledge.

Silverstone, Roger (2007) *Media and Morality: On the Rise of Mediapolis*. Cambridge: Polity.

Tomlinson, John (2008) '"Your Life—To Go": The Cultural Impact of New Media Technologies.' In Hepp, Andreas; Friedrich Krotz; Shaun Moores and Carsten Winter (eds,), *Connectivity, Networks and Flows: Conceptualizing Contemporary Communications*. Cresskill, NJ: Hampton Press.

Virilio, Paul (1986) *Speed and Politics: An Essay on Dromology*. Semiotext(e) Foreign Agents Series. New York: Columbia University Press.

PART I

Everyday Intersections

In the late-modern world, networked media practices have changed character from the extraordinary to the mundane in just a decade. A variety of platforms are today used, depending on social needs and desires, for establishing and maintaining contacts, or for more open-ended self-expression, creativity and voyeurism. Embedded in pre-established social routines and materialities, the territories that unfold through online networking attain a number of never before seen intersections. This regards, in particular, the intersections of private and public, the intimate and the formal, which may on the one hand extend individual opportunities for liberation and self-emancipation, but on the other hand raise issues regarding integrity, privacy, addiction and various forms of abuse. As the chapters in this section will illustrate in various ways, the social impetus of new media forms must always be analyzed through their situated appropriations, which means, for instance, that online activity may change the way we experience the everyday and our lived spaces, rather than necessarily making us more mobile and more "global." What will also be considered here are the various forms of social resistance that are somehow inherent to the sedimented character of everyday life, and may be expressed through alternative forms of media use, or through a questioning *and* subversive appropriations/celebration of mediatization as such.

In the opening chapter "The Everyday War: Iraq, YouTube and the Banal Spectacle" Christian Christensen addresses everyday mediations of war. While a great deal of research within media and communication studies has addressed coverage of military conflict by focusing on the unusual, the outrageous, the offensive or the entertaining, relatively little work has been done on how military conflict and occupation is presented to members of the general public as mundane, routine or even boring. In short, focusing on the "spectacle" of warfare, or, what Baudrillard has called "war porn." With the advent of video-sharing sites such as YouTube, however, soldiers serving in Iraq and Afghanistan have had the opportunity to upload images of military service that are at odds with mass-produced representations of war and occupation. In this chapter, Christensen discusses how, throughout history, and through various forms of media such as letters home, poetry, music and photography, troops have represented "the everyday" of military life. In that sense, there is nothing new in soldiers documenting the banal moments that can often dominate during conflict, a practice that strikes a dissonant chord with non-military members of the audience who often assume banality to be antithetical to war.

In Chapter 2 ("The Domestication of Online Pornography: How Cyberporn Found a Home in the American Home"), Jonathan Lillie draws on cultural theory and research to explore some of the everyday contexts of the home that shape peoples' encounters with online pornography. Since the popularization of the web in the mid-1990s, the explicit content available on the Internet has too often been characterized as a separate, yet easily accessible, world full of either sexual delights or perverse demons. Such narratives perhaps draw in part from Western civilization's social and physical compartmentalization of sexuality. Lillie argues that what is called by some "the cybersex panic of 1995," can be understood as a society's anxiety attack brought on by the perception that an untamed space of sexual expression had suddenly been opened to the masses without any rules for what can be done where and by whom. He further suggests that this "panic" did not last long and a regime of social discipline, what Michel Foucault termed a technology of sexuality that was already operationalized within domestic spaces, was quickly brought to bear on the untamed sexualities of cyberspace. More particularly, Lillie seeks to understand how the science of sexuality in a sense domesticated online pornography by creating narratives and guidelines for how this houseguest, who was both wanted and unwanted, could be tolerated.

The third chapter of the book ("Fans Online: Affective Media Consumption and Production in the Age of Convergence"), written by Cornel Sandvoss, the issue of fandom is tackled. As Sandvoss writes, being a fan has come a long way in the Internet age. Far removed from the days in which the word "fan" carried connotations inviting social stigma and pathologization, in the contemporary world of social networking being a fan has become one of only two fundamental modes of engagement with the world around us: on Facebook, we can link to others either by being their friend or their fan—the former, mirroring Thompson's analysis of fandom, requiring the reciprocal approval, the latter being a non-reciprocal statement of affect. This visibility of fans online—populating news groups, online fora, video sharing portals and fan created websites—has led some inside and outside academia to misinterpret contemporary fan practices as a consequence of technological change. Aiming to assess the interplay between (fan) audiences and their practices with the diffusion and subsequent transformation of the Internet, this chapter documents the different forms of fan productivity online, and in juxtaposing these with offline fan activity argues for a revaluation of audiences' impact on the shape and structure of the Internet.

In the past fifteen years, gambling has become an increasingly popular Internet application. Casino games, poker, sports betting, and lotteries are now major moneymakers on the Internet, sometimes legally, but often illegally. Disagreements between national governments have occurred over the operation of offshore Internet gambling websites. In chapter 4, "The Place of Internet Gambling: Presence, Vice, and Domestic Space," Holly Kruse looks at how interactive technologies have changed the places and spaces of gambling and social ritual, from public space to private space, and to spaces in between by focusing on the case on pari-mutuel horse race wagering. As Kruse observes, although Internet gambling accounts for a substantial portion of online financial transactions, it has been at best a marginal research topic in the field of Internet studies. Through an examination of Internet gambling and how it enables and influences forms of interaction, she addresses the socioeconomic implications of the relocation of horse race wagering and other forms of betting first to remote sites, and then into homes with Internet access. Such implications include the diminishing number of face-to-face interactions among individuals that take place at betting locations, the importance of forms of presence other than physical co-presence, and the changing notions of what should take place in public and what is allowed in domestic space.

In the final chapter of this section, "Spamculture: The Informational Politics of Functional Trash," Kristoffer Gansing (Chapter 5) argues that on the net we are all, in one way or another, working as subjective cyberneticians, constantly sending, sorting and filtering messages according to some (often collectively) imagined logic of order. Yet, Gansing contends, what actually counts as meaningful communication will always be a site of negotiation in relation to a complex set of political predispositions and interests, interwoven with spatial, cultural and social realities. This chapter attempts to examine such political aspects by looking at a borderline phenomenon that defies conventional communication categories. In today's online communications some informational objects have become definable as trash in the sense of appearing as immediately disposable to the user. An example of such a "trash" category of informational objects is the electronic communication phenomenon known as "spam." Being disposable as trash, however, only forms one of several possible functions for a spam object: at the same time most spam mails have a concrete function such as selling products, installing viruses, mal- or spyware, "phishing" for credit data, "junking" the receivers computer, etc. They are in effect "functional trash." In the course of this chapter, Gansing explores this functional ambiguity of spam as trash, as well as other related wasteful web-based communications phenomena, with the aim of developing a theory of the politics of contemporary spamculture.

CHAPTER ONE

The Everyday War

Iraq, YouTube and the Banal Spectacle

Christian Christensen

In the opening line of the book *Tales of the South Pacific* (later to become the play/musical *South Pacific*), author James Michener—in the form of a letter to a girl back home written by a crewman aboard a US navy vessel in the South Pacific—wrote, "I wish I could tell you about the sweating jungle, the full moon rising behind the volcanoes, and the waiting. The waiting. The timeless, repetitive waiting." Michener's reflections upon the boredom and ennui of warfare stood (and continue to stand) in stark contrast to popular discourses within which theaters of war are places defined by constant action, individual honor, collective bravery, unspeakable violence, physical pain and horror. The often stifling boredom of warfare discussed by Michener and others does not, however, negate or counter-balance the real horrors of military conflict. As will be discussed in this chapter, efforts on the part of troops to both illustrate and alleviate boredom through the construction, reproduction and transmission (via channels such as YouTube) of "normal" or "everyday" activities such as eating, sleeping, singing, drinking or joking in and from fields of battle only serve to reinforce the actual brutality of war. These rapidly constructed and easily distributed online reproductions of the offline everyday should also lead us to consider the relationship between the affordances of technology and potentially shifting notions of post-9/11 media "spectacles." These clips should also encourage researchers to expand analyses of the visual representations of warfare and conflict so as to include those everyday aspects which, according to many who have served in the armed forces, come to define their time in the services.

War on a Sliding Scale

This chapter is the third part of a trilogy of articles in which I examine the representation of the occupations of Iraq and Afghanistan on YouTube through material posted primarily by troops on the ground, but also by the official military hierarchy. In the first article on the subject (Christensen, 2008), I examined how US troops would upload images of their own violent, anti-social and occasionally criminal activities—firing at unarmed civilians, humiliating Iraqi children or exhibiting near-sexual pleasure during heavy combat—in Iraq and Afghanistan to YouTube, and how these images were juxtaposed with the positive, upbeat and highly propagandistic material—donating footballs to Iraqi children, cooperating with grateful locals and freeing captured hostages—produced and uploaded by the US military to the same YouTube system. The extreme variations in the two presentations of the occupations of Iraq and Afghanistan through these clips, and the fact that both forms were available side-by-side on the same website, I argued, generated a form of "dissonance" whereby propagandistic, nationalistic popular discourses on the values motivating the US occupations (often using buzzwords such as "democracy," "justice," and "freedom") were challenged by the violent and disturbing images shot and uploaded by the very troops sent to defend these values.

While in the first study I examined what one might call the more "conventional" aspects of warfare such as violence, action, jingoism and propaganda, in a subsequent article (Christensen, 2009b) I examined a more unconventional mode of presenting warfare: videos of various aspects of the occupations—from combat to funerals—set to music, produced and distributed by troops themselves. In this article I connected the music videos made by soldiers in the field and then uploaded to YouTube to the World War I poetry of Wilfred Owen, and how Owen's "memorialization and documentation of warfare through verse is but one example of how soldiers throughout history have used a variety of media (from print to photography to music) to record their experiences" and how, "setting videos of military activity to music and uploading them to YouTube is an extension of Owen's method of documenting warfare" (pp. 164–5). For many of the troops, the music videos were a form of documentation—or, in the case of videos dedicated to fallen comrades, memorialization—meant to show to the outside world the sacrifices made by those in battle, and also to lay down a digital marker, proving that what had happened was real (p. 174). The videos could also be seen, I argued, as simply a digital variant of the "letter home" sent by soldiers on the frontlines to family and friends.

One could say that, by including the current chapter into the equation, my three pieces on Iraq and YouTube address everydayness on a sliding scale: from the very non-everyday material (in the civilian sense of everydayness) of battle and military propaganda posted by the troops and government; to the more everyday music videos created by troops in which images of battle set to music are blended with footage of troops sleeping, joking and travelling; to, finally, the current chapter, in which I will discuss videos uploaded by troops showing everyday activity, devoid of not only the traditional spectacles of battle (gunfire, death, heroics, horror), but also of the formulaic and generic tactics (such as the use of music, rapid editing or image manipulation) that can lift the mundane into the realm of the spectacular. As noted, while the representation and reproduction of everydayness in these YouTube clips serve to highlight the spectacle of war through a veiled but overwhelming juxtaposition between horror and the mundane.

Everyday-ness and War[1]

It is an apt point of departure to discuss the notion of the "everyday" in relation to war, as the military enterprise has influenced how theorists have conceptualized the relationship between military life, state violence and what can loosely be called "civilian" activities. Lefebvre (1987) writes that, "the everyday has always been repetitive and veiled by obsession and fear" and that the everyday is marked by two forms of repetition: the cyclical (i.e. nature) and the linear (i.e., processes of rationality). He continues:

> The everyday implies on the one hand cycles, nights and days, seasons and harvests, activity and rest, hunger and satisfaction, desire and its fulfillment, life and death, and it implies on the other hand the repetitive gestures of work and consumption. In modern life, the repetitive gestures tend to mask and to crush the cycles. The everyday imposes its monotony. It is the invariable constant of the variations it envelops. The days follow one after another and resemble one another, and yet—here lies the contradiction at the heart of everydayness—everything changes. But the change is programmed: obsolescence is planned. Production anticipates reproduction; production produces change in such a way as to superimpose the impression of speed onto that of monotony. Some people cry out against the acceleration of time, others cry out against stagnation. They're both right. (10)

In the linear, rational and repetitive everyday proposed by Lefebvre, modern civilian life can easily be seen as an extension of militarist structure and epistemology. In her excellent study on the presentation of the everyday in relation to war in Romantic literature, Favret (2005) notes Foucault's (1975/1995) contention that "the military practices of the eighteenth century were the spawning ground of a new type of everydayness," and that "the European military helped organize the ordinary individual within a system of rules, routines, and practices, thereby fashioning not only a disciplined individual but also the concept of ordinary life as an affair of daily timetables, charted motions, performed duties" (pp. 613–4).

Importantly, the notion of linear everyday life as a by-product of military rationality is but one understanding of how the everyday developed in relation to professional soldiering. A related, but somewhat different conception of the everyday vis-à-vis war, Favret (2005) notes, is the argument made by a number of historians that,

> the concept of the everyday arose when the military enterprise of the early modern nation-state "faded as the pre-eminent activity of elites" and court life replaced camp life as a social and political ground. The ideal of military heroism, so much at odds with the everyday, yielded to the ideal of leisurely gentlemanliness. "There was a place now...to notice how pleasant ordinary occupations might be." (p. 610)

Although recognizing the connections between structured everyday life and militarism as discussed by Foucault, it is the latter conception of the everyday in relation to war that informs this chapter: the epistemological and psychological disconnect between "military heroism" and "leisurely gentlemanliness" and the social construction of space for the enjoyment of "ordinary occupations." While the vast majority of everyday activities of soldiers in Iraq and Afghanistan shot on digital cameras then uploaded to YouTube are, of course, far from the pursuits that would have constituted "leisurely gentlemanliness" of the 19th century, they are nevertheless pursuits clearly separated from the "ideal of military heroism" and/or stereotypical military activities.

The construction, reproduction and presentation of the everyday in and from theaters of war should be understood also within the context of not only the motivations for such activity, but also the affordances of the technologies used to record and relay the events. In terms of motivation, the historian Fussell (1989) wrote that, "waiting itself and nothing else becomes a large element in the atmosphere of wartime" and that soldiers throughout history have searched for ways in which to alleviate the boredom of waiting:

> Some antidotes to boredom took literary form. In relatively inactive theaters like India, bright British soldiers kept copious diaries, which they composed carefully, even artfully, writing sometimes 2500 or more words per day. That such diaries were strictly forbidden added to the pleasure. One man stuck in Syria wrote in a diary "to prove," he said, "that I am alive." (p. 78)

Addressing updated versions of the Victorian soldier's diary (or the war poetry of Wilfred Owen), Cammaerts and Carpentier (2009) discuss the rise of soldier blogs during the occupation of Iraq, and note how troops use these blogs to relay not only the stress of military life, but also "the sheer banality and boredom of war." Cammaerts and Carpentier note that these blogs (such as "A Day in Iraq" written by a soldier called Michael) contradict "the heroisation of American soldiers in Iraq by narrating the ordinariness of their activities" (12). In a particularly telling example from "A Day in Iraq," Michael notes how, upon their return from completed duties, fellow troops would sarcastically respond, "Oh, the horror" (playing off of Marlon Brando's infamous line from *Apocalypse Now*) when asked how things had gone. In truth, the duties had been mind-numbingly boring, leading Michael to note that the tedium of guard duty, "kills you from the inside out, eating away at you like a cancer" (in Cammaerts & Carpentier, 2009, p. 12). It is precisely at this point of intersection between warfare, boredom/ennui and technology that we might consider the representation of everydayness on YouTube.

Unlike the forms of media available to soldiers in earlier conflicts, digital cameras and websites such as YouTube give troops serving in Iraq and Afghanistan the possibility of producing relatively sophisticated audio-visual clips in a short period of time, and then upload these clips instantaneously. While I have argued that the videos produced by US soldiers can be seen as extensions of earlier forms of troop communication from the frontlines—letters home, photographs, poems, songs—the affordances of contemporary technologies have shifted what we might call "war communication" in such a way that the physical and temporal limitations (in terms of both, use in the field of battle and subsequent distribution of completed materials) of pen, paper, automatic cameras or audio cassette recorders are lifted in favor of instantaneous recording and relatively rapid distribution. This access to both, cheap technology and a global distribution system, should also be understood in relation to an increasing interest and participation in "confessional culture" in the form of reality programming and talk shows (e.g. Aslama & Pantti, 2006) as well as blogs and video-sharing, in addition to the practice of constant filming on the part of troops (Carruthers, 2008) and a sophisticated knowledge of the generic forms and intertextual codes of contemporary popular culture (Christensen, 2009b, pp. 166–7). The presentation of "everyday" activities from Iraq and Afghanistan is, therefore, a complex phenomenon, and more than just the happenstance, *ad hoc* recording of the everyday or the mundane; or, to put it another way, it is more than just the simple representation of the analog offline in the digital online.

The Everyday in Iraq/Afghanistan on YouTube: Capturing, Staging, Adapting

When examining the material posted by troops to YouTube, what becomes clear is that the videos representing activities that are "other" than regular military fare—the "everyday"—can be markedly different, falling into defined categories, thus illustrating the complex interplay between the online and the offline. For the purposes of this chapter, I have selected ten videos uploaded to YouTube that illustrate the variation in how everyday activity is documented.[2] The first category of clips representing/reproducing everyday activity has material with the feel and tone of the spontaneous: as far as can be gauged, the videos *appear* to have been shot without forethought or planning, there is minimal "intervention" on the part of the person doing the recording or others, and thus are closest to notions of the "documentation" or "recording" of "reality." These are clips that can be placed under the rubric of *Capturing the Everyday*.

Three clips serve as good examples of this type of material. The first, entitled *Soldier Sleeping*,[3] is a simple, 69-second clip showing an unidentified soldier sleeping on the top bed of a bunk. There is no voice-over or commentary in the video: simply noise of the soldier snoring. The clip ends with the camera moving down to the bottom bunk, where another soldier is sitting, smiling, and making hand gestures indicating a sawing action (making fun of his bunkmate's snoring). The second clip, *Bored Soldier in Iraq*[4] is a humorous video showing a large soldier "marching" a line of birds down a road. The incident appears to be spontaneous, and the humor of the piece is found in the fact that the birds appear to be following the orders of the soldier: walking in a straight line, moving left and right, and, eventually, leaving the area. The soldier also plays the part by giving orders in a typically loud, military fashion. The final video shows a soldier in a tank slowly eating a banana (*Eating a Banana in Iraq*)[5] while in an armored vehicle. The description of the video by the uploader reads:

> My gunner "Ace" taking a break to eat a banana. Shortly after a rocket flew over our heads, one of my soldiers got the sudden urge to eat bananas, well now you get to see it! woohoo.

His friends joke with him, offering him 20 dollars if he can get the whole banana into his mouth (thus simulating oral sex), but the soldier refuses, slowly eating the banana, exclaiming how good it is (with soldiers laughing during the filming). The video was only viewed a total of 750 times, and the following comment written by a viewer under the clip (the only comment, in fact) is either incredibly poignant, or an attempt to fool viewers:

> omg the one who was laughing [in the clip] is my big bro he died wed i still cant believe it

The second category of clips is one that we might call the *Staged Everyday*. In these clips the troops record everyday life on their bases or out in the field, but, unlike the "capturing" variation, these videos show troops engaged in activities that appears to have been staged, to some degree, for the camera. In *Bored Troops*,[6] a female soldier and her male colleagues play a game where the loser has to crawl through the legs of fellow troops ("the gauntlet") while they slap the loser's behinds. In this clip, as the female soldier runs the gauntlet, one of her fellow troops (the owner of the YouTube channel to which the clip is posted) drops his trousers slightly, causing the entire group (including the female soldier) to break into laughter. The *Staged*

Everyday also includes a fair amount of material in which troops play jokes on one another, or dare each other to endure physical pain for the amusement of themselves and their friends. Two examples would be *Soldiers Bored in Iraq*[7] in which a soldier lying on a bed volunteers to take a karate chop to the stomach, and the more elaborate, *Boredom in Iraq*,[8] in which a soldier stands spread-eagled against a large concrete wall while a friend shoots a ball at his genitals from a slingshot. In the second clip, despite the considerable distance between the shooter and the "target," the ball from the slingshot manages to hit the soldier squarely in the genitals, causing him to fall to the ground. In both clips, the witnesses—and, surprisingly, the victims—all break into waves of deep laughter following the painful events.

The final two examples of the *Staged Everyday* come from the same source: a soldier named Betty whose YouTube channel had a fair degree of success (over 160,000 views) despite the simple, everyday nature of her uploads. In the first, Betty provides a vegetarian soldier's view of eating while serving in Iraq (*Vegetarian Soldier's Eat in Iraq*).[9] In the 1-minute-40-second video, Betty shows the food that she (a vegetarian) eats, discusses what options are available to her as a vegetarian, and talks about food with a few of her fellow troops. In the second video, *"The Things They Dont Show You in Iraq,"* Betty put together a compilation of clips in order to show viewers what life is like in the military. As she put it in the description of the video:

> These are just a bunch of things that I put together of my unit in Iraq. Most of it showing what we did with our off time. Just for added info, this is not the only thing that happened. What I added is showing you what happened in addition to all the other crap that's going on over there.

The 3-minute video is very much a compilation of the everyday, and shows troops engaged in various activities such as playing football and baseball, working in an office, feeding kittens, playing video games and making jokes. Betty shows herself lounging in bed on her day off, and comments how power outages are a common occurrence in Iraq, forcing the soldiers to go outside and amuse themselves by playing sports.

The final "genre" of everyday clips is one that I have labeled *Adapting the Everyday*, in which scenes appear to be generally spontaneous, but where there is a degree of "intervention" on the part of the person recording the event, or those around him/her. These interventions do not alter the overall nature of the everyday act, but usually serve to inject an element of humor and/or commentary. The first example of such a video is the simple clip, *Drunk Soldier in Iraq*,[10] in which a heavily inebriated soldier is filmed sitting on a chair. The soldier is very young, and clearly on the verge of unconsciousness, and is questioned by the soldier shooting the video about how he feels, what he has been doing, and so on. The drunk soldier's standard response (other than several garbled sentences) is that he is drunk and "fucked up." The owner of the YouTube channel in question (who also appears to be the person who shot the video) posted responses to comments made to the clip by viewers. The following were three comments posted by viewers:

> Pathetic that you can't remain sober in a combat zone…probably under 21 too
>
> It is pathetic…I really dont understand why you posted this. You guys get to live like you are back in the States.
>
> you have some nice barracks for iraq

To which the uploader of the clip replied:

> And for the comment about how its pathetic that we cant stay sober in a combat zone…. Bitch, its pathetic that you wont go to a combat zone. If we have the fucking balls and nerve to make the ultimate sacrafice for our country so that assholes like you can have a decent life, then we have the right to have a few drinks.
>
> First of all, our barracks werent nice at all! They were called CHUS which are containerized Housing Units that were made of the cheapest materials possible from some third world country! Every other day something would fall apart on the damn things.

The subject of the clip, the drunk solder, also received comments from viewers, one of whom wrote that the boy was "cute" and wanted to know if he was single, to which the subject of the clip responded:

> that is me in the video and yes im single…this is highly embarassing we bought 2 half gallons of johnnie walker black off the iraqi army that night!!!

In a similar video, *Drunken Soldier*,[11] a drunk soldier is offered a dollar to spin on the floor 20 times, and does so much to the amusement of himself and the friend shooting the video (who is also drunk). Once again, what makes the clip particularly interesting is the following comment made by the uploader of the video, which was viewed 29,000 times (and which garnered a great deal of comments from viewers), in response to one individual who critiqued members of the military and their behavior:

> damn lots of arguing going on in my video comments…are you all for real? we were just having fun, yea were soldiers, but were still just normal people underneath that uniform. would this video have been viewed differently if my buddy on the floor had been doing circles in Dickie's and a tank top?

A final example of the adapted clip would be *Funny Soldiers in Iraq Talk about Hermaphrodites*[12] in which a group of soldiers have a long conversation about sex toys, having sex with hermaphrodites, taking money for giving oral sex to another man and other highly sexual subjects. The conversation is interspersed with a great deal of laughter, and is highly "macho" in nature, with repeated references to homosexuality and lengthy discussions on the "boundaries" between hetero- and homosexuality. The conversation is highly crass and vulgar, and one particular comment posted by a viewer (a comment raked as the "best" by other viewers), perhaps goes the furthest in highlighting the mediated everydayness of the soldiers' conversation:

> There grandfathers probably acted the same way. It just wasnt documented.

Representations of Warfare: Alternatives to the "Always Online"

The three "genres" discussed in the previous section were meant to serve as (brief) illustrations of the varying forms that representations of the "everyday war" might take on YouTube. The clips ranged from the poignant to the dull to the crude, but they all present versions of warfare and military life that are rarely seen by the general public. There have, of course, been myriad representations within popular culture of soldiers getting drunk or speaking crudely, but these representations have, for the most part, been relayed through fiction. The fictional representation of the everyday (eating, sleeping, swearing, drinking) is usually within the context of a clear storyline, and usually presented via characters with whom viewers/readers/listeners have some form of established relationship. The short, anonymous, often *ad hoc* nature of the YouTube clips, on the other hand, supply comparatively unvarnished representations of the trivialities and banalities of everyday life, and do so through individuals with whom viewers have, it is likely, little or no established relationship or emotional ties. While recognizing the problems with the term as discussed at length by Ellis (e.g., 2005), the "documenting" of the everyday in Iraq and Afghanistan via YouTube provides viewers with an archive of material in which military life is presented by soldiers whose relatively raw presentation styles (in relation to slick Hollywood productions) perhaps adds to feelings of empathy, disgust or sadness.

As noted in the introduction to this chapter, the online representation of the everyday activities of troops serving in Iraq and Afghanistan via YouTube should encourage researchers to reconsider the nature of journalistic "war reporting" and the representation of warfare throughout popular culture in general. The term "online" first came into use in the early 1950s, and is defined as the state of being "connected to a system" (in contemporary discourse, usually understood to be the Internet). In contextualizing war reporting over the years, it might be fruitful to consider the practice of being "connected to a system" in more general, non-digital terms. In many ways, specific aspects of war have always been "online," always connected to a system. In news reporting and popular culture (films, books, music, television), dominant representations—bordering on the hegemonic—of military activity have been constructed: representations containing action, blood, violence, valor, cowardice, machinery and death. As Griffin (2010) writes,

> Images of war do seem to have an inherent attraction. The attention paid to war-related news photography and video, the reproduction and sale of large numbers of war photography books, the long popularity of the war movie genre, the success of cable television channels devoted exclusively to military documentaries, and even the popularity of contemporary war-themed video games, confirm a widespread public fascination with depictions of warfare. Undoubtedly, this has something to do with the fact that war images offer viscerally exciting and voyeuristic glimpses into theaters of violence that, for most viewers, are alien to everyday experience. (p. 8)

Those who produce these streams of representation (both visual and rhetorical) have found ready and willing systemic conduits within all branches of the media from multinationals to small, family-owned local outlets. As such, sensational representations of military life, conflict and warfare have always been "connected to a system" (mainstream media) and, thus, are "always online" and always plugged in. The extent of the commodification of war—with an intertwining of news, information, propaganda and entertainment—is such that Andersen (2006) developed the term "militainment" to crystallize the complex overlapping of these areas.

Taking this admittedly flexible definition of "online" one step further, what can we then say of the non-digital "offline" in relation to military life? If particular representational forms of warfare have, historically speaking, been fast-tracked for exposure and packaged for commercial gain via media systems, which are the forms disconnected from these same systems? Despite claims by personnel that they constitute the temporal majority of military life, ennui, boredom and other facets of the everyday are aspects that have rarely found their way into larger-scale representations (with occasional popular culture productions such as HBO's *Generation Kill* and Hollywood's *Jarhead* as the exceptions to the rule). Without these systemic outlets, the military everyday has been relegated to the *ad hoc*, disconnected and (relatively) uncommodified offline, to be found in forms such as letters sent home, poems, idiosyncratic photographs, sketches and the like. To a great extent, the limited volume of material presented in this chapter, together with the material discussed in my previous pieces on YouTube and Iraq (Christensen, 2008, 2009b), are examples of what was previously in the domain of the offline. US soldiers engaged in war crimes, making music videos or simply sleeping and/or eating have now found a space within the online territory of YouTube.

Andersen (2006) wrote that the representation of war is essentially about who possesses sufficient "storytelling power," and it is clear that, to date, storytelling power in relation to military conflict has remained in the hands of the few: from journalists to governments to filmmakers. The online representations of everyday acts on the part of soldiers serving in Iraq and Afghanistan not only raise questions regarding a recalibration (albeit it extremely small at this point) of this storytelling power, but also how very particular notions and forms of spectacle and the spectacular have come to dominate both popular and academic discourse regarding warfare. As an example, in his voluminous work on post-9/11 US politics and media, Kellner (2004) took his cues from Debord (1967), but narrowed the notion of spectacle from Debord's initial, "overarching concept to describe the media and consumer society, including the packaging, promotion, and display of commodities and the production and effects of all media" down to research looking at specific, large-scale events:

> My main interest (…) is in the megaspectacle form whereby certain spectacles become defining events of their era. These range from commodity spectacles such as the McDonald's or Nike spectacle to megaspectacle political extravaganzas that characterize a certain period, involving such things as the 1991 Gulf war, the O. J. Simpson trials, the Clinton sex and impeachment scandals, or the Terror War that is defining the current era.

The work of Kellner and other scholars on these issues (many of which have become staple foci for media research) has added a great deal to our understanding of mediated warfare. Yet, the emphasis (in both popular media and media research) on larger-scale "visual spectaculars" such as the collapse of the Twin Towers in New York or the bombing of Baghdad only serves to reinforce the hegemony of militainment. An examination of the everyday activities of the soldiers involved in these conflicts, on the other hand, brings us back to a notion of spectacle within which the act of viewing is perhaps reinforced as a central element of defining the spectacle (we watch soldiers eat *despite* the banality), as opposed to the megaspectacle, within which the event itself is the key factor (we watch a building explode *because* we find it stimulating).

Of course, there is the danger here of a technological-determinist understanding of the presentation of the everyday war: the presentation and dissemination of everyday activities by troops is possible simply because of technology, and people watch simply because it is available through this technology. In short: troops make the material because they can, and people watch because they can. In the introduction to this edited volume, however, Christensen, et al. (2011) note that the concept of "Online Territories" is rooted in a deeper understanding of the interrelationship between the affordance(s) of technologies, the socio-cultural, economic and political factors influencing the material conditions of their access and use and the material generated. While lightweight digital cameras and broadband access enable soldiers to shoot and upload footage of activities from firefights to fellow troops playing games, and enable citizens in the US and worldwide to view these clips, the presentation of the banality of troop boredom cannot be separated from either the context of this banality (members of an occupying military force killing time while waiting to take part in violent engagement) or the deeper socio-political meaning behind the presentation of the everyday.

In a recent article (Christensen, 2009a) on the re-presentation of the everyday lives of Muslim families in two post-9/11 documentary films (*USA vs. Al-Arian* and *My Country, My Country*), I argued that the political power of the films was to be found in the very everyday-ness of the two families in question and the focus of the filmmakers upon the seeming banal aspects of their daily existence (making tea, birthday parties, driving, haircuts, eating, and the like):

> There is a socio-cultural *gestalt* at work where the totalizing image of the Muslim fanatic is expanded in order to absorb those within his (and it is almost always a 'his') private sphere. And herein lies the potential Brechtian power (as defined by Gaines 2002) of the presentation of the everyday in *USA vs Al-Arian* and *My Country, My Country*. The 'mere acts' of showing the Al-Arian and Aladhadh families engaged in 'mere acts' of everyday normalcy serve to undermine hegemonic Orientalist discourses. The dissenting nature emerges not simply through the content or direction of these acts [...] but also through the very *act* of participation in everyday life. The dissent, in these two cases, is not only against Islamophobia, however, but also against a US geo-politics that has transformed ordinary, everyday life for the Al-Arian and Aladhadh families into the extraordinary.

In the context of the occupation of Baghdad and the US War on Terror, the everyday served as a form of dissent, showing that, in the midst of the horrors of war and occupation (*My Country, My Country*) and the trauma of being accused of being a terrorist (*USA vs. Al-Arian*), the making of tea and the playing of games served to (at least in part) re-humanize individuals whose supposed inhumanity had formed the ontological basis for their very repression.

What, then of the soldiers on YouTube who present themselves engaged in everyday activities? Are their acts of eating, sleeping, playing and drinking also transformed from the ordinary into the extraordinary, from the trite and trivial into the dissenting? There can be no doubt that the clips uploaded to YouTube provide those who both support and oppose the US occupation of Iraq and Afghanistan with intellectual and emotional conundrums. On the one hand, the images remind the viewer of the youth and naïveté of many of the soldiers ordered to do the bidding of those who sit in Washington and London—re-injecting class into the equation of war—thus re-humanizing a group that had remained faceless cogs in the totalizing discourses of megaspectacles and militainment; on the other hand, images of soldiers playing, laughing and drinking while hundreds of thousands of Iraqis die serve to reinforce, in the most blatant of fashions, the senseless brutality and neo-imperialist overtones of their occupations.

Notes

1. Portions of this of this section also appear in Christensen (2009a).
2. Note that spelling and grammatical errors made by the viewers have not been corrected or noted with the word "sic." Errors have been left as they appeared.
3. http://www.youtube.com/watch?v=y00b9utTSuA&feature=PlayList&p=54FA5CDE379AC295&index=9
4. http://www.youtube.com/watch?v=uPXMgDb_CJc
5. http://www.youtube.com/watch?v=0VuAumiJH84
6. http://www.youtube.com/watch?v=SabTfaYsbxg
7. http://www.youtube.com/watch?v=GUZa0Bzj3I0
8. http://www.youtube.com/watch?v=HtZUd95C33w&feature=related
9. http://www.youtube.com/watch?v=j3VUmAOihqI
10. http://www.youtube.com/watch?v=R4VCRgvEoCU
11. http://www.youtube.com/watch?v=y5NTOorG-wY&feature=PlayList
12. http://www.youtube.com/watch?v=NfqSfiMXxrc

References

Andersen, R. (2006). *A Century of Media, a Century of War.* New York: Peter Lang.

Aslama, M. & Pantti, M. (2006). 'Talking Alone: Reality TV, Emotions and Authenticity,' *European Journal of Cultural Studies*, 9(2), pp. 167–184.

Cammaerts, B. & Carpentier, N. (2009). 'Challenging the Ideological Model of War and Mainstream Journalism?' *Observatorio (OBS*) Journal*, 9, pp. 1–23.

Carruthers, S. (2008). 'No One's Looking: The Disappearing Audience for War,' *Media, War & Conflict*, 1(1), pp. 70–76.

Christensen, C (2008). 'Uploading Dissonance: YouTube and the US Occupation of Iraq,' *Media, War & Conflict*, 1(2), pp. 155–75.

Christensen, C. (2009a) 'The Everyday after 9/11: Cycles and Details, '*Studies in Documentary Film*, 3(3), pp. 233–244.

Christensen, C (2009b).'"Hey Man, Nice Shot": Setting the Iraq War to Music on YouTube,' In P. Snickars & P. Vondereau (eds.), *The YouTube Reader.* Stockholm: Swedish National Archive; London: Wallflower.

Christensen, M., Jansson, A. & Christensen, C. (2011). 'Globalization, Mediated Practice and Social Space: Assessing the Means and Metaphysics of Online Territories.' In, M. Christensen, A. Jansson & C. Christensen (eds.), *Online Territories: Globalization, Mediated Practice and Social Space.* New York: Peter Lang (pp. 1–11).

Debord, G. (1967). *Society of the Spectacle.* Detroit: Black & Red.

Ellis, J. (2005). 'Documentary and Truth on Television: The Crisis of 1999.' In J. Corner & A. Rosenthal (eds.), *New Challenges in Documentary.* Manchester: University Press (pp. 342–60).

Favret, M. (2005). 'Everyday War,' *ELH*, 72:3, pp. 605–33.

Foucault, M. (1975/1995). *Discipline and Punish: The Birth of the Prison.* New York: Random House.

Fussell, P. (1989). *Wartime: Understanding and Behavior in the Second World War.* Oxford: Oxford University Press.

Gaines, J. M. (2002). 'Everyday Strangeness: Robert Ripley's International Oddities as Documentary Attractions,' *New Literary History*, 33, pp. 781–801.

Griffin, M. (2010). 'Media Images of War,' *Media, War & Conflict*, 3(1), pp. 7–41.

Kellner, D. (2004). 'Media Culture and the Triumph of the Spectacle,' *Razón y Palabra,* 39. URL: http://www.razonypalabra.org.mx/anteriores/n39/dkelner.html (viewed December 12, 2009).

Lefebvre, H. (1987). 'The Everyday and Everydayness,' *Yale French Studies*, 73, pp. 7–11.

CHAPTER TWO

The Domestication of Online Pornography

How Cyberporn Found a Home in the American Home

JONATHAN LILLIE

INTRODUCTION

My best friend, Casey, found a stack of *Playboy*s and *Hustler*s in his father's closet when we were both 10. A few years later we discovered a stack of hardcore magazines wrapped in a plastic bag in the woods. These initial encounters with pornographic texts were unlooked for and haphazard, but secondary exposures to those same texts were quite intentional. Like pirates from Hollywood movies, we hoarded these treasures in secret hiding places (tree houses and basement nooks), where the mysteries of sex were revealed in glorious detail by the adult magazine industry.

In comparing those times to the Internet era of today, it is hard to resist repeating the obvious platitudes of how different things are. One might proclaim that pornography is easier to access today than ever. It is true that with only a little effort I could pull up Web sites offering copious amounts of explicit material in a stunning variety of genres, much of it for free. But where exactly could I do this? Under what circumstances? At home? My wife knows I research Internet pornography, but my three-year-old son and his cousins are constantly running around the house. The oldest of them certainly can read the words "porn" and "sex," which makes even just reading literature on the subject something best done behind a closed door or else I'd have a lot of unpleasant explaining to do with the in-laws. I could read articles on the subject in my university office, but I would not want to be caught checking in on the state of the online commercial porn industry by a student or colleague, or violate any university guidelines for the proper use of technology. If I worked for a public university in the State of Virginia (U.S.A.), I could be guilty of violating a law on the proper use of equipment by state employees.

These examples illustrate that society's relationship with pornography might not be so different today than before capitalists and enthusiasts began populating the new cultural spaces of the Net with explicit materials and new modes of sexual interaction at-a-distance. A group of adolescent boys at a sleepover still need to craft a plan to sneak into the parents' office at midnight or cheat the blocking software on their own laptops to bask in the glow of the contemporary to my childhood's dirty porno mags.

This chapter seeks to contextualize potential everyday encounters with Internet pornography in the home. When telecommunication and computer companies brought the Internet into American homes in the mid-1990s, the press aggressively identified the threat posed by "cyberporn" to domestic harmony and safety. But far from being defenseless, the home as it turns out is well protected by several long-established regimes of social discipline—Michel Foucault called them *technologies of sexuality*—that were quickly brought to bear on the untamed sex of cyberspace. The stalwart 19th- and 20th century champions of family health, the medical professions and the state, rushed in to retool these technologies of sexuality that they created centuries ago to discipline sex and its representations. They helped to extend the science of sexuality of the American home and by doing so in a sense domesticated online pornography by creating narratives and guidelines for how this houseguest, who was both wanted and unwanted, could be tolerated, in some cases becoming a permanent resident. As digital media production tools have become easy to use and ubiquitous in middle-class homes many porn consumers have become amateur porn producers publishing snippets of their personal sex lives on public social media Web sites. In this second generation of online pornography, "WebPorn2.0," people can see other's sex and show their own sex, while publicly discussing and reacting to anything they see with text comments and reaction videos all from the comfort of home sweet home.

The Domestication of Cyberporn

Pornography, typically defined as sexually explicit media designed for arousal, was born in the lude illustrations, libertine memoirs, and risqué novels of late 17th-century Western Europe. Materials of this kind were first officially banned by the U.S. government in the Customs Act of 1842 as obscenity,[1] but for its first several hundred years of existence porn was largely the province of the secret collections of elite men (Kendrick, 1996). The 19th and early 20th centuries witnessed religious and state efforts to curtail obscene literature, as James Joyce's *Ulysses* was often defined during this period, so explicit novels and photos were still often kept secret by dealers and collectors alike. Mass production of crude periodicals and nude photos in the United States was stimulated by the enlistment of large numbers of young men in World War I, but porn was still a mostly veiled commodity due to continued prosecution of explicit materials under obscenity laws through the 1960s. A series of Supreme Court rulings on obscenity cases in 1957, 1964, 1966, and 1973 gave First Amendment protection to almost all literature and later to a wide array of, though not all, pornographic images and films. Thus, "porno mags" such as *Playboy* were able to reach a mass, mainstream audience of men (though women and children certainly sneaked plenty of peeks) in the 1960s and 1970s. In the 1980s commercial pornographers found new inroads to the American home in the form of videocassettes and cable television. By the time people started accessing nude photos and other content with dial-up modems in the early 1990s, pornography had been a staple in many homes for a long time, albeit an often hidden one.

"The cyberporn scare" in the United States lasted roughly from the mid-1990s to the turn of the millennium, as journalists, politicians, educators, and parents grappled with the question of what to do about kids' access to online pornography. A 1995 *Time Magazine* cover showing the confused, horrified face of a young child looking at a computer monitor is the most iconic representation from the scare. That *Time* cover article, "On a Screen Near You: Cyberporn," is typically cited as inciting the media and governmental outrage over what was said by many moral crusaders and politicians to be a dire threat to domestic innocence.[2] Religious leaders condemned the "cybersmut;" Congress sought to regulate it "to help parents who are under assault;" (Grassley, 1995, p. S9017) and software companies, encouraged by the Clinton White House, developed computer applications to block children from accessing offending content.

The cyberporn scare is the most recent, but not the first, moral panic over pornography's potential penetration into the home (Marwick, 2008; Potter & Potter, 2001). Only a little more than a decade earlier, for example, there had been considerable uproar in some areas over the inclusion of "adult" channels on community cable systems. In Mount Pleasant, Michigan, Cablevision had to delay its planned 1982 addition of the Playboy Channel due to public protests (Hofbauer, 1983, p. 140). Father Mortan Hill, founder of Morality in Media, Inc., went on a national talking tour "warning that 'pornography will be downstairs instead of downtown unless a stand is taken now at state and local levels' " (Hofbauer, 1983, p. 140). Numerous states and local municipalities did attempt to enact laws to limit where and how cable companies could carry sexually explicit content. The cable-porn and cyberporn scares can both be understood as technopanics, which typically have revolved around children's use of new technologies (Marwick, 2008). John

Springhall's research (1999) indicates that new information and communications technologies (ICTs) are sometimes defined as a threat to youth, often spawning initiatives by adults to censor or regulate. He concludes that, "Each new panic develops as if it were the first time such issues have been debated in public and yet the debates are strikingly similar" (1998, p. 7, quoted in Marwick, 2008).

The cableporn and cyberporn panics were both preoccupied, at least in part, with children's access to pornography via ICTs, newly introduced into millions of American homes. It is perhaps the coming together of the two often hotly contested terrains of pornography and technology that helped to create these moral panics. Though separately the incorporation of media technology and even sometimes pornography into the home has frequently occurred within much more mundane, routinized processes, especially when children are not identified as being threatened. It may not be too surprising then, though unfortunate, that research on both porn use and the uses of ICTs, particularly the Internet, has by and large ignored the spatial and place-based contexts of media use, such as the home.

The Pornographic Home

In *At Home with Pornography* (1998), Jane Juffer considers how women have over the years "carved out spaces" for the consumption of pornography within the gendered routines and boundaries of everyday life. "In the oscillation between public and private spheres, in this struggle for control," she argues, "we must analyze the conditions of access, which are determined not only by textual content but also through the publication, distribution, circulation, and reception of texts" (p. 5). When media messages and technologies are brought into the home, they are incorporated into the daily routines, domestic culture, and social relations particular to each family. Silverstone et al. (1992) map out several phases that cultural products transact within what the authors call "the moral economy of the home." *Appropriation* happens when a product is brought into the household. Many porn panics over potential threats to the innocence of women and children and the corruption of men in the United States have reacted to new pathways of appropriation of pornography into the home. The *objectification* of a media object has to do with how a household or individual identifies with it through display. Some households have television sets prominently displayed in almost every room, while others hide a single set behind cabinet doors. Storing a copy of the popular heterosexual how-to book, *Joy of Sex*, on a bookshelf in the den and explicit images saved in a password-protected folder on the office PC are both forms of objectification. The *incorporation* phase is concerned with individuals' symbolic and physical access to an appropriated media message or a technology such as the TV or computer. How is it to be used and by whom? A laptop computer for instance might be available to all members of the household when it is on the kitchen table, but only to adults or one particular adult when it is moved to the home office. Technologies in certain spaces, like bedrooms, may be off limits to some family members. Incorporation is of particular concern to Juffer's (1998) study of how and when women are able to access erotic media in the home. Incorporation is also a key element to many public discourses about families' abilities to limit and monitor children's access to certain types of media texts, such as pornography. The last phase, *conversion*, occurs when family members take public the meanings they have negotiated from cultural products appropriated into the private sphere of the home. Conversion often happens through conversations with peers, and in the case of pornography with sexual identity peer groups. With a product like pornography, which is often considered taboo and clandestinely appropriated into the home, conversion is a particularly important part of meaning-making and identity-making processes where we find out what others think about porn and how they use and experience it.

Well before 1995, both pornography and personal computers were highly regulated commodities within the domestic moral economies of many homes in industrialized regions. In the 1980s, British scholars developing the emerging field of reception studies applied extensive ethnographies of domestic media use and observed the social negotiations around a new ICT known as the "micro," or personal computer. Murdock et al. (1992), for example, observed that the authority structure and the variation in computer skills within a family help to determine parents' approaches to controlling their children's PC use (p. 157). The researchers speculated that by the mid-1980s, "There were at least four major discourses around home computing, offering competing definitions of its potentialities and pleasures:" a discourse of "self-referring practice" wherein computers offer spaces for problem-solving and creative activities; a discourse of "serious applications" associated with schooling and income-oriented work; discourses of games and entertainment; and a discourse of "righteous concern for the welfare of the young" (p. 157). Together these discourses provided the symbolic narratives that parents and their kids negotiated and struggled over through their home computer use.

Foucault's Technologies of Sexuality

The cyberporn panic of the mid-1990s did not so much introduce a new item to the list offered above as it did project these discourses of home computing more fully into the realm of childhood and adult sexuality, while also articulating a whole host of discourses and institutions that concerned themselves with sexuality to issues of home Internet use. Many of these institutions, such as the state and the medical sciences have been concerned with family sex since the 18th century. In the first volume of Michel Foucault's seminal work *The History of Sexuality* (1978), he posits that for three centuries Western society has developed a variety of institutional discourses and techniques —technologies of sexuality—designed to observe, know, and tell everything about our sex:

> Rather than the uniform concern to hide sex, rather than a general prudishness of language, what distinguishes these last three centuries is the variety, the wide dispersion of devices that were invented for speaking about it, for having it be spoken about, for inducing it to speak of itself, for listening, recording, transacting, and redistributing what is said about it: around sex a whole network of varying, specific, and coercive transpositions into discourse. (p. 34)

Foucault describes how the solitary institutional discourse of penance for excesses of the flesh in the Middle Ages (and the peasantry's everyday practicality about sex) was steadily joined by a variety of other distinct discourses of sexuality within the emerging sciences of medicine, biology, psychology, psychiatry, and demography, as well as philosophical movements within pedagogy, ethics, and political criticism. The emergence of the "population" as a specific concern of Western European states over wealth and manpower in the 18th century, for instance, required the development of demographic observation of birth rates, marriage licenses, genetic diseases, and fertility as well as medical observation of family health, the biology of sexual reproduction, and psychological disorders that were thought to cause deviance from the limited boundaries of procreation-related sexual intercourse. In particular, the sexuality of children and women were increasingly the subject of study, observation, and control. For example, Foucault explains how the healthy sexuality of boys in boarding schools became "a public problem" over the course of the 18th century that enlisted the disciplinary efforts of not just doctors, schoolmasters, and parents, but also dormitory architects and philosophers of pedagogical theory (pp. 27–29).

Sexuality for Foucault is a historical construct, a highly contested and multiplicitous modern apparatus, that rather than being forced on society by some outside power or ruling elite has developed from a bourgeois obsession over the effects of sex: procreation, health, disease, desire, and pleasure. "With this investment of its own sex by a technology of power and knowledge which it had itself invented, the bourgeoisie underscored the high political price of its body, sensations, and pleasures, its well-being and survival," he writes (p. 123). By the end of the 19th century, most of the institutional discourses of bourgeoisie sexuality, such as the medical and legal control of "perversions," had been extended throughout the entire social body for the sake of the society as a whole (p. 122). The cyberporn scare was a continuation of this discourse bent on the discipline of middle-class sexuality. The cast of characters who emerged in the mid-1990s to tame the threat of cybersmut includes the usual suspects from Foucault's accounting of the modern technologies of sexuality of the 18th and 19th centuries. Although discourses of the cyberporn scare came from a variety of institutional origins, there were two that had the most effect in redistributing techniques of sexuality in the home, (1) the state/press and (2) the behavioral/medical sciences.

Taming the Cybersmut

At first glance, the U.S. federal government's attempts to protect families from the horrors of cyberporn appear to be a series of failures. *Time's* cyberporn article was cited as evidence in congressional speeches supporting the first legislative solution to the problem, the Communication Decency Act (CDA) of 1996. But both the CDA and its successor, the Children's Online Protection Act (COPA) of 1998, were found by federal courts to be violations of the First Amendment rights of children and adults for being overly broad and vague in their attempts to make it difficult for children to be exposed to online porn.[3] Both laws in effect made most explicit media accessible via the Internet illegal, creating criminals out of those responsible for operating pornographic Web sites. For students of obscenity law, these rulings were not a surprise. In *Miller v. California* the Supreme Court had adopted a specific test for defining obscenity in 1973, which extended clear First Amendment protection to a wide array of explicit media. Anticipating the courts' reaction to the CDA in 1996, the Clinton White House started working on a solution to the cyberporn problem that was both pro business and pro family. In July 1997, the president and vice president hosted an event at the White House with business leaders and representatives from parents groups. They unveiled "a 'virtual toolbox' that would 'empower' parents to make the Internet safe for their children" (Montgomery, 2007, p. 58). The toolbox included software for computers used by children that blocked access to sexually explicit Web sites. Although the toolbox was endorsed early on by many civil liberty groups as a welcome alternative to the heavy-handed censorship of the CDA and COPA, critics of filtering-software products pointed out that almost all of them blocked sites with important educational content, such as those discussing safe sex practices and birth control. Encouraged by the PR success of the digital toolbox, Congress put filtering software at the center of its next attempt in 2000 to protect children from cybersmut with the Children's Internet Protection Act (CIPA). CIPA required public schools and libraries to install filtering software on computers.

The mainstreaming of filtering-type software is one of the most direct effects of the cyberporn scare. Today many parenting-focused Web sites and news articles recommend that parents use such software, at least in some capacity, to structure their kids' Internet use. Some applications allow parents to monitor what sites children are browsing rather than prohibiting access. This functionality speaks to the now cliché parental complaint: "I have no idea what my kids are looking at online." Surveillance of the kind facilitated by monitoring software is at the core of how Foucault theorized the production of power within many modern discourses, particularly discourses of sex where knowledge is directly tied to the researcher, therapist, doctor, or parent's ability to observe the sexual behavior of patients or children. Blocking/monitoring software can also facilitate what Foucault described as panoptic power. A panoptic power arrangement is one where the observee knows she is being monitored by authority and therefore modifies her behavior accordingly to avoid being disciplined. The physical and social structure of many institutions, such as schoolrooms, applies panoptic power arrangement, as do certain situations in the home. The use of baby monitors in the bedrooms of young children and "open bedroom door" policies for older children entertaining guests are just two examples. Foucault's description of panoptic power has been used by many scholars to critique the potential use of ICT hardware and software to surveil people's private lives (see, for example, Engberg, 2006), but long before NetNanny and SurfWatch panopticism has been a prime tactic within the moral economy of the home, perhaps since the emergence of the nuclear family and single-family home. Filtering and monitoring softwares address the rhetoric of the cyberporn scare that bemoaned a loss of control or knowledge about the daily media diet of kids, even though the type of power these digital sentries produce has been in the home's toolkit all along (even if some parents don't use it).

While legislators and software programmers were designing solutions to cyberporn, medical practitioners of psychology and psychiatry started to apply their knowledge of sexual psychoses to those who were afflicted with "problems" associated with viewing online pornography. Since the 18th century therapists had engaged in defining and treating the different deviancies, and in the mid-1990s their contemporaries sought evidence of different types of cyberporn addiction (see Cooper et al., 2000; Schwartz & Southern, 2000). As they did 200 years ago, the medical sciences sought to reveal the mysteries of sex. Psychology researchers initiated studies to extract knowledge of how much time people spend online, what types of sites they are looking at, and how consumption of online pornography affects their sexual identity and behaviors. Cyberporn addiction groups such as No-Porn.com began to offer online support forums that advocate a variety of behavior-avoidance techniques including self-disclosure and setting up blocking or monitoring software on home computers. In fact, expert advise for coping with cyberporn addiction shares many of the same techniques offered for protecting children such as keeping Internet terminals in non-private rooms; discussing browsing experiences and exposure to pornography as a family; discussing sex-related issues as a family; and installing blocking or monitoring software on all Internet terminals. These techniques of observation and self-confession are designed to produce knowledge, and through knowledge sex can be controlled and disciplined.

HOME WITH WEBPORN2.0

In *At Home with Pornography*, Juffer (1998)writes that, "We are now certainly at the point where we can move beyond reacting to anitpornography feminists and articulate a politics of pornography that does a more comprehensive job of addressing issues of production, access, and consumption, one that includes but is not limited to the questions of the subversive powers of individual consumers" (p. 14). This quote provides a good segue to consider how online pornography has changed since the quieting of the cyberporn scare. The emergence of Web2.0 has pushed consumers to the forefront of online culture as popular producers. Some Internet researchers, including those interested in porn and its subcultures, have been quick to examine this phenomenon which seems tailor-made for media studies' preoccupation with the endless production-consumption cycle of cultural products and meanings. Similar to the work done so far on Web2.0, much of the popular and scholarly discussion of WebPorn2.0 hypes the potential for user empowerment through personal publication and notoriety via social media sites. Two popular WebPorn2.0 sites include the social media site YouPorn, where users upload, watch, and comment on explicit videos, and the social networking site AdultFriendFinder.

Some academics and activists have sought to carve out a particular genre of WebPorn2.0 called "NetPorn" (see Attwood, 2007; Jacobs, 2007; and the *C'Lick Me Netporn Reader*, 2007), which they define as, "as grass root level activities, gift economies, and performative exchanges—or, as the editors of the *C'Lick Me Netporn Reader* put it, 'alternative body type tolerance and amorphous queer sexuality, interesting art works and the writerly blogosphere, visions of grotesque sex and warpunk activism'" (Paasonen, 2010, p. 1298). Susanna Paasonen (2010) argues that while NetPorn-focused scholarship[4] has lauded (often critical and political) amateur pornographic expression/dialogue that has emerged through the use of Web2.0 tools, some of this research attempts define NetPorn by differentiating it from "mainstream" online pornography. Noting that many of these binary-like differences are false or oversimplified, Paasonen explores "how the conceptual divisions drawn between amateurs and professionals, the non-commercial and the commercial, the alternative and the mainstream are played out in relation to definitions of netporn, alt porn and amateur pornography, Web 2.0 platforms and their participatory cultures" (2010, p. 1299). She uses the site *Beautiful Agony* (http://www.beautiful agony.com) as an example of "artistic erotica" attempting to detach itself from the visual codes of the commercial porn industry. The site invites users to upload videos clips showing their faces while they reach orgasm, thus drawing attention away from porn's favorite body parts. The site's content is not free, however, charging a monthly membership fee to watch the videos, and is owned by Gmbill, an Australia-based company that owns and runs several mainstream, commercial, fee-based, pornographic Web sites. Still, unlike the cyberporn of the 1990s, hybrid commercial/amateur sites like *Beautiful Agony* and *Suicide Girls* (http://suicidegirls.com), which features amateur and semi-pro 'alternative' nude female model photos, rely heavily on user-submitted content mostly shot and produced in these submitters' homes. In a critical analysis of *Suicide Girls* Matthew Wysocki (2010) argues that the site "is pornography made by women that believe they are free to choose how to express their sexuality....They are doing so because they believe that what they are creating is, in fact, a statement of their alternative nature" (p. 13). If the home was not a safe, or at least available, space to explore such creativity, it is doubtful that these Web sites could exist.

Reaction Videos, Web Sex Chat, and Conversion

While NetPorn and much mainstream online commercial pornography have different histories and very different ideological agendas, they can be understood as somewhat similar technologies of sexuality. Each is in its own way preoccupied with making sex reveal its various "truths" (though they may disagree on what these truths are). This is also the case for many non-activist Web2.0 users. NetPorn activists and erotic artists represent only a small fraction of the large population of people who have used Web2.0 tools to upload explicit photos or videos of themselves or their friends, offering up their own bodies to join countless others in the public archives of sexual knowledge. As Paul Booth points out: "It is no exaggeration to say that the production and distribution of online amateur pornography has revolutionized the porn industries across the board" (2010, p. 12). Rather than being ascribed as an explosion of crude exhibitionism, such behavior is better understood as attempts to negotiate, make meaning of, and rearticulate the confusing array of discourses of the sexual body. Silverstone et al.'s (1992) concept of conversion helps to explain how home Internet users negotiate their own understandings of the pornographic media they consume by having online discussions about porn and sex and by publishing their own creative expressions via Web2.0. As a cultural and social, more so than a technical, phenomenon, Web2.0 has opened up a huge space not just for archiving sex but also talking about it.

Although most legacy Web2.0 sites like YouTube and MySpace prohibit publishing sexually explicit multimedia content, users still find plenty of ways to talk about sex and porn. For example, many of the most popular "reaction videos" posted to social media sites like YouTube show individual's and group's reactions to shocking online porn videos that have gone "viral," meaning they have become very popular and have been copied and spread all over the Net. A series of works by artist Nia Burks explores the cultural phenomena of shock porn group reaction videos on

YouTube. Her video art pieces combine six of the group reaction videos at a time so that they play simultaneously aligned side-by-side (Figure 1). Burks (2010) notes that in her pieces, groups of young men "are judging that which they have sought out and gained access to. Each set of boys have also hit record on their recording devices and are aware of what they are doing; a performance of the male gaze."[5] Not unlike the group viewing of found porno mags from my childhood, the young men in the videos are also performing their sexual identities to one another, and in these moments of performance (at several levels) they are negotiating their individual and group understandings of sexual media. Social performance after all doesn't just show others our identity of the moment. It also helps us to understand, maintain, or alter our identity, in this case as it pertains to sex and sexual media.

Figure 1. Nia Burks' video art piece, "Reaction 1" (2010).

In late 2007, YouTube users started posting reaction videos to an online trailer for a Brazilian scatt-fetish film that came to be known as "2 Girls 1 Cup." While the trailer is quite grotesque by most standards, intellectually it is not nearly as interesting as the reaction videos or the pop culture foment that they inspired. The trailer and reaction videos were discussed on many popular blogs and in the mainstream press. George Clooney was asked to watch and comment on the trailer by *Vanity Fair*, and the YouTube videos have been spoofed in many venues such as the animated cable-TV comedy series *Family Guy* (see Figure 2). On YouTube you can watch several thousand individuals' and groups' reactions to watching the trailer. Some of the more popular "2 Girl 1 Cup" reaction videos on YouTube have been watched over 5 million times. As in Figure 1, you only need to watch a few of these videos to note the striking similarity in most responses. You'll see many disgusted, shocked, and disturbed faces. These mini commentaries on pornography are not as sophisticated as most NetPorn media, but they still offer a socially significant message: Watch me watch porn. Watch me as I give a somewhat scripted yet also genuine performance of mainstream sexual values. These performances, though not themselves shocking or surprising, validate a personal right and interest to watch, react to, and talk about sex, and to do so in a very public form. They also perform social rituals of disgust at socially taboo sexual behaviors (there are several in the trailer, including two young women eating feces). Those reaction videos that show the responses of groups of people also validate group (and perhaps also individual and private) viewing and negotiation of pornography. For example, in "My Chinese

Parents Watch 2 Girls 1 Cup," a YouTube producer filmed her parents' reaction to the trailer. Reaction videos here become an accepted and constructive form of conversion, in this case inter-generation acknowledgement, consumption, and negotiation of explicit media.

Figure 2. Brian watches Stewie watching "2Girls 1 Cup" in the *Family Guy* episode "Back to the Woods," which first aired February 17, 2008 on Fox.

Conversion processes within sexual identity peer groups, regardless of whether they are virtual or face-to-face groups, certainly mediate individual's future appropriation, objectification, and incorporation of online pornography. In the Web2.0 era, even those sites and services catering to more traditional, single-viewer, masturbatory practices encourage participants to communicate through comments, text chat and other forms of interaction. Online Web Sex Chat (WSC) services, for example, allow users to watch live video of amateur or professional sex performances while simultaneously using text chat to converse with other observers. In a recent study of WSC, Paul Booth (2010) notes,

> More so than merely receiving visual stimulation, the viewer of the WSC can see and understand the instructions and musings of other people watching the same WSC around the globe. This form of interactive communication helps to develop what Henry Jenkins (2006)...described as an online 'participatory culture,' in which consumer/users can not only enjoy texts, but also create their own texts to share with others. (p. 4)...By looking at Web Sex Chats, we can have a unique insight into the mechanized way sexuality functions in

> online pornography, and the way that performance and participation are inextricably linked. For, while there is a clear-cut delineation between the performer and the audience, the interactive nature of the chat belies a more complicated relationship between the two. (p. 14)

In most WSC environments, observees can use text chat to ask the live video performers to perform certain acts. Observess can also chat with other observees or just read the other observees' interaction with performers and their reactions to the performances. Similar to the reaction videos, WSC involves discussion and performance of sexuality at several levels. Social and individual negotiation are meshed into the same experience with performance. A WSC session in the home bedroom or office could involve appropriation, objectification, incorporation, and conversion of a media text all at once, though these are processes that typically would not happen together. Thus WebPorn2.0 is not only a "frenzy of the visible" as Linda Williams (1999) described over the top hardcore porn videos, but also a frenzy of texts, a frenzy of negotiations of meanings, and often a frenzy of conversion. Adult video sharing site like YouPorn, which at first glance does not seem as interactive and hypermediated as WSC, allows users to post comments on each video. So even if a person sitting at home browsing YouPorn does not post, he or she can still see what Catplay21 and DirtyNik think of this or that video. Through Web2.0, the connected home becomes an Internet Porn Café where people can connect to their favorite site, listen in on what others are saying, or talk directly with others about what they are doing.

Conclusion

Booth (2010) argues that services like WSC are technologizing sexuality. Here again we see sex in the home being disciplined through techniques of observation and control, not unlike blocking/monitoring software for kids and cyberporn addicts (I go into more detail on this in Lillie, 2004). But Web2.0 publishing tools and social media networks also facilitate a technology of sexuality in a different way, one not completely unlike the libertine memoirs of 18th-, and 19th-century Western Europe, wherein horny, erudite gentlemen of privilege would recount their clandestine encounters with prostitutes and other sexual escapades. Such memoirs could only be published at significant expense for a limited, elite audience. Today, however, it has never been easier to talk about your sex through online forums, or let your sex talk for itself though uploaded pictures and videos. As in the reaction video and WSC examples, people also use Web2.0 to talk about others' sex. It is possible that none of this could occur if the cyberporn scare did not place online pornography on the mainstream cultural map. The message might have been: It's here, and it's bad! But in the drawn-out process of letting everybody know about it, online porn became somewhat normalized. As van Doorn (2009) argues: "pornography has been involved in a 'mainstreaming' process over the past decade…[S]imultaneously, the public discourse on sex and sexuality has grown exponentially." Foucault observes how sundry discourses of sexuality espouse a veil of silence and prudishness towards sex while at the same time positioning people to seek knowledge about it, observe it, and talk about it. The rhetoric of the cyberporn scare asked society to wall up and hide pornography, but ended up forcing people to accept it and engage it more directly, whether it is by talking about it, joking about it, actively seeking it, or actively avoiding it.

Moral panics such as the cableporn and cyberporn scares can spawn periods of reconfiguration, or redeployment, of various technologies of sexuality, often as discussed earlier in the form of scientific analysis and panoptic scrutiny. Desires and techniques to speak about sex can also emerge as new social and cultural spaces are opened up for use by certain groups. The technologies of sexuality deployed during the cyberporn scare helped to define what could be done in these spaces and by whom, and how it could be controlled or at least better understood and accepted (in certain homes, at certain times, in certain rooms, with certain types of media texts). As opposed to the early and mid-1990s, when WebPorn2.0 came along several technologies of sexuality of the home were already available to help deal with this newcomer. While there are no quantitative studies on the subject, anecdotal observation suggests that the widespread use of Web2.0 social media networks and easy-to-use digital media production technologies have encouraged many people to display themselves online in their birthday suits. Only a decade after the decline of a moral panic over easy domestic access to online porn, the home is now not just the scene of so much consumption of pornography, but also a popular site of amateur production and publication of (mostly non-commercial) sexually explicit media. As Juffer (1998) points out, in some cases "domestic bliss can perhaps best be achieved through the proliferation of information to sexual pleasures and practices. The pornographic home may well be the real site of family values" (p. 39). Juffer's pornographic home has moved beyond merely carving out safe spots and techniques for porn consumption by different family members to invoke domestic media production as a form of creative expression and dialogue via the Net.

Web2.0 tools facilitate social critique (and meaning negotiation) of pornography and sexuality. The reaction videos function through a technology of sexuality that has perhaps been revamped or redoubled through the Net apparatus, but which also invigorates older techniques of observation of the body. Through reaction videos we observe ourselves and invite others to observe us observe the explicit bodies of others. These videos constitute a forum for public critique and analysis on pornography that seems much more productive and grassroots than so much of the media hype and congressional rhetoric. In *Formations of Pleasure* (1983) cultural theorist Fredric Jameson reminds us that moments of pleasure (*jouissance*), whether mundane or sublime, as well as our reactions to our own pleasure or the pleasure of others, always have political significance. He argues that we must "make it impossible to treat [jouissance] except as a response to a political and historical dilemma, whatever position one chooses (Puritanism/hedonism) to talk about that response itself" (1983, p. 9). Responses to pornography can span a great gamut between extremes of Puritanism and hedonism, but always have political contexts and significance. The reactions to "2 Girls 1 Cup" on YouTube of are often quite strong, but they seem to object more so to the taboo behavior depicted in the trailer than to the fact that it is sexually explicit. The first few seconds of the trailer, showing a woman taking off her shorts, does not solicit strong reactions; it's what happens next that gets people stirred up. Some sexual taboos still seem to have high cultural capital, but for most as shock entertainment rather than as moral outrage.

The contexts of the home are evident in the reaction videos depicted in Figure 2 as well as many amateur video uploads to sites like YouPorn. For many, though not all, the home has become the safest actual space of sexual exploration because its moral and sexual architectures are typically well known to its occupants. This safe, private place is now linked to the external, public sphere of the Internet, which for many people offers a sense of anonymity as well as intimacy since it can be accessed from the confines of the home. Many people do things on the Net that they would not do within other public spaces they frequent, like the front lawn or grocery store. The secreted porn collections of adolescents and adults, as producers or consumers, can now be stored in the seemingly safe zone of the Net, which, in its own way, seems to promise privacy, safety,[6] and publicity at the same time.

The prevailing discourse of the pornographic home has changed considerably since the beginning of the cyberporn scare. The dominant narrative of 15 years ago is well represented by the *Time Magazine* cover. The young boy exposed to cybersmut is alone, confused, scared, damaged. He is also lost in cyberspace. Unlike most of the reaction videos or the *Family Guy* skit (Figure 2), the physical contexts of home and family are not shown or available to him. Of course the *Time* boy is not real. Even without filtering software and congressional lawmakers to protect them, most real children have parents, siblings, friends, teachers, and community members who can (we hope) arbitrate encounters with sexually explicit media. *Time's* narrative contrasts with the fictional depiction of *Family Guy's* child character, Stewie, being exposed to pornography. Although the producers of *Family Guy* do not claim to or attempt to represent reality in their comedic portrayal of American family life, their faux "2 Girl 1 Cup" reaction scene is a deadpan spoof of the YouTube videos. It is also much closer to actual reality, where people encounter pornography in the moral economies of the home, and they make meaning of these encounters through conversion within face-to-face and online social negotiation. Brian, the *Family Guy* dog, is neither totally innocent nor totally malevolent, when he shows the trailer to Stewie. He seems to want to talk about it, to see Stewie's reaction, and to enjoy the shared, extreme mediated experience. Brian's role in this scene then, is not unlike that of Casey, my best friend when I was a child, who introduced me to the adult magazine genre. Moments of exposure to sexually explicit media are always charged, but they are also firmly grounded in the contexts of reception. In some homes, the context might not have changed much from decades ago when adolescents (and adults) in small groups or alone engaged secret collections behind locked doors. Today, however, we don't just collect media objects. We mash them up into new compositions, and sometimes we produce digital media documenting our private lives. And then some of us opt to press the "publish" button. Thus, the current era of WebPorn2.0 does not so much represent, as some contend, the sexualization of culture, as it involves the centuries-old process wherein we as a culture and as individuals continue to find ways to mediate, remediate, and talk about our sex.

Notes

1. As a legal term in the United States, obscenity is defined as explicit materials deemed to lack the qualifications for enjoying First Amendment protections.
2. The *Time* article drew heavily on a study by Marty Rimm about pornography on Internet bulletin boards published in early 1995 in the *Georgetown Law Journal*. The study was widely criticized for very poor methods and gross over exaggeration of the number of explicit images on the Net. The *Time* article and the Rimm study, however, caused a media uproar (138 magazine and newspaper articles were published on the topic between June 1995 to July 1996, Panepinto, 1998) and directly influenced congressional action leading to the passage of the Communication Decency Act in 1996.
3. See Marwick (2008) for a short history of the CDA and COPA.
4. Much of the scholarly foment around "NetPorn" emerged from two conferences held in Amsterdam, "The Art and Politics of Netporn" (2005) and C'Lick Me" (2007). Many of the papers from the second conferences where published in the *C'Lick Me Netporn Reader* (2007).
5. While Burks' "Reaction" series focuses on reaction videos with groups of young men, it is important to note that many reaction videos show groups of young women, groups of men and women together, and multigenerational groups.
6. It is interesting to note that most of the current concern over criminals befriending children online has to do with kids being lured away from the safe zone of the home.

REFERENCES

Attwood, F. (2007). 'No Money Shot? Commerce, Pornography and New Sex Taste Cultures.' *Sexualities* *10* (4), 441–56.

Booth, P. (2010). Participatory Porno: The Technologization of Sexuality. *NMEDIAC: The Journal of New Media & Culture, 7* (1).

Burks, N. (2010). Reaction 1. *NMEDIAC: The Journal of New Media & Culture*, 7 (1).

Cooper, A., Delmonico, D., & Burg, R. (2000). Cybersex Users, Abusers, and Compulsives: New Findings and Implications. In A. Cooper (Ed.), *Cybersex. The Dark Side of the Force* (pp. 5–29). Philadelphia: Brunner Routledge.

Engberg, D. (2006). The Virtual Panopticon. Retrieved from http://128.122.253.144/impact/f96/Projects/dengberg/

Foucault, M. (1978). *The History of Sexuality, Volume I: An Introduction*, translated by Robert Hurley. New York: Vintage.

Grassley, C. (1995). Cyberporn, *Congressional Record, 141* (26 June), pp. S9017–S9023; from the Congressional Record Online via GPO.

Jacobs, K. (2007). *Netporn: DIY Web Culture and Sexual Politics.* Lanham, MD: Rowman & Littlefield.

Jameson, F. (1983). Pleasure: A Political Issue. In Fredric Jameson (Ed.), *Formations of Pleasure.* London: Routledge.

Jenkins, H. (2006). *Convergence Culture: Where Old and New Media Collide.* New York: New York University Press.

Juffer, J. A. (1998). *At Home with Pornography: Women, Sex, and Everyday Life.* New York: New York University Press.

Kendrick, W. (1996). *The Secret Museum: Pornography in Modern Culture.* Berkeley: The University of California Press.

Lillie, J. (2004). Cyberporn, Sexuality, and the Internet Apparatus. *Convergence: The International Journal of Research into New Media Technologies, 10* (2).

Marwick, A. E. (2008). To Catch a Predator? The MySpace Moral Panic. First Monday,13(6).Retrieved from http://firstmonday.org/htbin/cgiwrap/bin/ojs/index.php/fm/article/viewArticle/2152/1966

Montgomery, K., C. (2007). *Generation Digital: Politics, Commerce, and Childhood in the Age of the Internet.* Cambridge, MA: MIT Press.

Murdock, G., Hartmann, P., & Gray, P. (1992). Contextualizing Home Computing: Resources and practices. In Roger Silverstone & John Hirsch (Eds.), *Consuming Technologies* (pp. 82–92). London: Routledge.

Paasonen, S. (2010). Labors of Love? Netporn, Web 2.0, and the Meanings of Amateurism. *New Media & Society, 12* (8), 1297–1312.

Panepinto, J. (1998). *Policing the Web: Cyberporn Moral Panics, and the Social Construction of Social Problems* (Doctoral dissertation, University of Massachusetts-Amherst).

Potter, R. H., & Potter, L. A. (2001). The Internet, Cyberporn, and Sexual Exploitation of Children: Media Moral Panics and Urban Myths for Middle-Class Parents? *Sexuality & Culture, 5* (3), 31–48.

Schwartz, M. F. & Southern, S. (2000). Compulsive Cybersex: The New Tea Room. In A. Cooper (Ed.), *Cybersex: The Dark Side of the Force* (pp. 127–44). Philadelphia: Brunner Routledge.

Silverstone, R., Hirsch, J. & Morley, D. (1992). Information and CommunicationTechnologies and the Moral Economy of the Household? In Silverstone, Hirsch (Eds.), *Consuming Technologies* (pp. 17–31). London: Routledge.

Springhall, J. (1999). *Youth, Popular Culture and Moral Panics: Penny Gaffs to Gangsta–Rap*, 1830–1996. New York: St. Martin's Press.

van Doorn, N. (2009). Keeping It Real: User-Generated Pornography, Gender Reification, and Visual Pleasure. Paper presented at the *Annual Conference of the Association of Internet Researchers*, November, Milwaukee.

Williams, L. (1999). *Hard Core: Power, Pleasure, and the Frenzy of the Visible.* Berkeley: University of California Press.

Wysocki, M. (2010). Alternative to What? Alternative Pornography, Suicide Girls and the Dominant Gaze. *NMEDIAC: The Journal of New Media & Culture, 7* (1).

CHAPTER THREE

Fans Online

Affective Media Consumption and Production in the Age of Convergence

CORNEL SANDVOSS

Being a fan has come a long way in the Internet age. Far removed from the days in which the word "fan" carried connotations inviting social stigma and pathologisation (Jenson, 1992; cf. Gray et al., 2007), in the contemporary world of social networking, being a fan has become one of only two fundamental modes of engagement with the world around us: on Facebook, we can link to others either by being their friend or their fan—the former, mirroring Thompson's (1995) analysis of fandom, requiring the reciprocal approval, the latter being a non-reciprocal statement of affect.

As much as the collection of long forgotten high-school class mates, dear colleagues, not-so-dear relatives, past and present acquaintances and former students rarely qualifies as who we label as friends in the offline world, becoming a fan of something or someone on Facebook is often a fleeting, ephemeral statement of simply expressing a like or dislike. Yet it also illustrates the degree to which being an affective and involved audience member—and sharing our thoughts on the media texts we consume—has become part of our quotidian routines. As mediated content multiplies, audiences' affective attachment to given narratives and icons is an ever more commonplace reading position. Dig deeper into friends' status updates on Facebook, and various references to the latest sports event or concert attended, a favourite song, excited anticipation of the airing of new episodes of one's favourite show, links to YouTube or newspaper articles can to be found. In few places is the permanent presence of (mass) media in our everyday life performances that

Abercrombie and Longhurst' (1998) describe as a condition of "diffused audiences" as visible as in the world of social networking.

This visibility of fans online—populating news groups, online fora, video sharing portals and fan created websites—has led some inside and outside academia to misinterpret contemporary fan practices as a consequence of technological change. Aiming to assess the interplay between (fan) audiences and their practices with the diffusion and subsequent transformation of the Internet, this chapter documents the different forms of fan productivity online, and in juxtaposing these with offline fan activity argues for a revaluation of audiences' impact on the shape and structure of the Internet.

As versatile, the converging medium, as wide is the range of different fan activities we find online. Beyond the "one-click fandom" on Facebook, the diversity of these practices necessitates the further qualification of how they cluster among different fan groups. Two definitions are thus required here: Firstly, the emphasis on fans' productivity in online spaces warrants further attention to the different types of productivity in "participatory culture" (Jenkins, 2006b). John Fiske (1992) distinguishes between three types of productivity among fans: semiotic productivity—the act of reading/watching/listening and thereby meaning construction on behalf of fans in encounters with the object of fandom and texts surrounding an object of fandom; enunciative productivity— which describes the interactions between fans surrounding their fan object such as fan talk, gossip, banter and speculation; and textual productivity—found in fans' creation of new texts such as fan fiction, music remixes, or fanzines.

These three types of productivity in turn correspond with the three groups of fan audience Abercrombie and Longhurst (1998) suggest: fans, cultists and enthusiasts.[1] "Fans" they argue (1998, p. 138) are "those people who become particularly attached to certain programmes within the context of relative heavy media use. They are individuals who are not yet in contact with other people who share their attachment." Their object of fandom is thus broader than the more specialised fan objects among cultists. Among football fans, for instance, fans thus have a general interest in the game or in particular competitions, whereas cultists follow a particular team (and watch little other football when their team is not playing) (Sandvoss, 2003). Given the specificity of their fan object, cultists' media use is more discriminatory and focused, often embedded in the social context of membership to a lose network of fellow fans. Such social organisation in turn plays a pivotal role for enthusiasts whose fandom focuses less on a mediated object but their own activity supported by highly specialised media use. Enthusiasts of particular television shows or novels thus take to writing their own sequels, music fans qualifying as enthusiasts record their own music, or sports fans focus on the home and away support of the a team in which the travel, chanting and choreographies become more important than the matches they attend. My argument is that these three types of different productivity (semiotic, enunciative and textual), which in turn reflect different clusters of fan behaviour, inform fans' participation in and appropriation of online spaces in the production and consumption of popular culture.

Textual Productivity

The heightened visibility of fans textual productivity online bears the risk that we too quickly gravitate towards technologically deterministic conceptualisations of the transformations of fan activity in which the Internet is identified as the primary factor in the rise of fans' textual productivity. In contrast to much of the "first wave of fan studies" (Gray et al., 2007) such as Henry Jenkins' (1992) study of fans of *The Beauty and the Beast* and *Star Trek*, Camille Bacon-Smith's (1992) parallel work on the popular American science fiction programme and much subsequent work focusing on forms of fan fiction and 'slash' literature, circulated and exchanged through fanzines, mailing lists and fan conventions, all demonstrate that these forms of fan productivity precede the proliferation of the Internet into a widely available household communication technology. Similarly, textual and enunciative productivity in sports fandom predates the rise of the Internet as Richard Haynes' (1995) study of fanzine culture documents. Fans' creative work—today often centring on remixed and re-edited video clips on YouTube—equally precedes digital media as fan-inspired artwork from paintings and sculpture, such as those produced by fans in memory of Elvis Presley (Doss, 1992) or songs (see Jenkins' [1992] discussion of filk). Yet, while the Internet is no premise to fans' textual productivity, it has nevertheless transformed its articulations and uses in three important ways.

Firstly, the Internet has made textual productivity visible to media professionals involved in the production of texts ranging from writers and producers of television shows via pop stars to athletes and other celebrities (who ultimately are authors of the fan text that is their public persona). Legally, this visibility has put online fan producers at the risk of litigation. As Rebecca Tushnet (2007, p. 60) observes "some copyright owners have […] taken an aggressive stance against fan creativity" threatening fans with lawsuits and "cease-and-desist letters." Yet, Tushnet equally notes, the greater visibility and availability of fan fiction has led to greater degree of normalcy, in which fan producers feel increasingly less threatened by copyright legislation.

More commonly then, fans' online textual productivity either exists in parallel to the original fan text, further diversifying the textual field within which fan objects are read (a point we will return to later) or facilitates a dialogue between fans and professional producers—a dialogue not without contention. As Williams (2008) documents, relations between fans of *Neighbours* (Seven/Network Ten, 1985–present) and a writer of the show on a fan forum turned acrimonious after the latter attacked fans who had criticised recent overdramatic storylines which failed to resonate with what fans regarded as *Neighbours'* narrative and stylistic canon. Similarly, Williams (2008) and Andrejevic (2008) both report the case of Aaron Sorkin, creator of *The West Wing* (NBC, 1999–2006), who "after a heated exchange with critical fans" on one of the most widely used television fan fora in the United States, Television Without Pity (TWoP), scripted an episode that "according to a *New York Times* account, portrayed hardcore Internet users as obese shut-ins" (Andrejevic, 2008, p. 27). The hostility between fans and producers in these cases thus contrasts with many instances of and interactions between fans and producers that is deemed beneficial by both parties. Lancaster (2001), for example, details one of the early instances of online exchanges between creator/producer J. Michael Straczynski and fans of his show *Babylon 5*. As in other cases since, fans offered welcome feedback to producers in their efforts to produce (commercially) successful texts and performances. Even among *The West Wing* fans on TWoP, Sorkin's attack was regarded as a sign of acknowledgement of their own textual work in relation to the show, similar to the case of "a recapper for a show on WB network called *Popular* [who] noted that in a backhanded gesture of recognition for his praise of the show, the writers named a character after him: a junkie who was killed in a car accident" (Andrejevic, 2008, p. 27).

However, to many television writers, actors and producers, perpetually threatened by the cancellation of their respective shows, a thriving online fan culture functions not only as a quasi-focus group, but also as a form of support in negotiations with studios and broadcasters (see Menon, 2007). In addition, as Andrejevic (2008, p. 27) argues, "market and production imperatives such as show promotion, mass audience appeal, and technical details are taken up in depth by TWoP posters, who, in elaborate postings suggest ways to more effectively tailor a particular show to its viewers."

Fans' textual productivity then, even if pursued without commercial motives or professional aspirations, constitutes as a form of labour. This contribution of fans' labour to the professional success of texts or artists is documented further in other areas of popular culture. Nancy Baym (2007), for instance, details the role that fans of Swedish independent musicians such as *Peter, Bjorn and John; The Knife;* or *José Gonzáles* play in the international dissemination of their music. Through lobbying, including reviews of new releases, social networking ranging from profiles and playlists on Last.fm that further publicise these bands to a wider public sharing similar genre preferences on Facebook, the participation in fan fora and classic textual productivity such as user generated remixes or videos, fans' performers serve as unofficial and unpaid publicists to these bands. According Baym and Burnett (2009, p. 434) "these fans serve as expert filters as they sift, sort, label, translate, rate and annotate a large, disorganised, and geographically remote set of cultural materials for international consumption." Tracing the international distribution of Swedish independent music, Baym and Burnett (2009, p. 434) argue that in the participatory culture of Web 2.0 "user generated content produced by a fans stands alongside professionally produced content in claiming audience attention" (cf. Benkler, 2006; Jenkins, 2006a) and note how both musicians and independent labels recognize and value the importance of fans' textual productivity in promoting them in the global—and thus to the small cultural producer also overwhelming—marketplace of online music. Baym and Burnett thus echo Jenkins' (2006a, p. 138) observation that "media industry is increasingly dependent on active and committed consumers to spread the word about valued properties in an overcrowded marketplace, and in some cases they are seeking ways to channel the creative output of media fans to lower their production costs." Given that fans' textual productivity thus becomes a vehicle of profit accumulation, Baym and Burnett—in reference to Hardt and Negri's (2000) and Terranova's (2000) early critical assessments of user generated content in online spaces as a form of 'immaterial labour' and thus the extension of principles of capitalist production in which consumers are increasingly unpaid labourers (a general tendency of late consumer capitalism, cf. Ritzer, 1996)—pose the question whether textual productivity online constitutes a form of empowerment or exploitation.

Following a similar line of enquiry Milner (2009) examines the labour contribution of fans of the computer role-playing game series *Fallout* in the programming and design of the third instalment of the popular title. Following a ten-year hiatus since the release of *Fallout 2*, the game licence had been purchased by a new software developer, Bethesda Softwork, who initiated an official *Fallout 3* forum during the development phase of the new game, utilising long-standing fans input in seeking to create a successful sequel of a title previously written by a different developer. Like the regular posters on TWoP or the enthusiast for Swedish independent music, the fans on *Fallout 3* forum constituted an unpaid but valuable workforce that could serve as "developer, public-relations specialist, focus group, technical support, journalist, and consultant all in one" (Milner, 2009, p. 497). And like other fan groups, *Fallout* fans contributed dedicatedly and enthusiastically, offering their labour with little or no expectation, nor, for the most part, for financial compensation. Enthusiasts in any realm of cultural production are thus confronted with the inescapable logic of the capitalist system: fans whose textual productivity holds commercial value to third parties, yet wish to opt out of the system of monetary exchange and preserve their fandom as non-commercial space, inevitably open themselves to the exploitative utilisation of their productivity by others. Only if they subscribe to the principles of capitalist exchange in the first instance, are fans able to avoid such exploitation, yet thereby erode the pleasures the fan–object of fandom relationship based on control and appropriation of fan texts that operate outside such principles and that cannot accommodate the material separation between fan and fan object manifested through monetary exchange (cf. Sandvoss, 2005a). Further qualifying Terranova's assessment, Milner (2009, p. 501) therefore adds that "I would have to amend the list with what proved to be fans' greatest pleasures: the best interest of the text they esteem and of the fan community they belong to."

This qualification that fans' productivity was "not for Bethesda's benefit […] These self-motivated knowledge workers were more loyal to *Fallout* than to Bethesda" (Milner, 2009, p. 504), and that hence the potential for conflict between the fans and Bethesda lay in the fans' feeling of not being co-opted sufficiently in the production process, I think, is lost in other assessments of the limitations of the emancipatory potential of fans' textual productivity. Andrejevic (2008, pp. 39–40) implies a false consciousness in television fans' (mis)recognition in and identification with media producers and by extension the culture industry as they use online fora assessing "continuity, plot and character development, in makeup and lighting, and even in publicity and promotional material":

> the danger, in a savvy era, is that the goal of self reflexive knowledge is not so much to reshape the media—to imagine how things might be done differently—as it is to take pleasure in identifying with the insiders. The next best thing to having power, on this account, is identifying with those who do rather than naively imagining that power can (or should) be redistributed or realigned.

While an important reminder that we cannot equate audience productivity and empowerment *per se*, Andrejevic's assessment is, I believe, problematic on three different counts: Firstly, Milner's observation that fans distinguish between text and producer seeking to enhance the former even when taking on *de facto* tasks of the latter, questions Andrejevic's equation of fans' affection for a given fan text with an identification with the producer and his or her goals. While some fans in Milner's (2009) study express the desire to eventually become professional game-developers, such desire is tied to fans' concern over preserving the quality of the text: in the words of one poster on the *Fallout 3* forum: "I don't like it. I could do better. Bethesda, hire me" (cited in Milner, 2009, p. 501). Jenkins's (2006a) rich explorations into various forms of fan activism online underlines this distinction between identification with a fan text and its producers: Jenkins reports how Lucasfilm when offering *Star Wars* fans free web space in 2000 met determined resistance by fans who were reluctant to translate their love of the *Star Wars* series into a willingness to surrender intellectual property rights and, no less importantly, creative autonomy to the studio.

Secondly, Andrejevic and Milner both overestimate the medium specificity of fans' textual productivity in their respective studies. Milner's (2009, p. 492) reliance on Drucker's (1998) concept of New Organisation in its emphasis on "the specialized labor of self-motivated 'knowledge workers,'" suggests that fan productivity was first utilised following recent technological change. Fan labour, however, have long played a central role in the production of a range of cultural texts. *In situ* events, in particular, are reliant on fans' participation. Since the rapid growth in television revenue following the deregulation of Western Europe media markets in the early 1980s, European football clubs have reconceptualised fans attending games not only as paying customers but vital performers in staging the 'media event' by creating its atmospheric quality (Sandvoss, 2003). Similarly the subcultural economy of rave and dance culture in the early 1990s Britain that Thornton (1995) documents, relies heavily on clubbers detailed subcultural knowledge and capital in attending and promoting events. To Andrejevic (2008, p. 40), in contrast, fans' online textual productivity, rather than its interactivity, is posing a "progressive challenge to a nonparticipatory medium;"

> offers to divert the threat of activism into the productive activity of marketing and market research. Interactivity turns out to be rather more passive than advertised. The drive of savvy viewers to make themselves seen (as nondupes) overlaps with the invitation proffered by interactive media: for audiences to reveal themselves in increasing detail to producers. The logic of a savvy site such as TWoP, which allows viewers to take pleasure in critiquing the programming within which they are immersed, seems to stage Zizek's (1999, 284) formulation of the drive as the ability to derive "libidinal satisfaction from actively sustaining the scene of one's own passive submission." This Lacanian formula applies to the savvy stance in general—and perhaps more generally to the version of interactivity being prepared for the viewing public by the promoters of the digital economy.

Andrejevic, I think, overestimates the impact of a medium-specific mode of interactivity on impeding the progressive and empowering forms of fan productivity, as he underestimates the continuities in fans' creative engagements with mass-mediated texts predating the proliferation of the Internet as household technology, (though the latter naturally lies beyond the scope of his investigation). A minority of fans—those who Abercrombie and Longhurst (1998) have described as enthusiasts—elect to radically appropriate given fan texts through the rewriting of the text in forms of textual productivity that reject key premises in the conventions of the source text as in the case of slash writing or other fan fiction (Hellekson and Busse, 2006; Cicioni, 1998). As I have suggested previously, the power relations between fans and producers and the emancipatory potential of fan consumption are therefore reflective of different types of fan productivity rather than fandom *per se* (Sandvoss, 2005a)—fans who are textually productive appear best equipped to question dominant or hegemonic cultural discourses through user generated content. Yet, such subversive forms of cultural production are no automatism, even among textually productive fans, and are always based on an external motivations and desires for cultural and social change which are expressed through the vehicle of fandom, such as the radical rewriting of gender roles in slash fiction or among fanzine writers in sports fandom whose productivity is motivated by a rejection of the commercial imperatives of modern sports. Where no such motivations rooted in the fans' wider belief systems and cultural experiences exist, fans' textual productivity remains inherently conservative and focused on the endeavour of preserving the given fan object from the threat of transformations by media industry forces, regardless of the medium through which such productivity is articulated: the letter writing campaign of *The Beauty and the Beast* fans to keep the show on air, and subsequently retreating to authoring alternative storylines in the face of a radically altered narrative in the show's third season (Jenkins, 1992), was as much aimed at the preservation of an original fan text in a pre-Internet era as the *Fallout* fans' participation in the Bethesda forum was driven by the desire to maintain what they regarded as the game's textual essence. Campaigns such as Viewers for Quality Television (Brower, 1992) followed as much a quasi-media producer logic as many who post on TWoP. Much of the textual productivity among fans of team sports centres on taking on the manager's role in choosing in the right players, tactics and formations. Fantasy manager games as a formal manifestation of such textual activity that today are commonly pursued online equally predate the Internet. Even where such textual productivity critically engages with existing structures of club ownership and leadership, these are often aimed at the reinstitution of a previous (though occasionally imagined) *status quo*. Victor Costello and Barbara Moore's (2007) assertion, based on their study of online television fandom spanning across fan sites of 86 programmes, that fan "activity produces the more enlightened, discerning, insightful consumer and a heightened viewing experience that yields little change in the external products of commercial culture" (p. 140) thus holds true for many fan cultures online and offline alike.

Costello and Moore (2007) also point us towards a third aspect of Andrejevic's critique that, I think, requires further clarification. They acknowledge that some enthusiasts appear successful in their attempts to influence writers and producers empowered by the medium: "Where pre-Internet fandom was largely decentralised and limited in mass, inhibiting the collective bargaining power of individuals and geographically dispersed fan consortiums, online fan communities have the potential to produce unified centres of resistance to influence the global industries of cultural production." Such resistance, admittedly, will only be aimed at the formal rational goal of improving the fan text—and hence of doing the job of the producer better and more efficiently than the current producer, rather than challenging the underlying principles and (neo-liberal) regimes of cultural production. As Costello and Moore, 2007, p. 140) note: "online fans are exerting a form of power, but the impact does not extend between the boundaries of the reading experience and the cultural byproducts of interactivity."

Yet there is a second, wider point regarding the potential empowerment of fans through their textual productivity: it is, as I have suggested elsewhere (Sandvoss, 2005a), precisely through heightened levels of engagement with fan objects and by extension the systems of production of mediated texts that fans, and none more than those who are textually productive, not only identify with but also critically question and reject regimes of late industrialism in the paradox of the one-dimensional society that Herbert Marcuse (1964/1991) describes. In contrast to Andrejevic I argue that fans' heightened identification with producers not only leads to seeming complicity with the production practices and ideologies of media, but also facilitates spaces of reflection, rejection and resistance. What's at stake in fans' online labour then may, to return to Zizek (1999), be less libidinal satisfaction of the self's passive submission, but rather what he describes, in an almost Brechtian tradition of enlightenment through *Entfremdung*, as the strategy of "overidentification": the overdrawn imitation of the dominant system that thereby lays bar its underlying, hegemonic structures. In this sense over-identification illustrates the two contradictory forces that Marcuse (1964/1991, p. xiv) identified as an inherent condition of advanced industrial society: that "advanced industrial society is capable of containing qualitative change for the foreseeable future" while at the same time "forces and tendencies exist which may break this containment and explode the society." While overidentification is based on the former, it facilitates the latter through the radical break with the system—what Marcuse describes as the 'Great Refusal'—resulting out of the emancipatory disillusionment through such closeness.

However, assessing whether the Internet has invited and facilitated such (over)identification, and thus offered textually productive fans new spaces and practices of empowerment through fans' textual productivity or entrenched fans' complicity with the *status quo* of cultural production, is further complicated but the fact that the medium in its capacity to bypass traditional gatekeepers, has begun to erode previously clear lines of demarcation between fans and media producers and performers. While a fair amount of academic attention has been directed at audiences' textual productivity in an online environment, in other words to the fact that audiences are also produces of (user generated) content, the fact that media professionals—musicians, athletes, artists and other cultural producers—are fans, too, is less commonly addressed. As Caldwell (2008) reminds us, media producers are always also media audiences whose practices are informed through their own and their friends' and families' media consumption. In contrast to the pre-Internet era, however, cultural producers' fandom not only informs their 'official' output in the form of textual references, tributes, pastiche and other references, but is also articulated through non-commercial and/or nonprofessional forms of fan productivity. Bertha Chin and Matt Hills (2008), for instance, examine the blog of former writer and producer of *Lost*, Javier Grillo-Marxauch on *Live Journal*. Javi, as he is known among fans, "participates in discussions that fans have on his posts, responds to comments and admits to adding readers to his friends-list, as well as reading their journals" (p. 266). While Chin and Hills (2008) acknowledge that Javi carefully distinguishes between the roles of television writer/producer and being a fan, he nevertheless performs parallel productivity as genre fan that allows him the "privilege [of] judging television as good/bad" by indulging in criticism of the work of Jim Carrey, Martin Scorsese and others (267). As Chin and Hills' parallel examination of the blog of Frank Spotnitz, former executive producer of the *X-Files*, reveals, it is not blogging *per se* that leads to Javi's dual role as fan and professional producer, but the content of its (fannish) media criticism.

Similarly, textually active fans who have moved from being enthusiasts to petit producers, media professionals or even celebrities often maintain high levels of reciprocality with their (formerly) fellow fans. Take the example of Cultural Studies graduate and pop star Little Boots alias Victoria Hesketh whose debut album *Hands* reached fifth-place in the UK album charts: Hesketh describes her extensive use of enthusiasts' fora dedicated to synthesisers and electronic music in her passage from part-time musician and pianist to full-time pop musicians,[2] Having achieved full-time professional status, however, Hesketh produces not only commercial content, but continues to produce "amateur" content in the form of videos in which Hesketh performs a mix of e-piano, synthesiser and Yamaha tenorion-based cover versions of her own favourite songs written by bands and artists ranging from *The Human League* to Miley Cyrus. These songs, distributed via YouTube, echo the informality of much user generated content: like other cover versions by fans' posted on YouTube, Hesketh performs these songs in informal settings, usually her bedroom, dressed equally informally. Her performances, while demonstrating the skills of a professional musician, are improvised and include occasional errors. Production values are distinctly non-professional recording her performances with a webcam and microphone undeterred by occasional background interferences such as her phone ringing. Other clips posted by Hesketh do not feature herself, but are captured on cameraphone during trips abroad or of other artists' performances at Glastonbury.

This seeming amateurism of course projects a sense of the most significant of values in aesthetic judgment spanning from official culture in Bourdieu's (1984) seminal analysis to various sub- and fan cultures (see Thornton, 1995): *authenticity*.

However, as in the case of other forms of fan labour discussed earlier, dismissing Hesketh's productivity or that of the semi-professional musicians in Baym and Burnett's (2009) study, as commercially motivated performances imposes an inescapable logic of commodification on all forms of cultural production. Naturally, in a capitalist order all such activity can be seen as aimed at monetary gain: even content that is free and not directly revenue-generating can form part of wider marketing strategies on behalf of professional textual produces. The further we move from more niche artists with specific and comparatively small fan basis to those with permanent mass media exposure, the more such seeming fan productivity may indeed be produced for primarily commercial considerations. Yet, what matters most in our understanding of the interplay between fans' textual productivity and the Internet is that professional or commercial goals cannot function as a meaningful distinction between user generated and industry generated content, as even content produced by fans and enthusiasts without past professional success can equally serve to aid professional aspirations. Baym and Burnett (2009, p. 445) observe that as one of the three types of rewards that fans gain from the unpaid labour in assisting Swedish independent bands "is for you to view their labour as an investment toward a future career that may eventually lead to inappropriate financial compensation." Whether Hesketh is motivated by commercial considerations in writing a blog on her website, in twittering and posting home made videos on YouTube or not, offers no meaningful distinction between her textual fan productivity and other user generated content. In online textual productivity the boundaries between industry produced and user generated content, between media institutions and their audiences, between fans and producers thus dissipate as fans operate as producers and media producers as fans.

Enunciative Productivity

This is not to say that fans do not distinguish between their own labour and that of paid professionals. As Baym and Burnett (2009, p. 44) note, a second strategy of fans' assessment of their labour is to dismiss their contribution as marginal and unworthy of compensation. To a third group of fans, their rewards are of a non-fiscal nature as they "position themselves as musicians' peers" and "see themselves as doing favours for people who either are or could easily become friends." Such interaction between fans and performers based on levels of either manifest or non-reciprocal intimacy, leads us to the second type of textual productivity among fans that Fiske (1992) identified and which has quickly become the most visible form of audience activity online: enunciative productivity—talk, chatter and discussion about the fan object. The boundaries between textual and enunciative productivity are fluent and on occasion ambiguous: fanzines, for example, are often both a form of talk about the fan text as and a type of fan text at the same time. Textual and enunciative productivity can equally constitute forms of fan labour. Much of the online fan labour that Andrejevic examines in this study of TWoP, for example, takes place through enunciative productivity in which fans evaluate the television shows they follow regularly. Both textual and enunciative productivity can function as a form of appreciation or criticism of the original fan object. The difference between the two is thus primarily one of form: textual productivity is often aimed at the creation of texts and artefacts that, broadly speaking, follow the stylistic and genre conventions of the original fan object/text. Enunciative productivity, in contrast, tends to take the conversational form of everyday life talk, providing a commentary and evaluation of the fan object as form of paratexts.

As everyday life talk, the transformative impact of enunciative productivity on contemporary audienceship lies not only in the relationship between fan and fan object, but also in the interactions between different fans. Online enunciative productivity as quotidian interaction between otherwise often dispersed audience members serves as a social cement that forms communities and audiences with other media consumers. The debates surrounding the question whether online interactions can offer the basis of something meaningfully described as community has been at the heart of early investigations of online fans and online communities. Menon (2007), in her study of the online message board of the US American niche show *Once & Again*, outlines two positions within these debates: the first (Doheny-Farina, 1996; Stoll, 1995; Humdog, 1996) disputes that forms of online interaction offer the basis of meaningful community relations, whereas scholars such as Baym (1995, 2000), Bird (1999) or Clerc (2000) have described the frequent visitors to spaces of online fan interactions as communities. These debates, I believe, can be controversial only to the degree that they subscribe to a misleading definition that conflates the concepts of 'community' and 'social group': If as Eisenstein (1979), Anderson (1991) and many others have highlighted, the cultural formations of the modern nation state, i.e. the concept of national community, was rooted in the rise of mass communication, and concretely, the shared textual and symbolic resources and corresponding practices of consumption and performance, then fans' enunciative productivity online equally offers the symbolic resources for the formation of new and alternative communities. The duality of participation in blogs, fan fora and other online spaces populated by fans and the parallel development of a shared and often ardently policed set of beliefs, conventions and values (Menon, 2007; Jenkins, 2006a), facilitates the formation of online fan communities.

Naturally, these online fan communities vary in size just as much as national communities do. In smaller and subcultural fan cultures communities emerge around continuous personal interaction between active fans: Yet personal interaction is as little a *conditio sine qua non* for online fan communities as it is for larger national imagined communities; in fan communities surrounding sport teams such as Manchester United, or internationally successful television shows such as *Lost* most enunciatively active members will never meet many, sometimes most of their fellow fans as they spread across different fan fora and spaces just as the citizens of a given nation cluster in different cities. Both sets of communities are based upon bilateral social interactions *and* imagined as a bond between strangers who will never meet but who live in each other's imagination as part of the community.

The more relevant question then concerns the cultural implications arising out of the formation of such online communities. Firstly, it is worth noting that the formation of fan communities, while an increasingly visible presence online, is not tied to any single communication technology and was a common feature across fan cultures before the proliferation of Internet. From Bacon-Smith's (1992) and Jenkins's (1992) seminal work on *Star Trek* fans, King's (1998) or my examination of football fan cultures (Sandvoss, 2003) to Cavicchi's (1998) study of Bruce Springsteen fans, a plethora of fan studies documents the formation of offline communities across different fan cultures. However, the crucial feature that distinguishes such communities from other forms of territorially bound communities, such as national communities, is heightened by the proliferation of fans' enunciative productivity online: in contrast to territorially bound forms of community and associated identity formations, membership to fan communities is voluntary, if though, as a number of studies of different fan cultures drawing on Bourdieu's (1984) sociology of taste have illustrated, far from coincidental. In a globalising world in which the economic foundations of nation states have largely been eroded (Ohmae, 1995) and a shift from Fordist to post-Fordist economic models alongside longitudinal (partly related) cultural and social transformation such as secularisation and mass migration have increasingly eroded previous referents for identity formation such as employment, religion and kinship (although this process is of course far from complete), the significance of these communities as spaces for the formation and articulation of group identities can hardly be overestimated. In the ephemeral world of liquid modernity (Bauman, 2000, 2005), community membership and identity are increasingly less predetermined but achieved and maintained through agency, including fans' textual, enunciative and semiotic productivity. However, as Bauman (2005) reminds us, we should not misinterpret the fluidity of contemporary communities and identities as an automatic form of empowerment of the individual, but rather to the many who have failed to benefit from the spread of global laissez-faire capitalism as a permanent challenge to their identity position. We must thus not overlook the profound material inequalities that frame participation in online spaces. Even beyond the digital divide, inequalities continue online. In the virtual worlds of multiplayer online role-playing games such as *World of Warcraft*, linguistic divisions reinforced economic ones, as (Western) players seek out and kill the virtual characters of "gold framers" who in Chinese gaming workshops convert their virtual labour paid in gold coins into dollars and yuan (Dibble, 2007).

Other territorially grounded social and cultural distinctions remain present in many online fan cultures. Persistent language barriers aside, the vestiges of the national patterns of broadcasting and the still largely nationally structured global television market mean that fan cultures surrounding television programmes tend to reflect existing divisions between broadcast territories. TWoP, for example, is as much dedicated to American network and cable shows as format television tends to attract largely nationally homogenous online fan cultures. Online sports and music fan cultures are more deterritorialised, as music is released and circulated globally and television sports is commonly broadcast live, offering meaningful participation in discussions for fans from across the globe. However, as Baym (2007) and studies of football bulletin boards (Ruddock, 2005; Sandvoss, 2007b) demonstrate, even in these online spaces physical or social proximity to the fan object is rated highly. Hence, online fan spaces offering focal points for fan communities are far from universal. Like all communities they are based on the necessity of Othering and distinction (cf. Morley, 2000): whether fans of quality drama construct their fan community in contrast to reality television fans, or sports fans rally around the common dislike of another team or athlete (Theodoropolou, 2007), enunciative productivity continues as the mechanism by which distinctions of taste are articulated and reinforced—and thus the structured and structuring structure of the habitus (Bourdieu, 1984) continues to shape identities and community memberships online as much as in an offline environment. However, such forms of distinction do not only operate between different fan communities but are also means of structuration and the creation of cultural hierarchies within online spaces. Forms of enunciative productivity thus serve as means to establish and articulate "subcultural capital" (Thornton, 1995) by demonstrating particular knowledge, awareness of a fan community's conduct and, often most importantly, either proximity to or influence on the fan object. Costello and Moore (2007) document how fans describe encounters and (regular) contact with their favourite actors as a privilege. Similarly, fans note that "having the opportunity to influence the storyline a bit by our instant feedback is exciting" (cited in Costello and Moore 2007, pp. 137–8).

Enunciative productivity in fans' online spaces, however, not only serves fans as a tool of community building, identity formation and cultural distinction, but is equally utilised by athletes, actors and producers as an alternative means of communication with their fans for a variety of purposes. Subcultural producers such as DJs (Hodkinson, 2006) utilise online fan communities to effectively advertise forthcoming events to their target audience, much in the same was as independent music bands use online spaces for international promotion (Baym and Burnett,

2009). Actor James Marsters, playing vampire villain Spike in the later seasons of *Buffy the Vampire Slayer*, uses online interactions with fans in order to participate in and impact on the reading formations of interpretative online communities (Hills and Williams, 2005; see below). Participation in fans' online discussions becomes an important and ever more frequently used strategy that allows producers and celebrities to circumvent traditional models of (mass) media gatekeeping by professional journalists as online fan fora present ready and largely unfiltered access to a captive audience: Poor's study of former Major League Baseball pitcher Curt Schilling's use of Boston Red Sox fan fora during a proposed trade from the Arizona Diamondbacks to Boston highlights attempts to bypass traditional media gatekeepers. Schilling, eager to communicate how keen he was on the potential trade, joined *The Sons of Sam Horn*, a Red Sox fan message board, disclosing information on the ongoing trade talks, much to the frustration of mainstream media outlets. Such communication was not without gatekeepers as Poor (2006, p. 50) reminds us: "fan sites are still run by someone, and that someone […] can choose to act as a gatekeeper by filtering comments or policing users." Yet, in Schilling's case this gatekeeping function remained limited to verifying his identity and ensuring that no impersonators could post under his name. To a sports star and an A-list celebrity such as Schilling, as Poor argues, this form of interaction with fans through online fan fora thus becomes an additional channel of communication as part of a wider media mix that allows to bypass traditional gatekeepers in the print and broadcast media but may be less effective in reaching a vast number of fans simultaneously.

Semiotic Productivity

In understanding the impact of fans' discussions and debates in online spaces we thus need to reflect not only on who is speaking but equally on who is listening, leading us to the third type of productivity Fiske (1992) identifies: semiotic productivity, the process of reading and negotiating texts. To every fan who is textually or enunciatively reproductive online, there are many who watch, listen or read; fan produced music videos, for example, reediting storylines of popular TV shows to condense the relationship between two characters in a given television show or to juxtapose characters from different programmes and films—so called "vids" (Gray, 2010)—commonly attract between a few thousands and tens of thousands of views. Vids that feature particularly popular characters such as "Buffy vs. Edward: Twilight Remixed" have been viewed in excess of a million times.

Similarly, fan fora attract a substantially wider readership than its regular participants. The aforementioned football forum *Werkself* was established in December 2003 and currently counts just over 5100 users. During the six years of its existence, these users have produced 720,000 posts, an average of 318 posts a day, with per day "views" between 60,000 and 150,000. As an open forum all content is available to non registered visitors and unregistered users (including registered users not logged into their account when visiting the site from a different IP address) who tend to outnumber registered users at a ratio of 1 to 1.5 to 2 at most times. Out of 5113 registered users, approximately two-thirds (3010) had not posted on the forum themselves by November 2009. A total of 1467 users (just over 8%) had written more than ten posts, 703 (13.8%) had posted more than a 100 times during their time on the forum; 178 users (just under 3.5%) each had more than a thousand posts, while nine users had in excess of 10,000 posts, accounting for 14.6% of all posts on the forum. Only 3.5% of users (those with 1000 posts or more) thus accounted for 489,151 posts, over two-thirds (68%) of all posts. Such data further substantiates my previous assertion (Sandvoss, 2005a; cf. Longhurst, 2007) that we best conceptualise the three fan groups outlined by Abercrombie and Longhurst (1998)—fans, cultists and enthusiast—in the shape of a pyramid in which textually active enthusiasts form the smallest group at the top, followed on by enunciatively active cultists, whereas the largest number of fans are primarily active semiotically. As these figures indicate, the proliferation of the Internet has not fundamentally changed the relative sizes of these fan groups. What it has changed, however, is the way in which fans encounter and read their fan objects.

The process of reading in popular media consumption is central to the interplay between audiences and popular media and crucial in our understanding of the cultural, social and technological impact of committed, emotionally involved media consumption. Since the first wave of fan studies scholars have emphasised the, in Jenkins' (1992) words, exceptionality of the readings of texts by fans, rather than attributing exceptionality to texts themselves. To Fiske (1989), drawing on Eco (1986), fans' reading formations are a form of semiotic guerrilla warfare that serves to empower audiences by subverting the might of the 'power bloc,' which, however, is not clearly defined as either "official culture" propagated by the bourgeoisie as Fiske's use of Bourdieu (1984) would suggest, or as the power of representation of the culture industry. Throughout the second and third wave of fan studies this normative weight has largely been lifted of semiotic productivity, yet contemporary fan studies still emphasise the centrality of fans' ability to appropriate and create meaning in our encounters with mediated texts. Approaches that have sought to conceptualise the intense emotional bond most fans maintain with their object of fandom are univocally—implicitly or explicitly—premised on the importance of fans' autonomy in the construction of meaning of such texts: whether Stacey (1994) and Elliott's (1999) use of Kleinian object relations theory to explain this bond as a series of projections and introjections, Harrington and Bielby (1995) or Hills (2002) explore the relationship between fan object of fandom as a transitional phenomena drawing on the work of Winnicott (1964), or whether fandom is based upon a self-reflective reading position in which fans become fascinated with their unrecognized reflection of self in the fan object as I have suggested (Sandvoss, 2003, 2005a), all such approaches highlight fans' capacity to attribute particular and personal meanings to fan objects and in the process make it uniquely theirs as textual possession that is experienced as belonging to, or even part of, the self.

In order to accommodate this wide range of diverging, sometimes opposing readings, fan texts are polysemic to an extent that I have suggested that they are better described as neutrosemic. The absence of intersubjective meaning here forms the extension of the multiplicity of readings, i.e. texts are polysemic to a degree that they are void of meaning in and for themselves (Sandvoss, 2005a). This neutrosemy facilitating fans' capacity to create particular meanings informed by their individual experiences and horizon of expectations (Jauss, 1982) is not based on fans' ability to negotiate a particular text alone but the simultaneous process of drawing textual boundaries in an inter-textual, multimediated field in the first instance (Sandvoss, 2004, 2005a, 2005b, 2007a). In other words fan objects ranging from given artists' work, via sports teams, musicians, television series, films to celebrities all are constructed out of a plethora of different texts in and across different media. A Manchester United fan, for example, will construct her fan object out of the mixture of primary texts and experiences such as attending games or watching on television, from reading newspaper articles and magazines about the team, listening to talk radio, exchanging gossip and news with friends and fellow fans both offline and online and possibly reading further "user generated content" such as fanzines. Similarly, a fan of a given television show will watch the programme itself, but is also likely to follow trailers, read interviews with producers or actors and fan posting's or spoilers about the show. The main impact of processes of media convergence on fan culture thus lies in a changing textual field, one that is different to an offline environments: the sheer accessibility of texts online, partly resulting out of the dialogical structure of the Internet, promoting the easy sharing and availability of fans' enunciative and textual productivity, facilitates the role of the fan/reader as an ever more active agent in the construction of textual boundaries.

The fact that fans read their object of fandom as relevant to them, allowing for a sense of enjoyment, belonging and self-reflection, is not only a matter of textual interpretation and appropriation, but also of textual selection. As the above examples illustrate, fan objects, much like other media texts, are constructed between different interconnected textual layers, lacking intersubjective textual boundaries. Some of these layers may appear more central, others more peripheral. Between the different fan groups Abercrombie and Longhurst (1998) identify, the centre of gravity between these different texts and textual fields varies: fans' object of fandom tends to be broader, and often focused on mass-mediated texts or genres, while the fan objects of cultists are more narrowly defined and potentially subcultural, meaning such fans will often resort to online media, seeking the interaction with other fans. For enthusiasts the object of fandom shifts from mass-mediated texts to their collective textual productivity. As audiences, aided by the heightened availability of texts through online, charter their way through such textual fields, not only the interpretation of the text is specific to each reader. To maintain a sense of the evolution of a text or a textual motif, if not textual hierarchies, it is useful to follow Gerard Genette's (1997) distinction between text and paratexts that Jonathan Gray (2010) develops in a comprehensive survey of media paratexts from trailers, posters, DVD bonus materials, and sequels to spoofs, spoilers, toys and games. Gray summarises (2010, p. 6) their dual and ambiguous role as follows: "a 'paratexts' is both 'distinct form' and alike—or, I will argue, intrinsically part of—the text." What we commonly describe as "the text" or the "source text" is only part of the text, with "the text always being a contingent entity, either in the process of forming and transforming or vulnerable to further formation of transformation. The text, as Julia Kristeva notes, is not a finished production, but a continuous 'productivity'" (Gray, 2010, p. 7). The forms of textual and enunicative productivity online discussed in the previous two sections are thus part of, rather than additions to, the fan text. As Gray observes (2010, p. 144) in reference to his and Jason Mittell's (2007) study of the spoilers—(informed) speculation about plot developments in yet-to-aired episodes—among *Lost* fans, "the spoilers as paratexts helped carve a more personalised route through the text." This matrix of source texts and paratexts forming the textual fields in which individual readers construct textual boundaries—privileging some while relegating other textual episodes—is facilitated to an unparalleled degree by the Internet. The Internet's hypertextuality is the natural *modus operandi* of intertextuality articulated through texts and paratexts. Those with access to the Internet have, with limitation of linguistic barriers, unparalleled access to professionally and user generated paratexts. While in our popular imagination the Internet as a medium is often envisioned as a space it primarily functions not as a *space*, but as an *archive* that allows to access to the multitude of texts and paratexts across time and space.

However, the example of Twitter points to the paradoxical nature of the medium as through its archival function dramatically enhancing fans' access to the textual fields surrounding the object of fandom, and thus dramatically expanding their power to determine textual boundaries on the one hand. The aforementioned Victoria Hesketh (iamlittleboots), for instance, updates her friends and fans with tweets not only about releases and concerts but recording sessions, club nights she attends, and mundane activities from shopping to eating out. Twitter as much as message boards and other online sources of information connect to an unparalleled

array of paratexts, ever-increasing the depth and multilayerdness of
al fields and thus requiring fans' active selection between different texts
in the construction of individual textual formations. On the other hand, Twitter
also fosters dialogue that undermines the non-reciprocality on which the polysemy of popular, mass-mediated texts, rests.

In a key contribution to the analysis of the relationship between fans and fan objects, John B. Thompson (1995) described the relationship between fans and media celebrities as a form of "mediated quasi-interaction," in which the star's appeal is based on fans' ability to appropriate their fan objects in ways that suited their individual self-narratives and life-worlds precisely because the communication between star and fan is nonreciprocal. Fans' interactions with stars in the pre-Internet era were limited to letter writing with often little expectation of a personal reply. Celebrities', producers' or athletes' communication with their fans through broadcast and print media in turn was inevitably bound to a collective mode of address, rather than dialogical interaction with specific fans. While stars could of course interact with fans in person, the growing socio-economic distance between, for instance, athletes and sports fans throughout the 20th century rendered such encounters at best spasmodic, not least given the trepidations of such personal encounters. Similarly, those attracting a geographically diverse subcultural following, and those best-known for their name and work rather than appearance, such as television producers, rarely meet fans face-to-face.

The Internet has eroded this nonreciprocality: as detailed above, media professionals and audiences now share commutative spaces offering sometimes conflicting paratexts, not all of equal status. Paratextual hierarchies reflect their perceived closeness to the source texts. Online fora in particular challenge fans' semiotic productivity when those closely associated with the production of a given text, and thus with the authorial intent behind the text, become part of interpretive discourses. Here the motivations behind the hostility with which the textual productivity of adversarial media professionals was greeted in different online fora become clearer. Dialogues between fans and those involved in the production of texts have introduced levels of reciprocality that, if fans accept professional producers' attempts to assert authorial control, dramatically limit the range of different readings of a given fan text. The stronger the emphasis on authorial authenticity in a given fan object—interestingly grouping fans of high culture and those of individual celebrities closely together—the more profound such challenges become. Interactivity and reciprocality are thus double-edged sword fans: while online interactions with those professionally involved in the production of texts potentially empowers fans to impact on the source texts through their textual and enunciative productivity, the converse textual and enunciative productivity of media professionals in online spaces potentially impacts on fans' autonomy of interpretation. Yet, as much as the various studies discussed in this chapter suggest that fans' influence on producers remains limited, so does producers' influence on fans' reading formations for three main reasons.

Firstly, polysemy is a key premise of a given text's popular appeal, as a host of research in media and cultural studies over the past three decades has demonstrated. Seeking to limit the degree of a given text's polysemy runs therefore contrary to producers' commercial interests. Secondly, many fan texts such as films, popular music, sports teams and none more than television shows, lack the single authorial figure of a novel, poem or, to a lesser degree, comic. The range of above-the-line, below-the-line and, if we include the important paratextual marketing materials that Gray (2010) analyses, nonsignatory workers involved in the production of television, renders authorial intent multifaceted and fragmented from the outset. This heterogeneity and consequential polysemy is illustrated in Hills and Williams' (2005, p. 357) study of James Marsters' participation on *Buffy the Vampire Slayer* fan sites such as *Bloody Awful Poets' Society* that attest a "semiotic/discursive tension between differently situated production staff, with writers and creators wanting to support a 'preferred reading' that may clash with actors' readings of 'their' characters, and with writer/creator hyphenates promoting TV industry meanings that stress their autonomy—versus actors oppositionally referring to fans' empowerment." Hills and Williams (2005) conclude that "this demonstrates that where media texts can be thought of as multiply-authored, extra-textual work has to be done to secure a clear 'preferred reading.'" Yet as their own study shows, extra-textual work on behalf of producers—confronted, I would argue, with the extra-textual work by fans and the creation of textual boundaries around their fan texts—remains largely futile.

Thirdly, it is precisely the authorial ambiguity of most media texts that has only been heightened by technological convergence and fans' textual and enunciative productivity that allows audiences to bracket such authorial intervention out of the processes of interpretation in their reading of fan texts. In the words of a fan cited by Hills and Williams (2005, p. 359), whose reading of *Buffy the Vampire Slayer* privileged a strong romantic bond between Marsters's vampire characters *Spike* and the female heroine *Buffy* that was challenged by Marsters's proclamation that their relationship was overwhelmingly sexual and Spike evil: "I respect [James Marsters] as an actor and as a human being. He has a right to his opinions […] But I believe in Spuffy [the romance between Spike and Buffy] and no matter what any of the *BtVS* or *AtS* cast or crew say."

Such defiance is supported by the way in which the Internet has facilitated this fan's ability to find paratexts to support, and discursive spaces to reinforce, this particular interpretation of the show with other fans. Online communication between professional producers and media audiences can thus never achieve the levels of reciprocality and communicative certainty of face-to-face interaction which restricts participants to narrowly defined spaces. In the vast textual universe of the Internet as a semiotic Tundra, fans can easily avoid, bypass and retreat from such authorial challenges to the particular reading formations and find shelter among those sharing particular routes through the vast polysemy of their fan objects. Consequently, rather than functioning as *interpretive communities* in which readings are contested and formed collectively, many online spaces thus constitute *communities of interpretation*: layers of (para)texts and spaces fans find and frequent precisely because they correspond with their reading of a given fan object, thus maintaining the self-reflective bond between fan and fan object.

In this sense the fundamental question that has shaped the field of media and communication research in the pre-Internet era—the struggle over who holds the power to represent and create meaning in mediated communication—has lost none of its significance. While the Internet has changed the game through forms of textual, enunciative and semiotic productivity articulated as affirmations of and challenges to media power, its basic rules and aims have remained the same.

Conclusion

Beyond discourses about interactivity and communities that have informed much of the study of the Internet, the analysis of different forms of fan productivity online across the spectrum of popular culture thus demonstrates that audienceship and fan activity—while they have become increasingly visible and found new forms of articulation—have not changed in their fundamental aims and motivations. The frame of analysis in this chapter was based on Fiske's pre-Internet era distinction between semiotic, enunciative and textual productivity—a distinction which still meaningfully classifies audience activity online. As studies of popular audiences have gravitated towards researching fan activity online for reasons of ease of access, time and cost efficiency—and possibly on occasion also the researcher's convenience—it is important to remember each of these types of productivity predated the rise of online media.

While much attention has been paid to how the Internet has shaped fan culture, this chapter therefore makes the opposite argument: that we ought to spend more time reflecting on how fan cultures have shaped the Internet. The continuities between offline and online productivity undermine the implicit (and occasionally explicit) technological determinism of popular and academic discourses on audience activity online. Rather, I suggest that online fan activity serves as an illustration of the social and cultural shaping of the medium: the contemporary face of the Internet and the shift towards Web 2.0 have not created new forms of fan productivity, but rather reflect the range of activities in which popular audiences already engaged before the diffusion of online media: fans' active role in constructing textual boundaries which draws on fans' engagement with a wide range of different texts relating to their fan object across different media, is not a mere result of the convergence of previous separate media such as television, radio, newspapers and user generated content, but this process of convergence is driven by transmedia *storytelling* and *listening* alike. Fans' consumption of fan texts across different media and their wish to share texts with fellow fans, in which message boards function as hubs of news exchange and news are shared on social networking sites, has contributed to the demand for offline content to be available online. Fans' highly specialised, discriminatory textual habitus, in turn has advanced the archival nature of video portals through both users' uploading and their consumption. Similarly, the affective quality of the bond between fans and fan object resulting in forms of dedicated and regular consumption of given media texts has driven fans' efforts to inform processes of production of their favourite texts as much as to debate such texts collectively. However, as the occasionally national and mostly transnational distribution of films, sports, television shows, books, comics and music has created territorially dispersed audience communities, the Internet reflects such deterritorialised audiences' desire for equally deterritorial spaces in which to debate their fan objects with fellow fans and media professionals. Thirdly, textual productivity and eroding boundaries between fans, enthusiasts and petite producers (some of who progress to become professionals) long preceded the Internet yet fuel the emergence of modes of online distribution (such as Live Journal or MySpace) that have deinstitutionalised initial steps towards enthusiast' professional careers as well as broadened the scope for fans' labour contributions in support of their fan object. While transformed and made increasingly visible by the Internet, all these forms of productivity have in turn shaped the structure, purposes, uses and technological development of the new medium.

As much as the rise of first print media reflected the social, economic, political and cultural transformations of early modernity and the industrialisation, radio the rise of totalitariansm and television the processes of suburbanisation and 'mobile privatisation' (Williams, 1974) predating, but accelerating, following the Second World War, the Internet and the way it has been shaped by fans reflect the condition of late consumer capitalism. In societies that are less and less about needs and more and more about wants, the affective habitus of being a fan is increasingly central to the formation of identities, communities, social networks, and forms of consumption and participation and thus in the shaping of the communication technologies facilitating these processes.

This, *pace* McLuhan (1964), is not to argue that the Internet is a neutral medium. The easy membership to deterritorialsed communities of interpretation, the parallel distribution of professional and user generated content, the increasingly eroding boundaries between the two and the accessibility of ever wider textual fields across time and space all profoundly impact on how audiences form, practice and articulate the affective bond with their object of fandom. Fans' online practices thus reflect the interplay between culture and technology, but there is a little evidence to suggest that the Internet in and for itself has empowered or disempowered audiences. Rather than having transformed the nature of fan cultures, the Internet has underscored the further proliferation of fandom as a dominant mode of media consumption and production and therefore deepened audiences' engagement in convergent media culture.

Notes

1. In the subsequent literature (Hills, 2002; Sandvoss, 2003, 2005a, 2005b; Longhurst, 2007) it has been noted that the terms do not fully correspond with their common usage or fans' self-descriptions (few sports fans, for instance, would describe themselves as cultists). Attempts have also been made to broaden and add to this taxonomy in the study of particular fan cultures .
2. See the Hesketh interview with Paul Morley, *The Guardian*, 16th June 2009, http://www.youtube.com/watch?v=tRsR6CQxz-Y&feature=related

References

Abercrombie, N. & Longhurst, B. (1998). *Audiences: A Sociological Theory of Performance and Imagination.* London: Sage.

Anderson, B. (1991). *Imagined Communities*, revised edn. London: Verso.

Andrejevic, M. (2008). Watching Television without Pity: The Productivity of Online Fans. *Televsion and New Media*, 9(1): 24–46.

Bacon-Smith, C. (1992). *Enterprising Women: Television Fandom and the Creation of Popular Myth*, Philadelphia: University of Pennsylvania Press.

Bauman, Z. (2000). *Liquid Modernity*. Cambridge: Polity Press.

Bauman, Z. (2005). *Liquid Life*. Cambridge: Polity Press.

Baym, N. (1995). The Emergence of Community in Computer-Mediated Communication In: Jones, S. (ed.) *Cybersociety: Computer-Mediated Communication and Community*. Thousand Oaks, CA: Sage.

Baym, N. (2000). *Tune in, Log On: Soaps, Fandom, and Online Community.* Thousand Oaks, CA and London: Sage.

Baym, N. (2007). The New Shape of Online Community: The Example of Swedish Independent Music Fandom. *First Monday*, 12(8), http://firstmonday.org/issues/issue12_8/baym/index.html

Baym, N. K. & Burnett, R. (2009). Amateur Experts: International Fan Labor in Swedish Independent Music. *International Journal of Cultural Studies*, 12(5): 1–17.

Benkler, Y. (2006). *The Wealth of Networks: How Social Production Transforms Markets and Freedom*. New Haven, CT and London: Yale.

Bird, S. E. (1999). Chatting on Cynthia's Porch: Creating Community in an E-mail fan group. *Southern Communication Journal*, 65:49–65.

Bourdieu, P. (1984). *Distinction: A Social Critique of the Judgement of Taste*. London: Routledge & Kegan Paul.

Brower, S. (1992). Fans as Tastemakers: Viewers for Quality Television. In Lewis, L. A. (ed.) *The Adoring Audience*. London: Routledge.

Caldwell, J. T. (2008). *Production Culture: Industrial Reflexivity and Critical Practice in Film and Television*, Durham: Duke University Press.

Cavicchi, D. (1998). *Tramps Like Us; Music and Meaning among Springsteen Fans*. New York and Oxford: Oxford University Press.

Chin, B. & Hills, M. (2008). Restricted confessions? Blogging, Subcultural Celebrity and the Management of Producer–Fan Proximity. *Social Semiotics*, 18(2): 253–72.

Cicioni, M. (1998). Male Pair-Bonds and Desire in Slash Writing. In Harris, C. (ed.) *Theorizing Fandom: Fans, Subculture and Identity*. Cresskill, NJ: Hampton Press

Clerc, S. (2000). Estrogen Brigades and "Big Tits" threads: Media Fandom On-Line and Off. In Bell, D. & Kennedy, B. (eds.) *The Cybercultures Reader*. London: Routledge.

Costello, V. & Moore, B. (2007). Cultural Outlaws: An Examination of Audience Activity and Online Television Fandom. *Television and New Media*, 8(2): 124–143

Dibble, J (2007) The Life of the Chinese Gold Farmer. *New York Times*, 17th June 2007.

Doheny-Farina, S. (1996). *The Wired Neighborhood*. New Haven, CT: Yale University Press.

Doss, E. (1992). *Elvis Culture: Fans, Faith and Image*. Lawrence, KS: University Press of Kansas.

Drucker, P. F. (1998). The Coming of the New Organization. In Harvard Business School Press (ed.) *Harvard Business Review on Knowledge Management*. Cambridge: Harvard University Press.

Eco, U. (1986). *Travels in Hyperreality*. London: Picador.

Eisenstein, E. (1979). *The Printing Press as an Agent of Change: Communications and Cultural Transformations in Early Modern Europe*. Cambridge, UK: Cambridge University Press.

Elliott, A. (1999). *The Mourning of John Lennon*. Berkeley: University of California Press.

Fiske, J. (1989). *Understanding Popular Culture*. Boston: Unwin and Hyman; rep. 1991, London: Routledge.

Fiske, J. (1992). The Cultural Economy of Fandom. In Lewis, L. (ed.) *The Adoring Audience*. London: Routledge.

Genette, G. (1997). *Paratexts: The thresholds of Interpretation*. Cambridge: Cambridge University Press.

Gray, J.A. (2010). *Show Sold Separately: Promos Spoilers and Other Media Paratexts*. New York: New York University Press.

Gray, J. A. & Mittel, J. (2007). Speculation on Spoilers: *Lost* Fandom, Narrative Consumption and Rethinking Textuality. *Participations* 4(1), www.participations.org

Gray, J. A. et. al. (2007). Introduction: Why Study Fandom? In Gray, J. A. et al. (eds.) *Fandom: Identities and Communities in a Mediated World*. New York: New York University Press.

Hardt, M. & Negri, A. (2000). *Empire*. Cambridge: Harvard University Press.

Harrington, C. L. & Bielby, D. (1995). *Soap Fans: Pursuing Pleasure and Making Meaning in Everyday Life*. Philadelphia: Temple University Press.

Haynes, R. (1995). *The Football Imagination: The Rise of Football Fanzines Culture*. Aldershot: Arena.

Hellekson, K. & Busse, K. (2006). *Fan Fiction and Fan Communities in the Age of the Internet: New Essays*. Jefferson, NC: McFarland.

Hills, M. (2002). *Fan Cultures*. London: Routledge.
Hills, M. & Williams, R. (2005). "It's all my interpretation": Reading Spike through the Subcultural Celebrity of James Marsters. *European Journal of Cultural Studies*, 8(3): 345–65.
Hodkinson, P. (2006). Subcultural Blogging? Online Journals and Group Involvement among UK Goths. In Bruhns, A. & Jacobs, J. (eds.) *The Uses of Blogs*. New York: Peter Lang.
Humdog (1996). Pandora's Vox. In Ludlow, P. & Godwin, M. (eds) *High Noon on the Electronic Frontier: Conceptual Issues in Cyberspace*. Cambridge, MA: MIT Press.
Jauss, H. R. (1982). *Toward an Aesthetic of Reception*. Minneapolis: University of Minnesota Press.
Jenkins, H. (1992). *Textual Poachers: Television Fans and Participatory Culture*. Routledge, New York.
Jenkins, H. (2006a). *Convergence Culture: Where Old and New Media Collide*. New York: New York University Press.
Jenkins, H. (2006b). *Fans, Bloggers and Gamers*. New York: New York University Press.
Jenson, J. (1992). Fandom as Pathology: The Consequences of Characterization. In Lewis, L. (ed.) *The Adoring Audience*. London: Routledge.
King, A. (1998). *The End of Terraces: The Transformation of English Football in the 1990s*. London: Leicester University Press.
Lancaster, K. (2001). *Interacting with Babylon 5*. Austin: University of Texas Press.
Longhurst, B. (2007). *Cultural Change and Ordinary Life*. Maidenhead: Open University Press.
Marcuse, H. (1964/1991). *One-Dimensional Man: Studies in the Ideology of Advanced Industrial Society*, second edition. London: Routledge.
McLuhan, M. (1964). *Understanding Media: the Extension of Man*. London: Routledge.
Menon, S. (2007). A Participation Observation Analysis of the *Once & Again* Internet Message Bulletin Boards. *Television and New Media*, 8(4): 341–74.
Milner, R.M. (2009). Working for the Text: Fan Labor and the New Organization. *International Journal of Cultural Studies*, 12(5): 491–508.
Morley, D. (2000). *Home Territories*. London: Routledge.
Ohmae, K. (1995). *The End of the Nation-State: the Rise of Regional Economies*. New York: Simon and Schuster.
Poor, N. (2006). Playing Internet Curveball with Traditional Media Gatekeepers: Pitcher Curt Schilling and Boston Red Sox Fans. *Convergence: The International Journal of Research into New Media Technologies*, 12(1): 41–53.
Ritzer, G. (1996). *The McDonaldization of Society*. Newbury Park: Pine Forge Press.
Ruddock, A. (2005). 'Let's Kick Racism Out of Football—And the Lefties Too!.' *Journal of Sport and Social Issues*, 29(4): 369–85.
Sandvoss, C. (2003). *A Game of Two Halves: Football, Television and Globalization*. London: Routledge, Comedia,.
Sandvoss, C. (2004). Technological Evolution or Revolution? Sport Online Live Internet Commentary as Postmodern Cultural Form. *Convergence: The Journal of Research into New Media Technologies*, 10 (3): 39–54.
Sandvoss, C. (2005a). *Fans: The Mirror of Consumption*. Cambridge: Polity Press
Sandvoss, C. (2005b). One Dimensional Fan: Toward an Aesthetic of Fan Texts. *American Behavioural Scientist*, 49(3): 822–39.
Sandvoss, C. (2007a). The Death of the Reader? Literary Theory and the Study of Texts in Popular Culture. In Gray, J. A. et al. (eds.) *Fandom: Identities and Communities in a Mediated World*. New York: New York University Press.
Sandvoss, C. (2007b). Public and Publicness: Media Sport in the Publics Sphere. In Butsch, R. (ed.) *Media and Public Spheres*. Houndmills, Basingstoke: Palgrave.

Stacey, J. (1994). *Stargazing: Hollywood Cinema and Female Spectatorship*, London: Routledge.

Stoll, C. (1995). *Silicon Snake Oil.* New York: Doubleday.

Terranova, T. (2000). Free Labour: Producing Culture in the Digital Economy. *Social Text* 18(2): 33–58.

Theodoropolou, V. (2007). The Anti-Fan within the Fan: Awe and Envy in Sport Fandom. In Gray, J. A. et al. (eds.) *Fandom: Identities and Communities in a Mediated World.* New York: New York University Press.

Thompson, J. B. (1995). *The Media and Modernity.* Cambridge: Polity Press.

Thornton, S. (1995). *Club Cultures: Music; Media and Subcultural Capital.* Cambridge: Polity Press.

Tushnet, R. (2007). Copyright Law, Fan Practices, and the Rights of the Author. In Gray, J. A. et al. (eds.) *Fandom: Identities and Communities in a Mediated World.* New York: New York University Press.

Williams, R. (1974). *Television: Technology and Cultural Form.* London: Fontana.

Williams, R.S. (2008). *Television Fan Distinctions and Identity: An Analysis of 'Quality' Discourses and Threats to 'Ontological Security.'* PhD Thesis, Cardiff University.

Winnicott, D. W. (1964). *The Child, the Family and the Outside World.* Cambridge, Mass: Perseus Publishing.

Zizek, S. (1999). *The Ticklish Subject.* London: Verso.

CHAPTER FOUR

The Place of Internet Gambling

Presence, Vice, and Domestic Space

HOLLY KRUSE

Since the introduction of the telephone and the phonograph into the home in the late nineteenth century, electrical media technologies have met with resistance when crossing the divide into domestic space. In earlier centuries, the home had been an important site of production, but industrialization largely removed explicit processes of production from the home. As home and work were separated from each other, the home came to be seen as a feminized realm and a retreat from the masculine world of work and public life (Spigel, 1992; Cowan, 1983). But home has also become, as David Morley observes, a place heavily "connected" to elsewhere through communication technologies like the telephone and Internet, and may "no longer function for [many people] as a haven of peace and tranquility" (2000, p. 57).

The movement of communication technologies into the home raises questions about the uses and gendering of the domestic sphere, and how its meanings have been discursively contested in recent decades. The arrival of practical home Internet connectivity in the late 1980s in the United States eventually made a number of activities once undertaken primarily in public, an arena traditionally gendered as masculine, part of the domestic life. One such activity is gambling. Although Internet gambling accounts for a substantial percentage of all online financial transactions, it has been at best a marginal research topic in Internet studies, even though online gambling's direct antecedents—betting via telegraph and telephone—are rich and important communication history topics.

Internet gambling today is a huge business, although because of the multitude of illegal online gambling sites, it is impossible know how much money is being bet online. There are, however, expert estimates and projections. Christiansen Capital Advisors (2004) a gambling consulting firm, estimates that almost $23 billion in Internet gambling revenue was generated worldwide in 2009 and over $24 billion is expected in 2010. A late 2008 *Washington Post* feature on online poker cites a smaller but still very large number, noting that Internet gambling revenue has "more than tripl[ed] over five years, to $18 billion annually, including about $4 billion from virtual poker" (Gaul, 2008a).

An examination of the ways in which this major global industry enables and influences cultural and ideological understandings has much to offer to the studies of both new media and communication history. By focusing on online betting and the discourses that surround and construct it, this chapter looks at how interactive media technologies have changed both the spaces and understandings of gambling and associated social rituals from public space to private space, and to spaces in between. Thus, this chapter also examines the shifting notions of what takes place in the masculinely identified public sphere and what is allowed in the feminized sphere of the home.

The Problem of Domestic Space

The ease of gambling from home has created anxiety among various groups and individuals in American society, in part because Victorian notions of the middle-class home as a haven from the stresses of modern life persist and have shaped much of the discourse about domestic communication technologies for more than one hundred years. While larger European and North American houses prior to the eighteenth century included semi-public spaces, by the nineteenth century these houses were increasingly divided into specialized rooms, culminating in designs that featured a more public front room for entertaining, with the rest of the house primarily private space (Frohlich and Kraut, 2002, pp. 4-5). Indeed, the Victorian home, as Lynn Spigel has noted, because of the division of space, embodied "the conflicting urge for family unity and division"; Victorians were able "to experience private, familial, and social life within conventionalized and highly formalized settings" (1992, p. 16). Adding more rigor to Victorian conventions of household design was the emergence of interior decorating as a profession in the late nineteenth century. Elsie De Wolfe, one of the first female professional decorators, argued for the importance of temperance in decorating. In her 1913 book *The House in Good Taste*, she wrote: "My business is to preach to you the beauty of suitability. Suitability!

SUITABILITY! SUITABILITY!" In decorating the domestic haven, women were expected to practice visual, if not financial, restraint, and to adhere to conventional ideas (Lynes, 1985, pp. 70, 346–347).

By the early-to-mid-twentieth century, popular discourses posited the home as a place of family unity (see Spigel, 1992), and houses were designed to include communal areas—the kitchen, the living room, the dining room—that opened onto each other and encouraged family togetherness. In the quest for in-home unity, media technologies were often touted as means for bringing people together. For example, early twentieth-century phonograph advertisements pictured families and friends gathering together around phonographs (Kruse, 1993). Radio too, after overcoming obstacles to its acceptance in homes—including the large battery that leaked acid onto living room floors and the lack of a loudspeaker—had by the late 1920s become a technology that the whole family could enjoy together (Spigel, 1992, pp. 28–29). Likewise, the advent of television prompted discussions in popular media of the ways in which families could grow closer around the new technology: discussions that were concurrent with the appearance of the "family room" as a center of household activity, and thus as a place where family members could easily come together in front of the television set (Spigel, 1992).

Similar hopes were raised with the introduction of the personal computer into the home in the late 1970s and the 1980s, illustrated by advertisements for brands like Commodore that showed family members (although interestingly and tellingly, usually not mothers) gathered at the computer. Not surprisingly, however, as with previous communication technologies, the incursion of the computer into domestic space led to worries about the effects of the medium on users' morals, on social interactions, and on family structures.

The personal computer became a multifunctional device in the 1980s, sold and discussed as a machine to be used to play games, do office work, prepare household budgets, word process, and keep grocery lists. With this limited range of perceived functions and the introduction of easy-to-use personal computers like the Apple Macintosh in 1984, by the mid-1980s the popular press was announcing that all those who wanted or needed personal computers had them and thus that the market for PCs was "dead." Clearly reports of the death of the PC were greatly exaggerated. In addition, from 1980 forward there was great concern in the popular media about where in the home to put the home or personal computer, how such computers fit into the décor, and whether one needed to create a separate space in the home for the computer. One architect affirmed in a 1985 *Washington Post* column that "most of family life takes place in the kitchen or living room, [so] a few families have wondered how to socialize the computer," adding that "It is not like adding a microwave oven to a crowded kitchen counter or a television set to a family room bookcase" (Ridley, 1985). The computer needed its own space at the center of home life, and the architect explained to readers how he was able to create a suitable space. Such discourse certainly mirrors the concerns expressed thirty years earlier about where in the home to put the early television set, and why.

Computer placement was indeed important, because research conducted by David Frohlich and Robert Kraut in the early 2000s as part of the groundbreaking Carnegie Mellon University HomeNet Internet-use project indicates that much family computer activity is social in nature. Computers needed to be in places where all family members could use them. In Frohlich and Kraut's (2002) study of families in the Pittsburgh area with computers, only one-quarter of computers could be found in the very private space of a parent's or a child's bedroom. More often computers were located in places like the family room. Frohlich and Kraut argued that, at least in these families, computers (which were most often used for checking email) brought families together and therefore had a sociopetal effect. A significant number of participants in the study reported that they engaged in joint computer use with members of their families, sometimes when playing computer games. Joint use also happened when one family member was using the computer and another family member was observing and gaining knowledge to be actively used in the near future. In particular, "Family members may enter into true collaborations with each other to operate a PC programme or Internet service together" (2002, p. 30). Such copresence is clearly facilitated when PCs are placed in a less private room in a house.

More recent research indicates that in families at least, copresence at a home computer is not unusual. In their ethnographic study of youth and gaming, Stevens, Satwicz, and McCarthy (2008), described several situations in which kids play games at the same computer at the same time in living rooms or family rooms, often collaborating to figure out how best to play games. Similarly, in her study of Silicon Valley families, Heather Horst (2009) found families that embraced playing online games together as way of bonding over an activity in which all members had an interest. Others got together around the computer to build websites, make videos, and edit photographs (p. 167). The 2008 Pew Internet and American Life Project report "Networked Families," reported that among Internet users who cohabit with a spouse and at least one child, 54 percent go online at the same computer with someone else a few times a week (Kennedy et al., 2008 p. 16).

But as mentioned earlier, and as with previous communication technologies, the incursion of the computer into the private sphere caused and continues to cause trepidation. In her discussion of television's increasing centrality in family life in the United States in the 1950s, Lynn Spigel linked fears that television would become a "man-made monster" inhabiting American living rooms to a history of alarmist discourse about new technologies' potential to grow out of control (1990, pp. 81–82). In the case of computers and the Internet, for instance, although Frohlich and Kraut found significant sociopetal effects in their study of family computer use, parents in the sample noted a tendency for their children to isolate themselves when using computers, and they also noted this tendency in themselves. One would expect that this tendency would only have become more widespread and intense in the past decade, with the proliferation of mobile personal information and communication technologies (ICTs) like smart phones and wi-fi equipped laptop computers, and the trend in middle-class families for family members to have their own media technologies, like televisions, in their personal household spaces. Indeed, Horst (2009) writes, "In many homes, the arrival of relatively affordable and portable media has solidified the importance of the bedroom as a space where one can use new media in these endeavors and assume individual control over one's own media world" (p. 156).

The struggle over placement of computers and other new media in the home is illustrated in the 2008 television documentary *"Growing Up Online,"* in which the mother who insisted that her children use the Internet at home only in the semi-public space of the kitchen area appears anomalous, including to her children. Yet Horst finds that many parents still opt to place family computers and other new media systems in the more public, less private spaces of their homes, often with the intent of monitoring their children's media use (2009, pp. 154–155). At the turn of the twenty-first century, parents in Frohlich and Kraut's study expressed fears that their children would encounter inappropriate material if left on their own to browse the Internet. Several parents saw allowing a child unsupervised World Wide Web access to be the equivalent of leaving a child alone in a public place (2002, p. 25). Those fears were still widespread by the end of the decade, as illustrated by this parent's statement about youth Internet use in *Growing Up Online:*

> The scariest, worst part for me is stalkers, is somebody becoming obsessed with one of my children. I have two very attractive daughters. You know, some guy that all of a sudden decides that, really, my daughter was meant for him—that kind of stuff scares me. Kids think, "I'm in my home. How could anything bad happen to me?" They don't realize that when they're sharing on that keyboard, it's, like, let 'em on in baby, because they're right here. (Dretzin and Maggio, 2008)

The dangers of masculine public space were invading the idealized, middle-class, domestic haven. Although fears of online predators were particularly prominent in the early twenty first century, the media and individuals espoused concerns about the infiltration of various forms of online vices into the private space.

The Rise of Online Gambling: Poker, the Home, and Technopanics in the United States

Although perhaps less than online pornography, and because of the ways that online gambling traverses the public and the private in, to many, a troubling and transgressive way, Internet gambling has become a particular site of struggle. In the past fifteen years, gambling has become an increasingly popular Internet application. Casino games, poker, sports betting, and lotteries are now major moneymakers on the Internet, sometimes legally, but often illegally. Using illegal gambling sites, wagering through legal sites but from jurisdictions where certain kinds of Internet betting are not legal, betting when under the age of 21, and failing to report winnings to the IRS are all "profane" forms of home Internet use (and they are uses that invite severe disciplinary incursions by the state into the "sacred" domestic sphere).

The act of betting in itself is viewed by many as morally debased and dangerous, even when it is legal, and bringing it into the home challenges the dominant tropes of domesticity. At the same time, many Americans engage in online shopping and other online financial transactions from the privacy of their homes in a relatively unproblematic way. So perhaps it is no surprise then that a familiar gambling card game, poker, was fairly widely accepted online when it gained popularity in the early to mid-2000s. Betting online on poker really got underway in 1998, with the launch of the Planet Poker, which tried to replicate on the Internet the in-person experience of playing poker in a casino (Matthews and Zamora, 2008). Online poker became very popular very fast. Most casinos were, and are, located offshore in the Caribbean and Central America to avoid being shut down by American authorities (Gaul, 2008b). Initially, many casinos accepted credit card bets, but actions by the United States Department of Justice discouraged casinos from offering that option, and credit card fraud had also become a problem. Many online casinos switched to forms of "e-money," accepting money transferred from users' accounts at sites like Neteller (Heffernan, 2009; Matthews and Zamora, 2008). Despite these obstacles, Internet poker continued to grow exponentially in popularity during the 2000s, and now brings in an estimated $4 billion a year to its providers (Gaul, 2008a).

The true boom in online poker began in 2003 when the appropriately named Chris Moneymaker won $2.5 million for his first place finish in the World Series of Poker after having prepared for the tournament by playing only Internet poker (Matthews and Zamora, 2008). The tournament aired on ESPN, and Moneymaker's win is credited with drawing thousands of players worldwide to online poker and creating a plethora of e-betting poker websites based in countries like Malta, the Isle of Man, Antigua, and Costa Rica (Gaul, 2008a). Today the Internet poker business is dominated by a few huge offshore sites like PokerStars and Full Tilt Poker (Heffernan, 2009). Offshore only, because in late 2006 President Bush signed into law a ban on using most kinds financial transfers to fund one's online bets. This in effect put out of business most online poker casinos that had located in the United States in hopes that the ambiguity of existing law would allow them to operate. Internet gambling opponent Robert Goodlatte, a Republican congressman from Virginia, contended:

> Virtual betting parlors have attempted to avoid the application of the United States law by locating themselves offshore and out of jurisdictional reach... These offshore, fly-by-night Internet gambling operators are unlicensed, untaxed, un-regulatable and are sucking billions of dollars out of the United States. (Quoted in Gaul, 2008b)

Offshore casinos are also likely to be open to cheating and fraud because if the lack of regulatory oversight. In a *Washington Post* series on Internet poker, reporter Gilbert Gaul notes of betting using offshore sites:

> Millions of the bets originate in the United States, where online poker and gambling sites are banned, forcing players to reach out across the Internet like modern-day bootleggers. Yet players have little way of knowing who is watching their bets or where their money is going. Often, owners hide behind multiple layers of limited partnerships, making it difficult to determine who controls the sites or to lodge complaints about cheating. (Gaul, 2008a)

Honest players can therefore find themselves losing large amounts of money to cheaters, with little recourse, although companies interested in maintaining customer loyalty have acted to compensate players who are the victims of cheaters. Gaul cites a case in which AbsolutePoker's software was hacked, and the company refunded $1.6 million dollars to players who had lost money to the hackers.

The perceived shadiness of the business of Internet gambling; the availability of the Internet and online gambling in the United States; the ease with which money can be spent through a Web interface; the morally questionable status of even legal gambling; the proliferation of home entertainment and information media; the fragmentation, specialization, and privatization of domestic space; and the presumed vulnerability of teens were all likely contributors to the increase in concern expressed in popular media in the mid-2000s over young people and online gambling.

To draw on Alice Marwick's (2008) work on cyberporn and online predators, discourses that raise significant concerns about teens, college students, and Internet betting qualify as technopanics. The idea of a technopanic is rooted in Stanley Cohen's (1972) notion of the moral panic, in which incidents of deviance—primarily youth deviance—become the focus of media coverage and public campaigns against a particular evil. Cohen argues that media responses to these incidents tend to present them as indicative of a far greater social danger than the public believes them to be (pp. 65–66). He also notes that the problematic behaviors are expressions of deeper structural problems that remain unaddressed (p. 204).

In a technopanic, the focus is on new technologies, including new media, and there is particular unease over the behavior of young people, which is pathologized. In the case of panics over the Internet, the purported facility of young people with computer technologies that baffle their elders contributes to the panic. Marwick argues:

> The trope of the teenager who possesses more technological knowledge than her adult counterpart and can program a VCR or set up a home computer is a powerful one. This image is furthered by movies from *Wargames* [sic] to *Hackers* to *Jurassic Park*, celebrations of young "techno–entrepreneurs" like Shawn Fanning and the wunderkinds of YouTube, and descriptions of the cultural competency demonstrated by teens as they blog, post digital pictures, talk to each other through instant messaging and interact through Facebook. (2008)

Troubling behaviors on the part of young people need to be stopped, and worrisome content restricted or banned, according to the logic of the technopanic (Marwick, 2008).

While there are often valid concerns at the root of a technopanic, such panics frequently exaggerate the extent of the problem, or suggest solutions that are more far-reaching than is justified by the reality of the problem. Still, it is important to keep in mind that the harms to victims of problematic behaviors are very real, and I do not want to diminish that. What I want to emphasize is the degree to which narratives of problematic Internet gambling by teens point to the struggle over meaning, and particularly in this case, over meanings of the private and the public. The protection of the family and private space is an important trope in dominant narratives of online gambling, even in early popular clinical discourses. A 2002 article in the American Psychological Association's *Psychiatric News* asserted a connection between Internet use among the young and online gambling, even without evidence presented of a specific causal link: "Adolescents use the Internet more than any other age group, and recent studies have found that about 3 percent of adolescents and 8 percent of college students have gambled on the Internet, according to the National Council on Problem Gambling" (Lehmann, 2002). In fact, the APA had flagged accessibility to young people in the home as a primary contributor to Internet gambling, emphasizing that lack of regulation meant that "there is no control on the hours of availability [or] age of participants," and that gambling sites would target children, who were most likely using computers within the confines of their homes, with offers of discounts and free items on sports, adventure, and action figure websites (2002).

By the mid-2000s, with the explosion in online poker's popularity, anxieties about Internet gambling and teens were widely expressed in the news media, in stories replete with horrifying anecdotes, almost exclusively about young men. A 2006 ABC News story described how Greg Hogan, Jr., president of the sophomore class at Lehigh University and the son of a respected minister, found himself robbing a bank in Allentown, Pennsylvania, as a result of his Internet poker losses. Hogan told *Good Morning, America*:

> I started playing with about $50... I would deposit money from my checking account with an account at [an] online gambling site, mostly at PokerStars.com or Sportsbetting.com. I tripled my money the first few times so it seemed like easy money, but then I lost $300 and just felt this rage. Eventually, I spent it all.

Online gambling made a good kid go bad. The assistant chief of police of Allentown underscored this interpretation when he said "The fact that he's a college student going, coming from an affluent family—he just does not fit your profile of someone robbing a bank" (ABC News, 2006).

The college press contributed to the spate of student gambling addiction stories. A 2005 feature in the California State University, Sacramento newspaper, for instance, described instances of college students who stole to support their gambling habits, who dropped out of school because of gambling, who lost relationships, who even contemplated suicide. While the story featured the views of college students who gambled for fun and who claimed not to be addicted, it also cited a recovered compulsive gambler who was skeptical that their gambling would remain problem-free (Gesuele, 2005). He also pointed out that it "makes sense" that an estimated 15 to 20 percent of college students had bet online, "because many older people are not Internet savvy. Students are smart, intelligent and have a better understanding of how to use the Internet" (Gesuele, 2005). Interestingly, a subsequent survey funded by the National Institute of Mental Health found a correlation between non-student status and frequency of gambling, non-student status and extent of gambling, undermining popular press assertions of a college gambling epidemic in the student population (Welte et al., 2008. p. 131).

Within the narrative that constituted the technopanic over Internet betting, college students were not the only tech-savvy young people at risk. High school students also fell prey to the allure of online betting. A Fox News report in 2006 asserted that gambling was on the rise not only among teenagers, but pre-teens as well, and claimed that "The trend can be attributed to a growing acceptance of gambling in American culture, an increase in accessibility because of the Internet and more betting shows on TV" (Donaldson-Evans, 2006). The story followed "Ross," who started betting online while a high school sophomore, had a bookie by the time he was a senior, began selling marijuana to support his gambling habit, and ended up $30,000 in debt before his parents learned of his problem and forced him into Gamblers Anonymous. Ross's story illustrates that bad behaviors–like gambling and drug use–are often linked, and that parents are often clueless about dangerous activities in which their children engage from the privacy of their homes. Readers are warned:

> Parents may miss the warning signs indicating the habit is getting out of hand because they could see gambling as a perfectly healthy way for children to spend their time.
>
> "This looks very benign," said [neuropsychiatrist Lawson F.] Bernstein. "They're home playing cards with their friends, they're not drinking or doing drugs. It all looks very harmless. But the problem is for a certain number of kids, it's going to be the addictive equivalent of pot in that they're going to get in trouble with it." (Donaldson-Evans, 2006).

Teen gambling is indeed far from benign, another set of researchers found. In fact, that "adolescent-onset gambling is associated with more severe psychiatric problems, particularly substance use disorders, in adolescents and young adults" (Donaldson-Evans, 2006). Ironically, the Fox News story indicts the glamorization of poker as presented in televised tournaments, even as Fox Sports has long held the broadcast rights to the World Poker Tour (Cypra, 2009).

The dangers of teen online gambling were spelled out in a personal way in a 2005 *Seattle Post-Intelligencer* column in which the mother of a former teenage gambler testified to the prevalence of teen betting, including online and in casinos, and complains about the lack of enforcement of existing gambling regulations. She described her efforts to bring attention to the problem, and explained her motivation:

> I am one of those parents of a teenage boy who played poker in high school and thought it was harmless fun. Ten years later, Ben's gambling addiction had such a grip on him that he lost his biggest bet of all.
>
> Poker, Internet gambling, and Black Jack became the sole beneficiaries of Ben's finances; maintaining his car was not a priority. The police report stated Ben died after losing control of his car due to mechanical failure. Actually Ben died after losing control of his life to gambling. It started as an innocent after-school poker game and ended with his car wrapped around a tree.
>
> When Ben talked about his struggle with gambling he often said, "Kids don't realize they are not only gambling with money, they are gambling with their lives." (McCausland, 2005)
>
> Were it not for gambling, including online gambling, Ben would have better maintained his car, and he would still be alive today.

In these accounts, the feminine domestic sphere is clearly under threat. Mothers are losing control of their families, even when their children are supposedly safe at home. The webpage of the Columbia University Medical Center Gambling Disorders Clinic warns that "easy access to computers and online gambling take away barriers to gambling in public. For teens, "online gambling is as close as their computer" (2009). Moreover, strangers, poker opponents who want to take teens' money, are imagined to be reaching through the computer into teens' home. Ill-meaning opponents are present in a disembodied way through the Internet, although the Internet is hardly the first technology to contribute to the disembodiment of presence. James Carey (1989) points out in his analysis of the electric telegraph that the device's invention and adoption in the 1840s were particularly notable because for the first time in human history, communication was no longer dependent on transportation, on a physical form of transmission. Jeffrey Sconce (2000) in his book *Haunted Media: Electronic Presence from Telegraphy to Television.* points to "the miraculous 'disembodying' presence" evoked by the telegraph (p. 44).

The telegraph was not, however, a home communication technology. At the end of the nineteenth century and early in the twentieth century the telephone and the phonograph emerged as in-home technologies that allowed disembodied presences, both familiar and unfamiliar, to penetrate the domestic divide. And later media technologies also evoked the idea that communication devices could separate being from body; television was lauded in its early years for allowing viewers to be present "in two places at once" (Sconce, 2000, p. 129)—presumably in one's home and, virtually, in the setting of the television show one was watching. Today wired and wireless devices allow users to simultaneously inhabit several spaces at once. One can be in one's home, as well as in the virtual space of the online poker game, and through the poker game into the homes, offices, and other locations of opposing players.

Conclusion

The middle-class American home is not, and never has been, the pristine separate sphere of civilizing feminine influence that dominant ideological constructions would have us believe and work hard to keep alive. It is only in the past 150 to 200 years that the home has been transformed from a primary site of economic production to a site primarily devoted to commercial consumption. In that relatively short time, much ideological work has been done to maintain the symbolic boundary between the domestic sphere and the public sphere, and more to the point, domestic space and public space. The case of Internet gambling is an especially instructive example of this struggle. Unlike with previous communication and entertainment technologies, the popular metaphor applied to the Internet says that it too is a space. Media technologies can be domesticated–phonographs could be placed in beautiful period cabinets, radios could be virtual hearths around which families purportedly gathered, high—definition televisions can be made thin and flat and hung on walls like vivid pieces of art, computers can be shrunk down or presented as sleek modern designs—but how can a technology that also constitutes a largely-masculinely-defined public space like the Internet be domesticated?

Traveling the Web means visiting sites, locations that exist in a public arena while one is within the physical confines of a feminized space, to a degree not afforded by other media technologies. Many of these places offer activities that were not easily accessible within the house, or even when away from home, like gambling. The dichotomies that have defined our understandings of what gendered space means, and what counts as private or public, are strained by phenomena as diverse as telecommuting and public breast-feeding, even as the contestation over activities, spaces, and meanings works largely to reinscribe traditional notions of what activities are appropriate for whom, and where. Online gambling is merely one phenomenon about which we can see active and even fervent symbolic work being done by popular discourse to maintain sense-making narratives in the face of a world in which boundaries between public and private are shifting and even disappearing, and forms of presence are multiplying.

References

ABC News. (2006, July 25). Student says he was driven to crime by gambling addiction. *ABC News*. Retrieved April 14, 2009, from http://abcnews.go.com/GMA/LegalCenter/story?id=2232427&page=1

Carey, J. W. (1989). *Communication as culture*. New York: Routledge.

Christiansen Capital Advisors. (2004). Internet gambling. Retrieved April 10, 2009, from http://www.cca-i.com/Primary%20Navigation/Online%20Data%20Store/Internet_gabling_data.htm

Columbia University Medical Center Gambling Disorders Clinic. (2009). Teen gambling. Retrieved April 14, 2009, from http://www.columbiagamblingdisordersclinic.org/teen.htm

Cohen, S. (1972). *Folk Devils and Moral Panics*. New York: St. Martin's.

Cowan, R. S. (1983). *More Work for Mother*. New York: Basic Books.

Cypra, D. (2009, February 17). Fox Sports Net to air world poker tour season VIII. *Poker News Daily*. Retrieved April 14, 2009, from http://www.pokernewsdaily.com/fox-sports-net-to-air-world-poker-tour-season-viii-1071/

Donaldson-Evans, C. (2006, May). Junior jackpot: Teen gambling on the rise. *Fox News*. Retrieved April 14, 2009, from http://www.foxnews.com/story/0,2933,195751,00.html

Dretzin, R., & Maggio, J. (Producers and Directors). (2008). *Growing Up Online* [Television documentary]. *Frontline*. Boston: WGBH/PBS.

Frohlich, D., & Kraut, R. (2002). The social context of home computing. Retrieved May 3, 2005, from http://basic.fluid.cs.cmu.edu/articles/frochlich02-FamComputing-2.pdf

Gaul, G. M. (2008a, November 30). Players gamble on honesty, security of Internet betting.

Washington Post. Retrieved January 22, 2009, from http://www.washingtonpost.com/wp-dyn/content/article/2008/11/29/AR2008112901679.html

Gaul, G. M. (2008b, December 1). Prohibition vs. regulation debated as U.S. bettors use foreign sites. *Washington Post.* Retrieved January 22, 2009, from http://www.washingtonpost.com/wp-dyn/content/article/2008/11/30/AR2008113002006.html

Gesuele, V. (2005, October 5). College students struggle with gambling addiction. *The State Hornet.* Retrieved April 14, 2009, from http://media.www.statehornet.com/media/storage/paper1146/news/2005/10/05/Features/College.Students.Struggle.With.Gambling.Addictions-2424613.shtml

Heffernan, V. (2009, March 6). Flop. *New York Times.* Retrieved March 9, 2009, from http://www.nytimes.com/2009/03/08/magazine/08wwln-medium-t.html?_r=2

Horst, H. A. (2009). Families. In Ito et al. (Ed.), *Hanging Out, Messing Around, and Geeking Out: Kids*

Living and Learning with New Media (pp. 149-194). Cambridge, MA: The MIT Press.

Kennedy, T. L. M., Smith, A., Wells, A. T., & Wellman, B. (2008). *Networked families.* Washington, D.C.: Pew Internet & American Life Project. Retrieved May 11, 2010, from http://www.pewinternet.org/~/media//Files/Reports/2008/PIP_Networked_Family.pdf.pdf

Kruse, H. (2002). Narrowcast technology, interactivity, and the economic relations of space: The case of horse race simulcasting. *New Media & Society 4(3),* 385-404.

Kruse, H. (1993). Early audio technology and domestic space. *Stanford Humanities Review 3(2),* 1–14.

Lehmann, C. (2002, August 2). Internet gambling alarms addiction experts, government. *Psychiatric News 37*(15). Retrieved April 14, 2009, from http://pn.psychiatryonline.org/cgi/content/full/37/15/4

Lynes, R. (1985). *The lively audience: A social history of the visual and performing arts in America, 1880-1950.* New York: Harper & Row.

McCausland, J. (2005, December 8). Teens are gambling with their lives. *Seattle Post-Intelligencer.* Retrieved April 14, 2009, from http://www.seattlepi.com/opinion/251179_teengambling08

Marwick, A. E. (2008). To catch a predator? The MySpace moral panic. *First Monday, 13(6).* Retrieved January 11, 2009, from http://firstmonday.org/htbin/cgiwrap/bin/ojs/index.php/fm/article/view/2152/1966

Matthews, C. & Zamora, A. (2008, December 1). Timeline: Internet poker and the law. *Washington Post.* Retrieved January 22, 2009, from http://www.washingtonpost.com/wp-srv/investigations/poker/time.html?sid=ST2008112902159&s_pos=list

Morley, D. (2000). *Home Territories: Media, Mobility, and Identity.* New York: Routledge.

Ridley, R. (1985, March 7). Bringing home computers out of the dark. *Washington Post,* Washington Home section, 29. Retrieved September 2, 2009, from http://0-infoweb.newsbank.com.library.utulsa.edu/iw-search/we/InfoWeb?p_product=AWNB&p_theme=aggregated5&p_action=doc&p_docid=0EB360DEBA95E1E5&p_docnum=29&p_query-name=1

Sconce, J. (2000). *Haunted Media.* Durham, NC: Duke University Press.

Spigel, L. (1992). *Make Room for TV: Television and the Family Ideal in Postwar America.* Chicago: University of Chicago Press.

Stevens, R., Satwicz, T., & McCarthy, L. (2008). In-game, in-room, in-world: Reconnecting video game play to the rest of kids' lives. In K. Salen (Ed.), *The ecology of games: Connecting youth, games, and learning* (pp. 41-66). Cambridge, MA: The MIT Press.

Welte, J. W., Barnes, G. M., Tidwell, M. O., & Hoffman, J. H. (2008). The prevalence of problem gambling among U.S. adolescents and young adults: Results from a national survey. *Journal of Gambling Studies 24(2)*: 119–133.

CHAPTER FIVE

Spamculture

The Informational Politics of Functional Trash

KRISTOFFER GANSING

On the net we are all in one way or another working as subjective cyberneticians, constantly sending, sorting and filtering messages according to some (often collectively) imagined logic of order. Writing about the modern consumer society in the late 1960's, Jean Baudrillard foresaw how this ordering was becoming a main characteristic of "technical civilization":

> ...if hypochondria is an obsession with the circulation of substances and the functioning of the primary organs, we might well describe modern man, the cybernetician, as a mental hypochondriac, as someone obsessed with the perfect circulation of messages.(Baudrillard, 2005, p. 29).

If we think about Baudrillard's idea of man's obsession with the perfect circulation of messages in relation to digital network communications, we can observe how some objects become definable as trash in the sense of appearing as immediately disposable to the user. An example of such a "trash" category of informational objects is the electronic communication phenomenon known as spam. Being disposable as trash, however, only forms one of several possible functions for a spam object as most spam mails also have a concrete function: selling products, installing viruses, mal- or spyware, "phishing" for credit data, "junking" the receivers computer etc. they are in effect part of a by-productional wasteland of "functional trash." In the course of this chapter, I will explore the functional ambiguity of spam with the aim of developing an understanding of the informational politics of contemporary "spamculture." This latter notion is here proposed as a field at the margins of online communication, which does not depart from a linear intentional or interpretative model where messages are first instrumentally encoded in order to be subsequently decoded and/or subversively appropriated but rather from "parodist communication" as the potential breakdown or obliteration of meaning. As I will explore further in this chapter, the communication breakdowns of spamculture also point to spatial ambiguities in the material culture of online communications. From Nigerian letters to malware attacks, the objects of spamculture occupy an in-between space, in which any clear distinction between online and offline territories are blurred—suggesting that while such categories emerge from differences on a technical scale, they are still haunted by the same social negotiations. Parodist by nature, the informational politics of spamculture goes against the normative power of networks and always reveals the socio-cultural (albeit technically mediated) negotiation of meaning inherent to communication flows.

A number of spamculture-related informational objects and practices will be examined, from the proto-spam of so-called "Nigerian letters" via malware and botnets to the compulsory flow of messages on Web 2.0 "social media" platforms. It is argued that the latter context calls for a critical re-evaluation of the meaning of spam, a shift if you will from a spamculture 1.0 to a spamculture 2.0, where the production of waste as communication takes on a new integrated function, migrating from the margin to the center of our understanding and practice of digital communications. This critical re-evaluation of spam as ambiguous waste product and excess of communication is drawing on what Hawkins and Muecke (2003) have called "the cultural economies of waste" working on symbolic, affective, historical and linguistic strata. What Hawkins and Muecke stress is the hybrid value of waste, describing how waste "matters" (2003, p. xvi) beyond being simply a negative value or a romantic recuperation of turning bad into good: "Far from signaling the elimination of value, or its end product, waste is always with us in its fundamental ambiguity" (Hawkins & Muecke, 2003, p. xvi).

History/Parody: Defining spam

Excavating the history of spam, trying to discern the origin of using the term in communication contexts reveals a rather muddled picture which in the end seem to make it impossible to bring a closed definition to the term. Just as in the original "spam" sketch by Monty Python which has given name to the phenomenon, the term spam becomes a mutable point of negotiation around what constitutes meaningful communication. In the Monty Python sketch we are spectators to a parody of communication in the sense of Georges Bataille (1985) and Pierre Klossowski

(1970) who saw in parodist forms the display of a radical expenditure, a wasteful tendency which disrupts the order of any utilitarian goal of the production/consumption dialectic: "In art, parody functions as a kind of perversion because it is a deviation from a standard and promotes the sheer pleasure of communication"(Silberstein, 2003, p. 11).

A couple seated at a restaurant asks the waitress (played by Terry Jones) for the options on the menu, but their request is only answered by a list of meals which all recursively contain the ingredient "spam"—the hybrid luncheon meat product which itself can be seen as a parody, of ham. The communicative negotiation departs from the waitress, absurd persistence of presenting a destructive "eternal return of the same," i.e. her insisting that her repetitive spamming discourse can be regarded as novel information, all the while the customer struggles to break through with her and request for menu items containing other ingredients. In the light of this never-ending state of negotiation over meaning, there can be no solution to the enigma of spam as such, and the search for origins of spam, as evidenced in the many attempts to locate the first instance of spam mails (Lynch, 2001; Templeton, 2003), could be treated as a psycho-pathological reaction to this impossibility: as a parody is precisely a hybrid form which has to always to be ready for change in order to produce new situations of disruptive and potentially wasteful communication.

In a similar vein to that of instrumental waiter-customer protocol of restaurant communication mentioned above, it is not at all uncommon that electronic newsgroups and mailing lists are terminated or become radically reconfigured due to being overflowed by trash communication in the form of spam (Broeckmann & Arns, 2001). A famous example is that of the Usenet electronic newsgroups which were starting to be inflicted by the first bulk transmissions of unsolicited e-mails in the mid-1990s, from then on increasingly contributing to the diminishing importance of the entire service (Grossman, 1997). Or at least this is how the story is usually told.

In the case of Usenet history it becomes clear that the negotiation of meaningful communication takes place from very early on and that spam is one significant focalizer of this discussion. For example, in Usenet jargon it is common to refer to a phenomenon known as "Eternal September" 2007, a term directly related to the pervasive desire for a pure and spamfree communication order. As Grossman (1997) writes, the month of September was by the early nineties already a nuisance for Usenet regulars, as this month always saw an influx of new and potentially disruptive users in the form of university students commencing their studies or returning from summer breaks. "Eternal September" inaugurated in 1993, as Usenet jargon for when AOL (America Online—at the time the most popular Internet Service Provider and portal in America) included Usenet groups as a service accessible by its vast and ever-growing amount of users. This meant that an enormous amount of new users were joining the Usenet newsgroups, which in effect were swamped by people who most likely did not know about the specific rules of conduct carefully built up in the Usenet community since the end of the 1970's , ch. 1).

The advent of Usenet's "Eternal September" is an example of how in the eyes of some, a meaningful communication channel goes to waste by way of an operation similar to spamming, in this case the depressing prospect of an endless introduction of new users and their uninitiated ways of communicating online, breaking established protocols. Note the ambiguous nature of the rule breaking going on here: the new, as the introduction of unknown meaning, is seen as the dissolution of meaningful communication. This "ambiguity of waste" as both new meaning and the dissolution of meaning hints at the paradoxical possibility that all communication have on the political level the potential of becoming spam-like, whether or not it is intended as specifically such.

Following this line of argument, according to cybernetic information theory, the attribution of labels such as "trash" or noise to the phenomenon of spam would ontologically be incorrect. This is because the mathematical theory of information does not try to account for the meaning of messages, but only that messages are consistent pieces of information that needs to traverse channels of more or less noisy character (Shannon & Weaver, 1998). The noise of cybernetics, however, cannot be reduced to "spam" since this would mean attributing a semiotic understanding to a mathematical structure which is inherently non-representational. Thus, it is not in the cybernetic order that we should look for the motivations for filtering out spam, even if this often seems to be the case in the pervasive tradition of instrumental analysis of (anti-)spam rhetorics which produces a "naturalized" discourse on "useful" and "wasteful" communication. This naturalizing hides the economical and ideological structures from which these analyses take their cue. By not taking these definitions for granted we can turn the stakes around and depart from the political level of the constitution of an always already politically negotiated "spamculture."

Spam/Anti-Spam: The Geography of Malware

- ~ 230 dead as storm batters Europe.
- ~ U.S. Secretary of State Condoleezza Rice has kicked German Chancellor Angela Merkel.
- ~ British Muslims Genocide
- ~ Naked teens attack home director.
- ~ A killer at 11, he's free at 21 and kill again!
- ~ Examples of e-mail subject lines of the 2007 Malware Storm "Dref" (Sophos, 2008, p. 6; Sophos 2007a)

A case of malicious weather hit Europe in the beginning of 2007, apparently both of the meteorological as well as of the digital kind. Simultaneously as storms were causing deaths across the continent, electronic spammers were quick to latch on to the public interest by sending out messages such as the ones quoted above—hence the 2007 "malware storm" was born, taking its name from the Storm worm known as Dref or Dorf usually distributed as an attachment to the message (McMillan, 2007). The electronic storm actually long outlived its meteorological counterpart and re-appeared throughout the whole of 2007 in an amazing series of 50'000 different variants, peaking at the rate of 1 out of every 200 e-mails infected (MessageLabs, 2007; Sophos, 2008).

The further life of this particularly inventive spam storm throughout 2007 shows the convergence of virtual and "real-world" events: American users were targeted next around St. Valentine's day with subject lines such as "A Bouquet of Love" and were later swamped by calls for Independence Day celebration around the 4th of July. Later in the year, the public concern for surveillance was exploited in mails posing to be written by a private investigator, who would claim to be tapping the recipient's telephone. The private investigator message was accompanied by a "fake" mp3 file purporting to be an actual recording of the surveillance, but which instead predictably revealed itself to contain a Trojan horse (Sophos, 2007c).

If we are to believe the numerous companies specializing in anti-spam research and development, this increasingly creative spam storm is not likely to come to an end any time soon; as Internet security firm MessageLabs warns us: "Stormy weather ahead: During 2007 a number of major new players began to dominate the threat landscape; cyber- criminals who may be perceived as inspirational to their more amateur peers. Responsible for one of the largest botnets in the world, the Storm botnet is an experienced and professional team which MessageLabs predicts will have further impact in 2008, through its own activities and the antics of new players attracted to the buoyant market." (MessageLabs, 2007).

In spam jargon the technological institution that makes a thing such as a malware storm possible is a sophisticated "botnet." This is basically a network of computers on which the spam software sending out the mass of messages operates, along with scripts for automating scams, identity "phishing," virus distribution and other illicit activities on web-pages, chat services, p2p networks or other susceptible media. A botnet is a distributed network (cf. Galloway, 2004), radically de-centralized, in that the "end-users" or should we maybe rather say "end-victims" of its activities are themselves made unknowingly complicit in the automation of the spamming activity. This is accomplished through software that when installed on a recipient's computer starts to run hidden processes, not only junking (slowing down) the existing computer processes but also using the computer as a host for sending out new spam. Such a hijacked computer is known as a "Zombie" and the automated practice of such "Zombiefication" on a massive scale, through spam containing Trojan horses, ensures that a botnet is potentially always increasing in size (Sophos, 2006, p. 1).

This hidden automation can be compared to geographers Nigel Thrift and Shaun French's notion of "The automatic production of Space" (2002) as a discussion of the "automatically reproduced backround" (Thrift & French, 2002, p. 309) of software, which by way of code is seen as operating in a space that is "...constantly in-between, a mass-produced series of instructions that lie in the interstices of everyday life..." (Thrift & French, 2002, p. 311). The authors further suggest that this unconscious "traffic between beings" that is not immediately accessible to most of us is bound to turn up as effects in material space: "Increasingly, therefore, as software gains this unconscious presence, spaces like cities will bear its mark, bugged by new kinds of pleasures, obsessions, anxieties and phobias which exist in an insistent elsewhere." (Thrift & French, 2002, p. 312).

Various (anti-) spam research sources estimate that out of all e-mail messages being sent between as much as 70% and 80% can actually be classified as spam. In September 2009, Denmark took the lead as the most spammed country in the world, with 95.6% of all e-mails sent classified as spam (Symantec, 2009). In this spammed information environment, the workplace emerges as a striking example of a space bugged by the new kinds of pleasures and phobias Thrift and French talks about. The immense and ever-growing field of anti-spam research and development is quite obviously occupied with the threat posed by spam and malware to working productivity. The numerous research articles and "white papers" from big field players such as Sophos, McAfee and MessageLabs are of course written with the prospect of making businesses invest in the security software products connected to their research. In "Workload models of spam and legitimate e-mails" Gomes et al. point out. "The impact of spam traffic on the productivity of workers of large corporations is also alarming. Research firms estimate the yearly cost per worker at anywhere from US$-50 to US$-1400, and the total annual cost associated with spam to American businesses in the range of US$ 10-billion to US$ 87-billion." (Gomes, et al., 2007, pp. 690–91).

As a consequence of this perceived threat the discourse employed by anti-spam corporations often provide ample evidence of the kind of communication "phobia" surrounding spam as infected information. Thus in "Stopping Zombies, Botnets and Other Email- and Web-Borne Threats" Sophos notes: "Zombies have been found in organizations of all kinds, from financial planning companies to universities and nursing homes. They cause business disruption, network damage, information theft and harm to an organization's reputation."(Sophos, 2006, p. 1). Note how, in this context somewhat offbeat, "nursing home" is used as a potential target, suggesting that not only entities engaged in more or less immaterial production are at risk, but also sites of "innocent" biological life itself.

However, it is arguably most often in the imaginary or biopolitical realm surrounding production in various institutions such as the nation, and work or gender relations that spam is most usually perceived as a serious threat. This is evident in the further adventures of botnets such as the aforementioned Storm, showing that national conflicts can play a part in the politics of spam. Having expanded rapidly throughout 2007, from 2,815 attacks in the year 2007 to 1.7 million attacks in June/July the same year, the Storm botnet fell under suspicion to have converted into a soldier of fortune, ready to be recruited in emerging "cyberwars." As quoted in a SecureWorks August 2007 press release, anti-spam researcher Joe Stewart stated: "We don't know the motive of the Storm author; however, one possible theory could be that the hacker plans to use the trojan for more malicious activity than sending spam. It could be that the hacker is rapidly building up the botnet so it can be leased to other hackers so that they can launch massive attacks against whatever target they choose: an organization, country, etc." (SecureWorks, 2007).

The dystopian sci-fi scenario evoked by the SecureWorks report seemed a reality shortly after as May 2007 saw a massive distributed-denial of service (DDoS) series of attacks on Estonian government and police web-sites. This practically meant that those pages and accompanying services would be unavailable to regular users, blocked by massive amounts of simultaneous requests—most likely caused by automated botnets (Sophos, 2006; Thomson, 2008; Tung, 2007).

Adding to the tinge of conspiracy theory in this story, the Estonian government and most sources reporting on the subject believe that this botnet DDoS attack was an act of retribution from the Russian secret service, responding to an earlier offense where Estonia had relocated a Russian war memorial; even though this theory has never been finally proved, and naturally denied by Russia as well. Nonetheless, the Asymmetric Threats Contingency Alliance (ATCA) went on line to say: "The attackers used a giant network of enslaved computers on 9 May, perhaps as many as one million in places as far away as North America and the Far East, to amplify the impact of their assault," (as cited in Tung, 2007). No matter if the accusations are actually true, this sensational story of "cyberwar" implicates how malware relates to the geo-political image of software production. As Sophos security consultant Carole Theriault noted, "it's interesting to see how malware varies depending on location, often exploiting current country-specific online trends." According to the Sophos "Security Threat Report: July 2009 Update, the top malware hosting countries after the US (39.6%) are China (14.7%), Russia (6.3%) and Peru (4.3%) (Sophos, 2009, p. 6), while USA and Brazil compete for the title of most spam-relaying nation in the world. (Sophos, 2009, p. 8).

The term "malware" itself implies a kind of negative mirror image of software, as a definition used to connotate an activity supposedly far from the classical high-tech centers of software development such as Silicon Valley, New York or London. As such it should be understood as a criminalizing term, distancing a certain type of software production from "legitimate" types, which similar to the relationship between spam and communication, displays the same tenuous ambiguities. Of course, malware is not a "non-software" and on closer scrutiny one would most likely find that the "illicit" techniques it employs are not so far from the ones to be found in the hidden processes of proprietary software such as Windows Media Player or ITunes, two products notoriously known for their non-transparent "background effects" (Dowdell, 2005; Doctorow, 2006). Seen in this context, malware would rather constitute a case of "bad mimesis," a term derived by Deleuze and used by Thrift and French (2002, p. 329) to illustrate the imperfect and always changing nature of software as the "not-quite-copy." This also hints at the origins of the contemporary usage of the term "spam," both as a hybrid ham-imitation product and as deployed in the infamous Monty Python sketch, as a parody of communication: a wasteful and always mimicking form of counter-cultural production. In my next case I aim to show how this approach might be fruitful when it comes to analyzing a specific category of spam, that of the infamous 419 letter fraud also known as the Nigerian letter. This includes looking at the terrains being crossed, from the criminalizing responses of anti-scam research and policies to the cultural production such as stories of actual scam victims, Nollywood movies dealing with scammers, digital vigilantes also known as "scambaiters" and other potential "stakeholders"—in short a veritable playground for the politics of spamculture.

Scam/Bait: Politics of Spamculture

> "Oigbo man, I go chop your dollar,
> I go take your money, disappear
> 419 is just a game
> you are the loser, I am the winner"
>
> Lyrics in Pidgin from Nigerian 2005 hit "I Go Chop Your Dollar" by Nkem Owoh (Owoh as cited in Lindqvist, 2005)

One of the spam "objects" that most e-mail clients' junk folders will be overtly familiar with is the "Nigerian letter." A curiously "analogue" form of Internet scam, the study "Genres of Spam: Expectations and Deceptions" even identifies the letter genre as the next common form of spam (Cukier, Cody & Nesselroth, 2006, p. 6). Who has not wondered at the politely addressed "Attn. Sir/Madam," "Dear Friend" or "URGENT REPLY" mails that from time to time make their way into our legitimate in-boxes? Or fantasized about the actual people behind the often elaborate stories revealed by opening these mails? Mails with amazing plots and characters involving recently deceased dictators or industrialists, their missing or suffering families, mysterious mistresses, esteemed "Dr.'s" and barristers and of course huge amounts of money soon to be making it your way. As with my earlier discussion of the Usenet spam politics, distinguishing Nigerian letters from "legitimate" communication might not be such a clear-cut affair after all.

This form of letter fraud has its roots in the classic turn-of-the-century scheme known as "The Spanish Prisoner" which is a so-called "confidence trick" where the victim is tricked into voluntarily giving up more and more money under the impression that he/she will eventually receive a huge amount of money (Spanish Prisoner, 2009). The contemporary successor to this scheme is the AFF "advance fee fraud" (Advance Fee Fraud, 2008) and this is where the so called Nigerian letter comes into the picture, also known as "419 fraud," taking its name from the section number of the Criminal Code of Nigeria dealing specifically with AFF (Wizard, n.d.). The shift to Nigeria as the main geographical point of reference for this type of fraud in itself reflects structural shifts in global culture and economy. Numerous sources claim that letter frauds emerged in Nigeria in the 1980s as an alternative source of income following the collapse of oil prices and the history of rampant corruption in the country (United States Department of State, 1997, p. 5; Wizard, n.d.).

It is not only the global geography of scams that has changed but also the media used in order to execute them. From postal letters and faxes in the 1970's and 1980's to e-mails and other electronic means of communication from the mid 1990's and onwards. According to one private consultant in the field, the array of different communication channels now used in AFF include "Mail, Fax, Phone, E-mail, Chat rooms, Dating web sites, Internet auction sites , Social and business networking sites , Mobile phone SMS , Internet phone and Internet gaming..."(Ultrascan Advanced Global Investigations, 2008). While certainly not all of the AFF's conducted in these media can be identified as coming from Nigerian "scammers," they still comprise such a large percentage that Nigeria, together with other sub-Saharan African countries, have come to represent the origin of this type of scam. As Ina Nolte of the University of Birmingham's Centre of West African Studies notes: "The availability of e-mail helped to transform a local form of fraud into one of Nigeria's most important export industries," (Nolte as cited in Andrew, 2006).

In Internet security reports, perpetrators of the 419 e-mail frauds are depicted as belonging to organized rings of cybercrime, with far-reaching connections to other types of criminality such as drug trafficking and even terrorism (Ultrascan Advanced Global Investigations, 2008; United States Department of State., 1997). There has been very little field work which deals directly with these supposed hardened third-world criminals. From another perspective, as Jenna Burrell has pointed out in a pioneering (field) study, West African scamming can be seen as a form of grassroots media production: "These efforts can be described as a form of grassroots media production because they involve content generation and distribution (often through collaborative efforts) outside of established media institutions by individuals with limited access to international media outlets." (Burrell, 2007,pp. 14–16). Her work points to the less than glamorous reality of young men in Ghana populating the local Internet cafés and who occasionally succumb to initiate 419 scams, thus "…compelled to work within a set of false perceptions and distorted archetypes that they viewed as alien and even openly resented." (Burrell, 2007, p.19). In Burrell's analysis, the "strategic misrepresentation" undertaken by African 419 scammers is ultimately regarded as a "problematic empowerment" in that their stories of corrupt regimes with poor suffering Africans in need of money and health care reinforce stereotypical Western views. The scammers are, according to Burrell, then playing the victim, using an inwardly projected "Western gaze" from which they would rather escape but are now forced to endlessly repeat.

At the same time it is clear that in Nigeria and other West-African countries, a pop-cultural counter-narrative with its own terminology and mythology have arisen around the 419 scamming. Internet cafés are thus transformed into sites of the economy of desire (cf. Ludovico, 2005) as young men spend ten hours a day, seven days a week engaged in "Yahoo yahoo," dreaming of becoming the next "Yahooboy"—referring to the popular free e-mail service, these are the local slang expressions for scamming and scam millionaires, respectively (Lawal, 2006; McLaughlin, 2005). The Yahooboys have even become the subjects of popular films from the thriving Nigerian film industry (Nollywood) with features such as *The Master* which in 2005 also gave birth to the "419 anthem": the comedian and movie star Nkem Owoh's hit song "I Go Chop Your Dollar," eventually banned in Nigeria (Libbenga, 2007). Even though, as Burrell states, scammers can be regarded as victimizing themselves in order to fit the Western gaze, the romanticization of scam culture in everyday life and pop culture also point to another post-colonial reading. In discussing "agency" and "counter-discourse" in the Nigerian video film industry, Chukwuma Okoye (2007) discusses the International and domestic negative reactions to the success of "Nollywood." Okoye emphasizes the transformative aspect of looking into one's own mirror, that even when Nigerian films are derivative of Hollywood or Bollywood models, "they are translations or appropriations rather than copies" (2007, p. 26) and that the Nigerian video film thus "functions as a postcolonial system of decolonization" (2007, p. 26). Taking this interpretation to the Nigerian letter, it is interesting to note how the elaborate plots of this genre display similarities to some of the standard elements of the Nigerian movie which are steeped in urban legends involving supernatural get-rich-quick schemes leading to the main character's downfall and ultimate redemption through the re inscription of Christian family life (Haynes, 2007). As Jonathan Haynes (2007) writes about the films of Kenneth Nnebue, one of the Nigerian video pioneers, they "(…) established Nollywood's essential themes: corruption, moral turbulence and pervasive anxiety of the post-oil-boom era; the garish glamour of Lagos; titillating and dangerous sexuality; melodramatic conflicts (…)" (Haynes, 2007, p. 30). Suggesting that it is not so easy to dismiss the Nigerian letter plots as the result of trying to exploit the Western image of African life but rather that these images are already part of a more complex post-colonial narrative.

However, if these pop- and counter-cultural expressions of 419 scamming establishes a kind of reverse master-slave "narrative" of first-/third world relations, this fantasy is troubled by the actions of so-called "jokemen." Better known as "scambaiters" in the West, jokemen are people who try to get back on the scammers by out fooling them, using the same means of deception (Tuovinen & Röning, 2006). Scambaiting is a kind of everyday vigilantism that prolongs the scam, playing with the performativity of the involved subjects. What we have here is a subversion of the already reversed master-slave narrative, often with a tragic-comic outcome. These parodic narratives are presented online in the "trophy rooms" and "halls of shame" which are common features on scambaiters web-pages. On the site "419 Eater" you can for example read of how infamous scambaiter "Shiver Metimbers" (real name: Mike Berry) fooled the unsuspecting Mrs. Joyce Ozioma and her would-be scamming colleagues into copying an entire *Harry Potter* book by hand. The case is presented on the site by the 419 Webmaster, Shiver Metimbers himself, in the following way:

> TITLE: Harry Potter & The Well of Scammers
> SCAMMER NAME: Joyce Ozioma
> SCAMMER LOCATION: Nigeria
> SCAMBAITER: Shiver Metimbers
> One of the golden rules of scambaiting is 'make your scammer do all the work.' Tie the guy up. Make his life miserable by dragging out the dealings for as long as humanly possible. What better way to keep a scammer busy than to make them copy an entire book by hand?"(Berry, 2006)

The presentation proceeds with the introduction letter from the 419 fraud stock character "Barrister Musa Issah" who asks the recipient (Metimbers posing as "Arthur Dent") if he would be interested in being the caretaker of a 27 million $ deposit, stemming from the house of the late Sanni Abacha, former President/Head of State of the Federal Republic of Nigeria. All he has to do is provide a telephone number for further arrangements. The mail itself displays many of the dead give-aways of the Nigerian letter genre, one of them being the somewhat archaic expressions such as "intimate me" or "If you will be interested to act…" (Issah as cited in Berry, 2006). As Douglas Cruickshank (2001) has further noted in an analysis of the literary qualities of this "crime genre" usually "no effort is spared in keeping the letter's recipient informed and reassured." (Cruickshank, 2001, 3) and in this case we can accordingly note how the author takes extra care in reassuring his own honest intentions by using phrases like "my humble self" and "thanks for your sincere understanding" (Issah as cited in Berry, 2006). Our scambaiter, now posing as Arthur Dent, PhD, director of "Singlesideband Systems," answers in a tone mimicking the official tone of the barristers mail, politely declining his offer, instead presenting his own company's activities: "a very important 4 year long research project on Advanced Handwriting Recognition and Graphology systems."(Berry, 2006). Would the barrister perhaps be interested in serving as a test person for this research, by providing handwritten samples for which the company pays a US $100.00 per page fee? Of course he would, and thus the scam has been turned around.

Further down the mail-exchange we find that the barrister has passed on the responsibility to his employee Mrs. Joyce Oizoma who proceeds with producing handwritten, page-by-page exact copies, of the entire second volume in J.K. Rowling's *Harry Potter* series, *"Harry Potter and the Chamber of Secrets."* The resulting 293 handwritten pages were scanned by Joyce and subsequently published at 419 Eater.com along with a recording of a brief telephone conversation as the increasingly desperate "copywriter" tries in vain to get her due payment.

Even more bizarre stories than the Harry Potter case are presented on scambaiters forums, among the most famous, one involves a hand-carved 1:1 replica of a Commodore 64 computer and another a hapless video-film re-enactment of the Monty Python "Dead Parrot" sketch (Rosenbaum, 2007). The tedious tasks and occasional journeys scambaiters compel their mostly African victims to complete are known as "Safaris" and they consistently refer to them as "mugu's," the Igbo word for "fool" originally used by 419 scammers to refer to their victims. Thus scambaiters shroud their triumphant counter-scamnarratives in a curious neo-colonial discourse, which shows how even a tactical approach to media production such as 419 scams can be re-appropriated by its supposed antagonistic culture.

"You be the mugu, I be the master," the phrases from the 419 "anthem" are endlessly re-iterated in the scam/bait game as the "black" and "white" man take turns in getting back on one another. What is disabled in this game is the more instrumental function of the economical motive behind the Nigerian letter, as well as that behind spam-filtering—the scambaiter could after all choose to ignore these mails altogether. What we are left with as observers could be likened to a kind of parody of communication in capitalist societies, stripped of its ability to function economically.

The parody form is used by 419 scammers in order to at all break through in the electronic communication order. As economically instrumental as their goals may be, their parodying still invades and indeed migrates into new configurations such as the scam/bait game where functional trash seem to turn into dysfunctional trash, temporarily disabling any rational communication protocols.

Prod/Using: Functional Trash Goes Womp!

Classifying what is spam and what is legitimate communication take on new dimensions of complexity in Web 2.0 services. In the context of new social media platforms we may ask, as the title of a spam-research paper suggests, "Is Britney Spears Spam?" (Donath & Zinman, 2007). Apart from the risk of receiving Viagra messages or other typical spam topics, users are supposed to expand their networks by sending out friend requests, thus "Differentiating between welcome and unwelcome communication is subjective, dependent on the taste and interests of the user." (Donath & Zinman, 2007, p. 1). For example, MySpace is a service built on the prospect of having as many friends in your network as possible. The service in a way encourages spamming as people compete to send out as many requests as possible, some of them from fake "bot" identities or celebrities who (like Britney Spears) have automatized this process through humans acting as "bots" for their message sending. But if you really did make friends with everybody in the MySpace network, how would you actually go on about communicating with them?

The success of Facebook could partially be explained by their different approach to the friend-making activity, as their service is built on a congruence between "real-life" and virtual identity, it encourages that all communication build from your already existing friend network. Paradoxically, at the same time as Facebook gives the impression of an isolated clean environment on the net, where supposedly people really are who they make themselves out to be, the mode of communication still adheres to the quick "social messengering" format. Thus MySpace as well as Facebook users are constantly spamming each other not only with friend requests but also with messages from their everyday: on their current location, where they will go out tonight, on who's hot and who's not etc. Spaces like Facebook have been portrayed as liberating consumers from their passive role (Leadbeater, 2007), but this view often overlooks the limited protocols of the communicative regime users are being liberated into.

As is made clear by the Facebook social ad system (Hodgkinson, 2008) it is not the media content of the users that matter but rather the exploitation of the relations built around this content. As a consequence, the communication and media content on these sites take on a form of immediate disposability. We can observe this in the endless remix and mashup culture of YouTube and in the short-message mode of communication favoured on Twitter, the "microblogging" service which limits messages (known as "tweets") to 140 characters.

The often referenced movement of users turning into producers is here rendered in a different light, as a production intimately tied to a new kind of modular consumer society. In *The Production of Space,* Henri Lefebvre used the term "spatial economy" for the kind of space where certain relationships between people are valorised over others, functioning by way of an economy in which the users are strangely silent since "It imposes reciprocity and a communality of use" (Lefebvre, 1991, p. 56). In Lefebvre's spatial economy, waste has a concrete function at the end of the production cycle, "giving it meaning and justification" (1991, p. 58) as, for example, in the expenditure of workers going on a charter vacation. These moments of waste are planned "with greatest care" by the consumer, and are encouraged by the mode of production of the 20th-century consumer society, in which the linear production cycle always seem to be restarted in a circular fashion. In the process based and networked digital consumer society, production doesn't seem to need to be restarted in this way, as production is modular and occurring simultaneously with the constant wasteful moment.

The production of waste in process-based "informational" capitalism then, might further be viewed as a development of what Jean Baudrillard saw as the "systematization of the inessential" in modern consumer societies (2005, p. 9). In Baudrillard's perspective on technology and society laid out in *The System of Objects*, this inessentiality can be viewed in the growing discordance between an object's technological reality and its function in the system (of consumer objects). This discordance is further used as a founding principle of how consumer society works by an "ideological function of integration" (Baudrillard, 2005, p. 9) which gradually impinges a differential subjective economy on the rational organization of technology (2005, p. 9, note 7). This leads Baudrillard on to the discussion of serially produced objects of mass consumption in furniture and interior design where "functional colour" does not any longer refer to any moral aesthetic regime but to purely "abstract values," which are not symbolic but only meaningful as "relative to each other and to the whole." (2005, pp. 34–36) (as in the whole system of serial and "functional" production). This could be compared to the flow of uploaded images on a Web 2.0 service like Flickr. (http://www.flickr.com). In the contemporary system of objects, the modular spatial economy, the images on a service like Flickr are not valuable because of their representational content, but because of the collective amount of data they are part of and its connected user relations, on the basis of which ad space can be sold. Value no longer needs to be serially hinged even on an already fleeting concept such as color but can be attached to the modular relations between data, leaving the content, by way of its instant disposability/recyclability/remixability, to take on the form of functional trash:

> ...'functional' in no way qualifies what is adapted to a goal, merely what is adapted to an order or system: functionality is the ability to become integrated into an overall scheme. An object's functionality is the very thing that enables it to transcend its main 'function' in the direction of a secondary one, to play a part, to become a combining element, an adjustable item, within a universal system of signs. (Baudrillard, 2005, p. 63)

In the case of the functional trash of Web 2.0 culture, the transcendence that Baudrillard spoke of, from the main to the secondary function has been conflated—as the integration into a system of production is re-inscribed also in the materiality of the digital object itself. In regards to the reciprocal logic of abundant online communication, Web 2.0 emerge also as a Spamculture 2.0, consisting of sites optimized for the capitalist exploitation of the production of waste. As spam-like communication/production is re-instrumentalized in a strategic business model, Spamculture 2.0 tries to "disambiguate" functional trash through the normative power of social networks. However, this technological integration and conflation of the production/consumption logic does not mean that the ambiguity of waste entirely disappears into a streamlined flow of smooth communication. On the contrary, disruptive phenomena such as malware and spam can only be expected to increase even more along with the growth of web-based social networks (cf. Sophos 2009, p. 17), proving Bill Gates irrevocably wrong in his 2004 claim that "Spam will soon be a thing of the past" (Gates as cited in Weber, 2004).

As a blogging service gaining popularity throughout 2009, Twitter became a particularly popular target of creative spammers who found web-based ways of exploiting its security gaps. A young hacker going by the name of Mikeyy Mooney repeatedly managed to implant a "worm" into Twitter, first causing mayhem through messages promoting his Twitter imitation service "StalkDaily"—a site that supposedly "follows the same functions as Twitter, except more advanced" (Kincaid, 2009). What StalkDaily promises to do is allowing users to "stalk" each other, meaning that they're notified each time their victims uploads images or post comments on other users (Kincaid 2009, Bialer & van Poppel, 2009). When Twitter users followed the link to StalkDaily, they were hit a by a XSS, cross-site scripting attack, which turned their accounts into spam relays for further messages promoting the site across Twitter. In a more advanced version of the worm, Twitter users simply needed to visit other infected accounts in order for themselves to also get infected.

> (...)I did this out of boredom, to be honest. I usually like to find vulnerabilities within websites and try not to cause too much damage, but start a worm or something to give the devel-

> opers an insight on the problem and while doing so, promoting myself or my website, (Mooney as cited in Bialer & van Poppel, 2009).

In a further development of the worm, Mooney hit Twitter again in May 2009 causing users to involuntarily spam each other with small aphorisms and jokes all containing the word "Womp": "Money is not the only thing, it's everything. Womp. mikeyy." and "'Your future depends on your dreams,' So go to sleep. Womp. mikeyy." (Cluley, 2009). In the online repository for vernacular language that is urbandictionary.com, looking up Womp results in no less than 34 entries, several of which state the ambiguity of this term: "Can mean ANYTHING, or is just said randomly at any given time." (urbandictionary.com).

As much as Web 2.0 services are built on appropriating waste, they do still generate their own waste products: new forms of spam that seem to be intensifying the kind of parodies of communication previously observed in the scam/bait games of Spamculture 1.0. In his book on living currency, *La Monnaie Vivante* (1970), Pierre Klossowski introduced a kind of parody of Utopian concepts of gift economies by suggesting that the capitalist economy could be replaced with one based on the exchange of pleasure instead of currency. As Silberstein (2003) suggests, it was a parody of Utopia because, as Klossowski also maintained, "the body cannot be fully appropriated by the other and resists exchange" (Silberstein, 2003, p. 187). Meaning that something will also go to waste in this exchange, a production of waste that in exposing the limits of utopia "shifts our focus to the immediate experience of pleasure" (2003, p. 20). Through the Klossowskian reading of parody and Utopia we might, as Silberstein suggests, approach the question of how to "understand the conditions for communication and exchange without creating an oppressive, totalizing discourse?" (2003, p. 212).

In the Spamculture 2.0, we see how spam-like and never-ending communication flows are integrated into a strategic business model, yet the ambiguity of negotiating between legitimate and illegitimate communication does not go away that easily. This is how spam works in that "form by camouflage" (Baudrillard, 2005, p. 61) is also re-inscribed back into its technological functioning, in the deceptive analogue methods of Nigerian letters as well as in the more advanced forms of "phishing" and message relaying through malware and Web 2.0 platforms. Annoying and destructive as these interruptions in our daily communication order might be, they still demonstrate the possibility that normative networks might be tamed to different ends. They do this by way of what Baudrillard observed as the continual "regression" of the consumer society, taking place along with its continual development. The result could be a contracting movement, as individual spammers are, and have been proven capable of, slowing down entire corporations (McWilliams, 2005) and in the context of social media this could be a Baudrillardian example of how technology turns against its own system and becomes fragile and dysfunctional. For Baudrillard this happens in the breaches and malfunctions of a system, allowing sexuality to revive the static systems "even if for only a moment, even if it takes the form of a hostile force (...), and even if its emergence in such circumstances means failure, death and destruction." (2005, p. 143). We have already seen some examples of how this might be put to further (un-)productive use in the world of social media, not only as new forms of web-based spam, but as waste products exposing the limits of communication through a parodying of the entire Web 2.0 system of communication.

References

Advance Fee Fraud. (2008, June 6). In *Wikipedia, The Free Encyclopedia.* Retrieved , November 5, 2009, from http://en.wikipedia.org/w/index.php?title=Advance_fee_fraud&oldid=217593834

Andrew, R. (2006). Baiters Teach Scammers a Lesson. Retrieved August 20, 2008 from http://www.wired.com/techbiz/it/news/2006/08/71387

Bataille, G. (1985). *Visions of excess* (A. Stoekl, Ed.). Minneapolis: University of Minnesota Press.

Baudrillard, J. (2005). *The System of Objects.* London: Verso.

Berry, M. (2006). Harry Potter & The Well of Scammers. Retrieved March 19, 2008 from http://www.419eater.com/html/joyce_ozioma.htm

Bialer, J. & van Poppel, M. (2009). 17-Year-Old Claims Responsibility for Twitter Worm. Retrieved November 05, 2009 from http://www.bnonews.com/news/242.html

Broeckmann, A. & Arns, I. (2001, November 13). The Rise and Decline of the Syndicate: The End of an Imagined Community. Message posted to http://amsterdam.nettime.org/Lists-Archives/nettime-1 -0111/msg00077.html

Burrel, J. (2007). Problematic Empowerment: West African Internet Scams as Grassroots Media Production. *Information Technology and International Development* Retrieved February 10, 2008, from: http://people.ischool.berkeley.edu/~jenna/jburrell_problematic_empowerment.pdf

Cluley, G. (2009, April 18). New Mikeyy Worm Makes Jokes at Twitter's Expense. Posted to http://www.sophos.com/blogs/gc/g/2009/04/18/mikeyy-worm-jokes-twitters-expense/

Cruickshank, D. (2001, August 07). I Cave Your Distinguished Indulgence (and All Your Cash). Salon.com. Retrieved August 20, 2008 from http://archive.salon.com/people/feaure/2001/08/07/9 scams/index.html

Cukier, W. L., Cody, S. & Nesselroth, E. J. (2006). Genres of Spam: Expectations and Deceptions. *Proceedings of the 39th Annual Hawaii International Conference on System Sciences—Volume 03,* 51.1. doi:10.1109/HICSS.2006.195

Doctorow, C. (2006, January 11). iTunes Update Spies on Your Listening and Sends It to Apple? Posted to http://www.boingboing.net/2006/01/11/itunes-update-spies-.html

Donath, J. & Zinman, A. (2007). Is Britney Spears Spam? *CEAS 2007—Fourth Conference on Email and Anti-Spam.* Retrieved February 10, 2008, from http://smg.media.mit.edu/papers/Zinman/britneyspears.pdf

Dowdell, J. (2005). Is Microsoft Using Windows Media Player to Spy on Users?. Retrieved April 10, 2008 from http://www.securitypronews.com/insiderreports/insider/spn-49–20050315Is Microsoft Using Windows Media Player to Spy on Users.html

Eternal September. (2007, February 1). In *Urban Dictionary*. Retrieved 13:15, November 5, 2009, from http://www.urbandictionary.com/define.php?term=eternal%20september

Eternal September. (2009, November 1). In *Wikipedia:The Free Encyclopedia*. Retrieved, November 5, 2009, from http://en.wikipedia.org/w/index.php?title=Eternal_September&oldid=323317305

Galloway, A. R. (2004). *Protocol : How Control Exists after Decentralization*. Cambridge,: MIT Press.

Gomes, L. H., Cazita, C., Almeida, J. M., Almeida, V., & Meira, W. (2007). Workload Models of Spam and Legitimate E-mails. *Performance. Evaluation,* (64, 7–8) (Aug. 2007), 690–714. DOI= http://dx.doi.org/10.1016/j.peva.2006.11.001

Grossman, W. M. (1997). *Net. Wars*. New York: New York University Press. Retrieved from http://www.nyupress.org/netwars/

Hawkins, G. & Muecke, S. (2003). *Culture and Waste: The Creation and Destruction of Value*. Oxford: Rowman & Littlefield.

Haynes, J. (2007). Nnebue: the anatomy of power. *Film International, 5*(28), 30–40.

Hodgkinson, T. (2008, January 14). With Friends Like These....The Guardian. No pagination. Retrieved February 15, 2008 from http://www.guardian.co.uk/technology/2008/jan/14/facebook

Kincaid, Jason. (2009, April 11). Warning: Twitter Hit by StalkDaily Worm. Posted to http://www.techcrunch.com/2009/04/11/twitter-hit-by-stalkdaily-worm/

Klossowski, P. (1970). *La monnaie vivante*. Paris: Éric Losfeld.

Lawal, L. (2006). *Online Scams Create "Yahoo! Millionaires*. Retrieved March 19, 2008 from http://money.cnn.com/magazines/fortune/fortune_archive/2006/05/29/8378124/

Leadbeater, C. (2007). *We-think: The Power of Mass Creativity*. London: Profile.

Lefebvre, H. (1991). *The Production of Space*. Oxford: Blackwell.

Lindqvist, N. (2005, Oct. 21). I Go Chop Your Dollars. Message posted to http://www.lindqvist.com/en/i-go-chop-your-dollars

Ludovico, A. (2005). Spam, The Economy of Desire. *Neural*, Retrieved March 10, 2008 from http://www.neural.it/art/2005/12/spam_the_economy_of_desire.phtml

Lynch, K. (2001). *Keith Lynch's Timeline of Spam Related Terms and Concepts*. Retrieved November 01, 2008, from http://keithlynch.net/spamline.html

McLaughlin, A. (2005, December 15). Nigeria Cracks Down on E-mail Scams. *The Christian Science Monitor*. Retrieved, February 15, 2008 from http://www.csmonitor.com/2005/1215/p07s02-woaf.html

McMillan, R. (2007). *Storm Malware Shapes Up as Worst 'Weather' in Years*. Retrieved January 10, 2008 from http://www.computerworld.com/action/article.do?command=viewArticleBasic&articleId=9008818

McWilliams, B. (2005). Spam Kings, The Real Story behind the High-Rolling Hucksters Pushing MessageLabs. (2007). Intelligence 2007 Annual Security Report: A Year of Storms, Spam and Socializing With the Enemy. Retrieved February 08, 2008 from http://www.messagelabs.com/mlireport/MLI_2007_Annual_Security_Report.pdf

Okoye, C. (2007). Looking at Ourselves in the Mirror: Agency, Counter-Discourse, and the Nigerian Video. *Film International, 5*(28), 20–29.

Rosenbaum, R. (2007, June). How to Trick an Online Scammer into Carving a Computer out of Wood. *The Atlantic Monthly*. Retrieved February 05, 2008 from http://www.theatlantic.com/doc/200706/cyberscam

Sandoval, G. (2008). MySpace Wins Suit Against 'Spam King.' Retrieved February 15, 2008 from http://news.cnet.com/8301–10784_3–9930977–7.html

SecureWorks. (2007). Bots Launching Storm Attacks Increase Dramatically Totaling 1.7 Million in June/July. Retrieved February 22, 2008 from http://www.secureworks.com/media/press_releases/20070802-botstorm/

Shannon, C.E. & Weaver, W. (1998). *The Mathematical Theory of Communication.* Urbana: University of Illinois Press.

Silberstein, L. (2003) *'Live Currency' Prostitution and Parody in the Works of Pierre Klossowski.* (Doctoral dissertation, Columbia University, 2003) Retrieved from http://app.cul.columbia.edu:8080/ac/handle/10022/AC:P:5292

Sophos. (2006). Stopping Zombies, Botnets and Other Email- and Web-Borne Threats. Retrieved March 4, 2008, from http://www.sophos.com/security/whitepapers/sophos-zombies2-wpus

Sophos. (2007a). Trojan Spam Storm Hits Inboxes, Races to Top of Malware Charts. Retrieved March 4, 2008, from http://www.sophos.com/pressoffice/news/articles/2007/01/malware storm.html

Sophos. (2007b). Sophos Security Report 2007 Reveals United States Is Worst for Malware Hosting and Spam-Relaying. Retrieved March 15, 2008, from http://www.sophos.com/pressoffice/news/articles/2007/01/secrep2007.html

Sophos. (2007c). Has Your Phone Been Wiretapped? The Dorf Trojan Wants You to Think So. Retrieved March 4, 2008, http://www.sophos.com/pressoffice/news/articles/2007/11/detective-dorf.html

Sophos. (2008). Security Threat Report 2008. Retrieved March 4, 2008, from http://www.sophos.com/sophosthreatreport08

Sophos. (2009). Security Threat Report: July 2009 Update. Retrieved November 5, 2009, from http://www.sophos.com/sophos/docs/eng/papers/sophos-security-threat-report-jul-2009-na-wpus.pdf

Spanish Prisoner. (2009, July 14). In Wikipedia: The Free Encyclopedia. Retrieved November 5, 2009, from http://en.wikipedia.org/w/index.php?title=Spanish_Prisoner&oldid=302064280

Symantec. (2009). Symantec Announces September and Q3 2009 MessageLabs Intelligence Report. Retrieved November 5, 2009, from http://investor.symantec.com/phoenix.zhtml?c=89422&p=irol-newsArticle&ID=1336240&highlight=

Templeton, B. (2003). Origin of the Term "Spam" to Mean Net Abuse. Retrieved November 5, 2009, from http://www.templetons.com/brad/spamterm.html

Terranova, T. (2004). *Network Culture : Politics for the Information Age*. London: Pluto Press.

Thrift, N. & French, S. (2002). The Automatic Production of Space. *Transactions of the Institute of British Geographers* 27(3), 309–335. DOI:10.1111/1475–5661.00057

Thomson, I. (2008). Russia 'Hired Botnets' for Estonia Cyber-War. Retrieved January 10, 2008 from http://www.itnews.com.au/News/NewsStory.aspx?story=53322.

Tung, L. (2007). Storm Worm Botnet Threatens National Security? Retrieved February 10, 2008 from http://www.zdnet.com.au/news/security/soa/Storm-worm-botnet threatens-national security-/0,130061744,339281305,00.htm

Tuovinen, L. & Röning. J. (2006). Scamming the Scammers—Viglante Justice in Virtual Communities. *European Conference on Computing and Philosophy*. Retrieved January 10, 2008 from http://www.anvendtetikk.ntnu.no/ecap06/program/Tuovinen.rtf

Ultrascan Advanced Global Investigations. (2008). 419 Advance Fee Fraud. Retrieved, March 19, 2008 http://www.ultrascan.nl/assets/applets/2007_Stats_on_419_AFF_feb_19_2008_version_1.7.pdf

United States Department of State, Bureau of International Narcotics and Law Enforcement Affairs. (1997). Nigerian Advance Fee Fraud. Retrieved November 05, 2009, http://www.state.gov/documents/organization/2189.pdf

Weber, T. (2004). *Gates forecasts victory over spam.* Retrieved March 10, 2008 from http://news.bbc.co.uk/2/hi/business/3426367.stm

Wizard, B. (n.d.). *Nigeria—The 419 Coalition Website.* Retrieved November 5, 2009 from http://home.rica.net/alphae/419coal/

Part II

Citizenship, Public Space and Communication Online

Social space is saturated with an abundance of media and it is not possible to conceive of physicalities without the contingencies of virtualities, and vice versa. As a result of new-media enhanced social mediation and pervasive mediatization, institutional and normative structures within which groups and individuals have operated for the purposes of democratic participation, communicative action and social/political networking are in a state of flux. New media tools (from online extensions of social movements to blogs, podcasts, social networks and other virtual tools) and citizen/user generated content have crucial bearings on the way individual and communal "presence" is experienced and social space is re/produced today. Politics and civic expressivity find new avenues of articulation and new opportunities of descaling and rescaling online—both in terms of spatial reach and scope of action—engendering new offline spatialities, new prosperities and new destitutions. Yet, despite the obvious permeability of online space to power, domination and inequality, by expanding immediate and spatially delimited social presence and (dis)order into spatiotemporal vastness, virtuality opens up new territories for social re/negotiation.

Such moments of social flux necessitates a reconsideration of our understanding of concepts such as identity, self and community, citizenship, social interaction, social control, rights and activism and social territory vis-à-vis processes of mediation and mediatization. This section aims to expand upon the conceptual scopes of such charged vocabulary and related theories and to offer nuanced, empirically informed analyses of the ways in which online communication today both afford *and* limit particular modes of social voice and presence on the whole.

In "Mediapolis, Human (In)Security and Citizenship: Communication and *Glocal* Development Challenges in the Digital Era" (Chapter 6) Thomas Tufte begins by assessing communication for social change from a citizen's perspective. Tufte then provides an example—taken from Tanzania in Eastern Africa—of a civil society–driven media platform which seeks to enhance processes of empowerment and ultimately influence good governance and promote social change. Using both Roger Silverstone's concept of 'mediapolis' to explore the mediated public sphere as a space for civic action and participation, and the concept of 'human (in)security' which refers to how cultures of fear and insecurity permeate public discourse and the public sentiment, Tufte discusses how to theorize communication for social change in a digital era, approaching it from a citizens perspective and connecting the contexts outlined in the chapter—human (in)security, mediapolis and citizenship—with the Tanzanian reality.

Chapter 7 ("The Rise and Fall of Online Feminism") is an investigation by Liesbet van Zoonen into how, during the early years of the Web, feminists thought that the lack of physical presence and the anonymity of the Internet would make it possible to experiment with diverse gender identities and even non-gendered identities. Such practices would reverberate in the offline world, undermine dichotomous gender discourse and liberate women and men from their fixed genders. Only ten years later, the authors of this chapter contend, anonymity has acquired a bad reputation due to anxieties about predators and bullies. Utopian prospects have been annihilated by a discourse of risk and fear that makes gender experiments suspect. It is a classic story, van Zoonen et al. write, of rise and fall: in this case of online feminism.

In "Social Movement Web Use in Theory and Practice: A Content Analysis" (Chapter 8), Laura Stein examines US-based social movement organization's (SMO) Internet use at one of its most visible points of access, the World Wide Web. Stein reviews and critiques the limited questions raised, and methodologies used, by existing communication and social movement scholarship aiming to study social movement communication and Internet use. And, drawing on alternative media studies, she develops a typology of communication functions central to social movements. This typology includes information provision, action/mobilization, interaction/dialog, lateral linkages, creative expression, and fundraising/resource generation. Stein extends previous methodological approaches by surveying a random sample of SMO websites to determine whether and to what degree they exhibit features or attributes related to these types. These sites are drawn from a more comprehensive list identified with six prominent movements, including: environmentalism; the gay, lesbian, bisexual and transgender (GLBT) movement; the anti-corporate globalization movement; the human rights movement; the media reform movement, and the women's movement. The survey results suggest that the majority of US-based SMOs are not utilizing the web to its full potential and posits a number of reasons why this might be the case, including organizational objectives, organizational resources, and resource sharing. Ultimately, the chapter aims to ground and refine studies of social movement web use in the actual practices found within this sphere.

David Phillips (Chapter 9, "Identity and Surveillance Play in Hybrid Space") approaches identity as the sharing, creating, and performing of socially meaningful relationships. As Phillips writes, identity negotiations occur in space, but spatial settings are themselves are negotiated and created along with the performances, the roles, and the identities that they support. The author identifies two historical shifts that are causing deep structural changes to the negotiations of space and identity. The first is the increasing prevalence of surveillance as a mode of knowledge production. Surveillance may be visual or actuarial, or a hybrid of these. The second is the development of mobile and "cloud" ubiquitous computing—sensors, computation devices, terminals, and responders embedded into mobile devices and stationary objects, and networked to mesh physical and data environments. Depending on how and to whom resources of surveillance and mobile computing are made available, they may facilitate the generation of spaces that are open and public, in that they sustain and are sustained through more or less egalitarian, emergent, collective activity. Or they may become mechanisms of normativity and the re-entrenchment of politically and economically useful identity categories. Phillips explores these issues, using location aware social media as the site of structured but messy interplay among modes of surveillance, interactions between data space and physical space, and performative play.

Addressing the field of music, in "Hacking, Jamming, Boycotting, and Out-Foxing the Commercial Music Market-Makers" (Chapter 10), Patrick Burkart uses the concept of colonization as a way to explore music and cyberliberties activism, media activism, and NSMs. Colonization, Burkart argues, joins commodification, spatialization, and structuration as an observable process of media rationalization in late capitalism. The "colonization of the lifeworld" is the theory of communicative action's interpretation of the reification thesis as introduced by Marx. Colonization is detectable in visible processes of bureaucratization, and commercialization of media cultures, and the cutting off of flows of music and other media. As Burkart outlines in detail in this chapter, the colonization thesis describes processes that contribute to the conversion of the Internet into a mass-media distribution platform for entertainment industry.

CHAPTER SIX

Mediapolis, Human (In)Security and Citizenship

Communication and *Glocal* Development Challenges in the Digital Era

THOMAS TUFTE

In a time when online territories are evolving as mediated practices and social spaces, how do we assess civic action and participation in social change processes? Moreover, how relevant is the debate about online territories when assessed from an East African reality where Internet access is minimal? These are some of the questions address in this chapter. In the first section of the chapter I discuss communication for social change in the digital era from a citizens' perspective. Secondly, I provide an example of a civil society–driven media platform which seeks to enhance processes of empowerment and ultimately influence good governance and promote social change. The example is from Tanzania in East Africa. Thirdly, I introduce Roger Silverstone's (2007) notion of 'mediapolis' as a conceptual approach to explore the mediated public sphere as a space for civic action and participation. In the fourth section I present the concept of 'human (in)security' which refers to how cultures of fear and insecurity permeate public discourse, public sentiment and ultimately civic action. Finally, I discuss how to theorize communication for social change in the digital era, approaching it from a citizens' perspective and connecting the contexts outlined in this chapter—human (in)security, mediapolis and citizenship—with the Tanzanian reality.[1]

Communication, Citizenship and Change in the Digital Era

Media development, with new media proliferating, is mostly seen as *either* revolutionizing our organization of time, space and social relations, *or* just seen as 'business as usual,' that is an extension of previous media and their role in society. I approach new media as a combination, being both an extension of established media and communication practices *and* a new development challenging established social order. In line with the overall approach of this book, this 'co-evolution' of new and old media is opening up for yet un- and underexplored appropriations of media and communication, appropriations which in diverse ways are articulating social and political change. In this context my research interest focuses on exploring and understanding these appropriations.

This chapter aims particularly at identifying and discussing conceptual and analytical entry points to understand these appropriations. My approach is colored by my experience of working in countries where the Internet still has very limited penetration, with Tanzania—where only 3% of the population have Internet access—being the case in point in this chapter. In such a reality, the very concept of online territories may seem a somewhat abstract scenario. However, the global trends in media development are also producing emerging research agendas for places like Tanzania. Some of these agendas will be discussed in this chapter.

Central to this discussion is a citizens' perspective to communication and social change. This perspective is based on a notion of citizenship understood as a social practice grounded in everyday experiences. As British development researcher John Gaventa argues, enhancing citizenship is about being 'the claimants of development' rather than the beneficiaries (Gaventa, 2005, p. xii). Citizenship is not merely a set of rights and responsibilities 'bestowed by the state,' but rather 'a multi-dimensional concept which includes the agencies, identities and actions of people themselves (ibid; Tufte and Enghel, 2009, pp. 14–15). Consequently, the foci of this chapter are the citizens conceived not only as receivers of, but equally as participants or activists in communication-based strategies for change. I conceive civic action as the active manifestation of citizens as claimants of development, a process in which agency, identity and action comes together in a deliberate action for social change.

Speaking to the relation between media and communication, citizens and change, let me share with you a quote from Manuel Castells, from his latest book, *Communication Power* (Castells, 2009). In reflecting upon social movements, insurgent politics and the new public space, Manuel Castells states that:

> in a world marked by the rise of mass self-communication, social movements and insurgent politics have the chance to enter the public space from multiple sources. By using both hor-

> izontal communication networks and mainstream media to convey their images and messages, they increase their chances of enacting social and political change—even if they start from a subordinate position in institutional power, financial resources, or symbolic legitimacy. (Castells, 2009, p. 302).

What is particularly noteworthy about Castells' new book is his recognition of the centrality of communication in society. With instances from, for example, Obama's presidential primary campaign in 2008 and from the longstanding social movement of environmentalism, Castells uncovers the strategic role communication—including the communication in and with the new social media—has played in articulating social and political change. He furthermore offers a rich pallet of conceptual tools to unpack the complex relationship between social movements, communication practices and change processes.

The above quote from Castells points towards a bottom-up communication strategy of social movements, and, furthermore to insurgent politics to enact social and political change. In the integration of horizontal communication networks with the use of mainstream media, Castells identifies a potentially powerful pathway for change.

As I read Castells, with a recognition of the need for horizontal communication, insurgent politics, bottom-up approaches and visibility in the public space, he speaks of some of the founding principles in communication for social change as we know them from the first generation of predominantly Latin American scholars in communication for social change. I think of Paulo Freire's emphasis on dialogue, Orlando Fals Borda's emphasis on participatory action research, Juan Diaz Bordenave's participatory communication in rural communities, Frank Gerace's call for horizontal communication also drawing from rural community experiences. And, we have indirect references to insurgent politics in Mario Kaplun's work on popular education, and in Rosa Maria Alfaro's works on popular culture and participatory communication (see original articles by these authors in Gumucio-Dagron and Tufte, 2006).

Some of these ideas, conceived 30–40 years ago, are clearly central still today. However, while the challenges to achieve a socially just, equitable and participatory development process remains the same, the technologies in the network society, as Castells argues, allows for another—a global—scale of dialogue, participation and horizontal communication, and then possibly also for another scale of impact.

Also, the new social media raise new questions regarding the nature of mediated practices and regarding the sites and spaces of civic action. Castells takes an overly optimistic view as to the potential for 'mass self-communication, social movement and insurgent politics.' A basic challenge to address in this respect is how citizens can obtain the power—or not—to enact change: how *can* ordinary people engage in development—through online media and communication practices and networks? How can they, or we, engage in and influence the social and political change processes that impact upon our own lives? My specific focus is on the role of civil society—NGOs, community-based organization and social movements. What particular opportunities do they have to ensure, facilitate or enhance citizen-driven change processes?

Consequently, the key question I am reflecting upon here is: how can civil society–driven media and communication initiatives—in the digital era—enhance processes of empowerment and ultimately good governance.

This question refers to the relation between four key elements in society: elements and relations which are the focus of my current research project taking place in East Africa and which I will briefly present in the next section. The four key elements are:

Civil society. Firstly, civil society, social movements and NGOs have gained tremendous influence and space in recent years—gaining visibility and enhancing both social change and political influence, often times using the new social media actively and creatively. In many countries, civil society represents today a new and insurgent space, a public sphere if you wish, of participation. The North American anthropologist James Holston—in his fascinating account from the Brazilian social movements in recent decades—calls this 'insurgent citizenship' (Holston, 2008). So, it is this articulation of citizen participation, this articulation of insurgent citizenship in civil society which is my first element of inquiry.

The citizens. Secondly, my focus is on the experience of ordinary people. While civil society analysis can tend to focus on some of the institutional dynamics and broader issues, the question to remember to ask is: do ordinary citizens *de facto* feel connected to and participants in the processes articulated by civil society organizations and movements? We should also ask: to which degree does civil society connect with the feelings, opinions and aspirations of their constituencies? And how do these mediated social practices unfold in developing countries where Internet still is incipient?

Government and decision-makers. Thirdly, government and decision-makers are the most common targets for civil society mobilization. These are the people NGOs and social movements wish to influence. The growing attention to 'good governance,' calls for transparency and accountability and speaks to the relation between citizens and their governments, including the role of (online) media and communication practices in establishing these relations.

Media and communication contents and outlets. Fourthly and finally, civil society organizations (CSOs) are increasingly themselves the producers, owners and disseminators of media and communication contents outlets. Online spaces are increasingly providing new opportunities, articulating visibilities, networking and allowing for new forms of political action. CSOs orchestrate campaigns and other forms of media and communication interventions at all levels—both locally, nationally, and in transnational networks—and with all forms of media and communication tools. And, importantly, they most often produce and control the contents of these initiatives. However, do these contents represent the feelings, opinions and aspirations of the relevant constituencies they are about?

The assumption guiding my own research, and central in this chapter, is that civil-society-driven and civil society–controlled media platforms, digital or not, are increasingly playing decisive roles in enhancing social and political change. This assumption is obviously supported by Castells' work on communication power. There is an already enormous and rapidly growing interest around the insurgent potential, the transformative dimension and thus the power of new social media in particular and the way social movements, individuals and civil society in all its diversity, are making use of these media platforms to articulate cyberprotests, network activists, re-claim space in the public sphere and drive mediated political action.

My point is not to deny this potential, a potential I acclaim, but to question the often times universal discourse and claims made about them. By putting the wished-for 'communication power' into the context of the reality of young Tanzanian women, my aim is to challenge the often celebratory attitude to both new social media and to the transformative potential of online mediated practices. By contextualizing the debate in a Tanzanian reality I wish to contribute to developing a more grounded understanding of communication for social change in the globalized digital era.

Let me now briefly introduce the case study I am presently exploring in Tanzania.

Media, Empowerment and Democracy in East Africa (MEDIeA²)

Democratic development in Tanzania, and in Africa in general, has been developing with particular intensity over the past 10–15 years. This is seen most clearly in the moves away from one-party to multi-party systems. It is also seen in the emerging civil society and the gradually more developed media sector. Increasingly free and independent media have been part of this process. When it comes to digital media, only about 3% of Tanzanians have access to the Internet at home, and while approximately 50% today have mobile phones—this has been an exponential growth in the last 2–3 years (TAMPS, 2009 in Femina HIP 2009).

Many African countries, including Tanzania, have furthermore experienced not just an economic de-regulation and liberalization, but a strong economic growth. It has been around 6% per year in Tanzania for the last decade (World Bank, 2007). As for major development problems, one of the most significant has been the challenge of HIV and AIDS. Today, about 6.2 % of the adult population has the HIV infection (HIVInSite, 2007).

In this context, the MEDIeA project is interested in the role of youth, particularly of young women, in the democratic development in Tanzania—looking into the degree to which the development process is socially inclusive of marginal young women. A major constraint for poor people and citizens of low-income regions or countries like Tanzania is the lack of an effective voice in public life, and particularly in regard to decisions on policies and laws which directly affect their livelihood. Media and information and communication technology *can* be powerful tools in promoting social inclusion. The question is how and to which degree civil society is facilitating this to happen. MEDIeA, therefore, wants to unveil how, if at all, civil society–driven media platforms are contributing to empower young women to speak out, participate in the public debate and in other ways engage in processes of participatory governance.

Our focus on youth is very deliberate. They have been and are severely affected by the HIV/AIDS pandemic and by the high unemployment rate. Challenges as these are leaving this new generation marginalized, something which is a contradiction when a youth generation is supposed to be the window of hope for a country and its development process.

In recent years, young people have, for good reasons, become the new focus of development policies of states and international donor agencies as well as among NGOs and CSOs. Not only are young people perceived as key to economic, democratic and socio-cultural development, but they are also understood worldwide as decisive agents with regard to peace processes and political stability on a local and global scale.

Media wise, the youth constitute the innovators. They are the generation of actors and (future) citizens who are increasingly exposed to and make use of media/ICT, both for entertainment *and* informational purposes, for social networking and mobilization, and for knowledge sharing. The recent years' mobile phone boom underlines the eagerness with which young Tanzanians, and Africans more broadly, seek to appropriate the new digital media, even under the constrained socio-economic conditions the majority of them face.

In this Tanzanian development context of democratic opening, civil society growth, HIV/AIDS and unemployment influencing the lives of the aspirational generation of youth, one particular NGO has had particular success in developing a strong and influential media and communications platform to articulate empowerment and social change. It is called the Femina Health Information Project, or Femina HIP. It was founded in 1999 as a health information project, servicing secondary school youth with a glossy magazine telling stories and providing information about sexual and reproductive health. The magazine, called FEMA, is distinctive in the sense that it is very embedded in youth culture, in the topics it takes up, in the colloquial language used, in publishing primarily in Swahili and not English, and thus in many ways putting a strong effort into connecting with the life-world of the youth.

In terms of the strategy of communication applied by Femina, they use entertainment-education as their primary communication strategy to engage youth. However, differently from most entertainment-education initiatives that use fiction, be it radio drama, TV-series or theatre, Femina uses real-life stories and written publications. They argue they are also cultivating a culture of reading.

While they began as a health communication NGO, they have today expanded their topical agenda significantly, and their media vehicle has also grown significantly. Today, they are engaged in the production of eight different types of communication activities. They produce the two largest magazines in Tanzania and a recent national survey showed that Femina, through their magazines, their radio drama, their 500 youth clubs in secondary schools, their interactive website, their TV talk show, their active use of mobile phones, and their community outreach program altogether expose almost 25% of Tanzania's 44 million inhabitants to their products. Thus, their use of online communication is one of eight elements used, and still in the margins of their communication strategy. However, they are developing a Femina Facebook group and they are exploring the possibilities to use blogging. From a co-evolutionary media and communications perspective, they are gradually exploring how to combine the possibilities the Internet, along with the booming mobile phone access, with their other and most important media and communication vehicle: print, radio and TV.

Viewed upon holistically, this conglomerate of Femina's media platforms constitutes what has become one of the strongest civil society–driven communication initiatives in Tanzania and in East Africa. Their aims are ambitious: to stimulate open talk, critical thinking and social change that will foster healthy lifestyles and positive, responsible attitudes toward sexuality, HIV/AIDS and democratic culture. Many things point towards Femina achieving this.

However, one issue is Femina's reach, which indeed is significant. Looking then at this NGO from the youth's perspective, what sense-making processes are, in reality, occurring, and what discussions are sparked by being exposed to Femina's media and communication initiatives?

To what degree can the youth identify with the characters in the comics, and do they relate to the topics brought up in the TV talk show or in the magazines? What elements of Femina's communication initiative spark them to send an SMS, and do they at all have access to the interactive website Femina has set up? What can they use a Facebook group for, and what role does the interactive website have, and for what social groups in particular? Does the Femina media and communication flow at all articulate public debate? Does it engage young women in any way, and if so, how exactly? And what do the young Tanzanians do themselves to influence Femina's agenda?

In other words, how can we, in the MEDIeA project, assess the ways and means whereby a media and communication vehicle as Femina contributes to enact social and political change in the country?

A first study we conducted, of letters sent by youth to Femina, reveals that the long-term, recurring nature of the Femina HIP media is vital for creating a gradual process of engagement resulting in groups and clubs taking on the agenda (Tufte et al., 2009). In this line of findings, the emerging online territories can potentially piggy-back on the brand value already established by Femina through its offline communication practice. The study also underlines how important it is for Femina HIP to work with community mobilization and to help ensure that groups and clubs are set up as these are a dynamic forum for learning, interpreting and taking action not just as individuals but as a collective (ibid). Having Femina clubs setting up their blogs—as has begun to happen—can be seen as yet another extension of offline dynamics into the mediated practice in online territories.

At another level of questioning, we are asking ourselves, what mechanisms Femina has developed to hold governments accountable and transparent? Do they at all engage explicitly with advocacy? And how far can civil society go at all in criticizing the government, and what influence do their listeners and readers have upon their formulation of social and political critiques? Finally, how does Femina ensure their own legitimacy and accountability to the groups of citizens they advocate on behalf of? It is questions like these that the MEDIeA research project engages with. As outlined, Femina HIP seems not to have any explicit distinction between online and offline technologies—the focus lies on the strategic use of whatever media and communication vehicle there is in order to tackle current youth development challenges in Tanzania.

Looking at the organization and its operations, one interesting and today very relevant element we wish to deconstruct is the interconnection between global, national and local discourses in their communication practice. It speaks to the reality of the network society and it opens up for the analysis of interconnections between public spaces at these different levels.

The fact of the matter is that both the director and other core staff of Femina HIP participate many times a year in international meetings abroad, mostly in South Africa and Kenya but also in Scandinavia and elsewhere, visiting and interacting with other experts from bilateral donor organizations, UN agencies, INGOs and elsewhere. Are the new Internet-based media playing some role in transnational networking efforts—for example, Skype conversations, emailing, Facebook networks, Twitter, etc. And, how does this interaction connect with the national agenda of Femina in Tanzania? Does it influence the way the organization is managed, the way communication strategies are designed, monitored and evaluated; does it influence the themes dealt with in the media and communication outlets like Femina has? Answering these sorts of questions can help us situate Femina's operations in the network society, in the transnational advocacy networks, and in the general intersections of global, national and local agendas.

The preceding brief outline speaks to two conceptual entry points that hopefully will help the MEDIeA project form the analysis of Femina's work in Tanzania. The first set of entry points is about *the interrelation between global and local discursive interaction, the notion of world development* and *the different notions of the public sphere*. The Tanzanian case will be situated in this broader debate of how development processes and discursive processes at different levels are interconnected. Roger Silverstone's concept of *mediapolis* will be used to help develop an understanding of the mediated public sphere. My particular interest is to conceptually explore the possibilities and limitations in cultivating the public sphere as a space for civic action and participation.

The other conceptual entry point is to engage with the concept of *human security*. This is especially relevant to help understand the lifeworlds of youth today. Human security, or rather insecurity, is increasingly an issue across the globe. A growing number of social scientists have identified how cultures of fear and insecurity have grown and increasingly permeate the public discourse and the public sentiment. My working hypothesis is that in the case of Tanzania, issues like HIV and AIDS, climate change and unemployment, may well contribute to the articulation of a strong sense of human insecurity amongst the ordinary citizen. However, this remains to be studied.

Mediapolis—World Development, Citizenship Disjuncture and a Call for Humanity

The Tanzania case speaks in part to the interconnectedness of development processes and not least the dynamic flow and interaction of ideas between local, national and international communities. As the Norwegian anthropologist, Thomas Hylland

Eriksen argues, globalization creates the conditions for localization, that is, various kinds of attempts at creating bounded entities as countries, faith systems, cultures or interest groups (Hylland Eriksen, 2005, p. 28). This interrelationship, between the global and the local, termed by Roland Robertson (1995) as 'glocalization' is a condition, a perspective upon development which I take on as a fundamental perspective in order to really understand processes of change.

To give an example: James Holston (2008), in his book *Insurgent Citizenship* refers to 'disjunctions of citizenship,' reflecting critically upon the way democracy, conceived narrowly as electoral democracies, has swept the globe as a universal norm for the organization of decision-making in nation states. However, a serious disjuncture, Holston argues, exists between this emphasis on the political project and the reality of the lived lives:

> This kind of political focus fails to account adequately, if at all, for precisely the sort of disjunctions of citizenship that I have analyzed in Brazil and that are prevalent among most emerging democracies—namely, the coincidence of democratic politics with widespread violence and injustice against citizens. This disjunction has become just as global a condition of contemporary democratization as free elections.' (Holston, 2008, p. 311)

In exploring the citizen's involvement in democratic development in Tanzania, this perspective will be kept in mind.

Within the debates of development, the grand paradigms of the 1960s (modernization) and 1970s (dependency) were followed in the 1980s and 1990s by a multiplicity of generally less assuming approaches, some of which radically questioned the very concept of development. As Dutch sociologist Jan Nederveen Pieterse (2001) has pointed out, there is an unholy alliance between the strong neoliberal perspective, associated with economic globalization and structural adjustment, and the radical post-development perspective, proposing local de-linking and resistance to globalization, in their common repudiation of 'development' as discourse and politics. But following the deconstruction of development, we can now witness its gradually emerging reconstruction as world development.

Development is no longer a process reserved for 'developing countries': all societies are developing as part of a global process, making the dichotomy of 'first' and 'third' worlds obsolete—at least in the geopolitical sense. The entire world is 'in transition' and development must therefore be rethought as a regional, transnational, global project (Pieterse, 2001, p. 45). The example of democratic and citizenship disjunctures has to do with this transition.

From the perspective of humanity, the Polish sociologist Zygmunt Bauman is another scholar who has explored the citizenship disjunctures when unveiling the social cost of economic globalization and when emphasizing the interdependency under which we live. Bauman has shed light upon the way many people—in times of cosmopolitanism, migration and mass tourism—have remained 'locals by fate rather than choice' (Bauman, 2000, p. 100); he has written about how the 'human waste' in our globalized society has been chucked aside, away from the public space, referring, for example, to refugees locked up in '*nowherevilles*' (Bauman, 2003, p. 142).

The interconnectedness between globalization and localization is discernable, and to understand a local process of social and political change, the *glocal* perspective and the notion of world development provides a relevant conceptual entry point.

In response to the many socio-economic and human downsides to these processes, Bauman argues in his latest book for 'humanity' as the required project for modernity: 'At its earlier stage, modernity raised human integration to the level of nations. Before it finishes its job, however, modernity needs to perform one more, yet more formidable task: raise human integration to the level of humanity' (Bauman, 2010, p. 69). Recognizing the interdependency of peoples and nations, Bauman argues for the creation of a global equivalent of the 'social state,' but looking not in a reinforcement of the UN or similar international governmental bodies. Rather, he argues that non-governmental bodies will take the lead in these processes:

> There is no decent way in which a single or a group of territorial states may 'opt out' from the global interdependency of humanity...I suspect that the vehicles likely to take us to that 'social planet' are not territorially sovereign states, but rather admittedly extra-territorial and *cosmopolitan non-governmental organizations* and associations; and those that reach out directly to people in need above the heads of and with no interference from the local 'sovereign' governments...(Ibid, p. 70, emphasis added)

The problem is, as Bauman argues, that "the state is today unable, and/or unwilling, to promise its subjects existential security ('freedom from fear,' as Franklin D. Roosevelt famously phrased it)" (ibid, p. 65). When the state acts in this way, the individual citizen is left to his own, unable to obtain ***existential security***, that is, unable to obtain and retain 'a legitimate and dignified place in human society and avoiding the ménage of exclusion' (ibid). This leaves the individual, for example, the marginalized youth in Tanzania, the *favelados* in Brazilian mega-cities or the immigrants in the suburbs of Paris to pursue life based on skills and resources of each individual on his or her own. And, it leaves these many individuals in a situation of 'enormous risks, and suffering the harrowing uncertainty which such tasks inevitably include' (ibid).

In response to the failure of the state in many places many non-governmental organizations have taken action. In the past two decades, parallel to processes of globalization having been at their most intense moments, we have seen an exponential growth in the number of NGOs and social movements in the world. This has become most manifest in the parallel NGO events at the large UN-summits and G8 summits, but most significant has subsequently been the large World Social

Fora, starting in Porto Alegre in Brazil in 1999, assembling many thousands of civil society activists. The World Social Fora have witnessed gigantic encounters of NGOs and other civil society organizations to deal with exactly the lack of government response to glocal development challenges. While not constituting any formal global governmental decision-making body, far from it, the social fora establish the space for hundreds and thousands of small NGOs to meet, learn, speak out and bring new ideas back home to the local struggle for social and political change. The global network of civil society has evolved into large transnational advocacy networks, e-networking being an instrument, alongside face-to-face encounters.

Consequently, out of this situation arise two analytical perspectives relevant to the Tanzanian case presented in this chapter, but also relevant to any localized study of citizen action. The first perspective is that of multi-level and interdependent ***agency***, exploring local agency as a social practice which is connected with others through multi-level and multi-nodal networks. Transnational advocacy networks both contribute to supporting local struggles as well as contributing to the configuration of transnational platforms for action and reflection. Thus, the second perspective is the creation and continuous development of transnational ***activist spaces*** at the global level: fora which, following Bauman's argument, establish elements of a global social state in recognition of the failure of the governmental initiatives.

These analytical perspectives connect well with what the late British media sociologist, Roger Silverstone, called 'The Rise of the Mediapolis.' This speaks to the rise of a mediated public sphere where civic action and participation has an opportunity to grow.

The Rise of the Mediapolis

In his last book entitled *Media and Morality*, Roger Silverstone (2007) developed the concept of mediapolis. The German sociologist Ulrich Beck has called the book 'a new cosmopolitan critical theory of the emerging global civil society and its contradictions.' What Silverstone does is to develop a new theory of the public sphere in which the logics, dynamics and opportunities of the media gain center-stage. The ideas put forward by Silverstone provide us with a conceptual framework to be able to situate and understand media and communication practices in the context of the globalized world. The mediapolis is, according to Silverstone:

> '...the mediated space of appearance in which the world appears and in which the world is constituted in its worldliness, and through which we learn about those who are and who are not like us. It is through communications conducted through the mediapolis that we are constructed as human (or not), and it is through the mediapolis that public and political life increasingly comes to emerge at all levels of the body politic (or not).' (Silverstone, 2007, p. 31)

Silverstone is concerned with the totalitarian dimensions of modernity, concerned with how mediated spaces represent or constitute public life and to what degree these spaces are inclusive or exclusive and whether they enable or disable public debate. Furthermore, he is concerned with the development of media literacy as a civic activity, a 'secondary literacy' (Silverstone, 2007, p. 179), where media literacy becomes a political project, just as media civics become a literacy project. Media civics and media literacy are interdependent.

Mediapolis, although embryonic and imperfect, is a necessary starting point, Silverstone argues, for the creation of a more effective global civil space. The mediated space of appearance is at best, a space of potential and of possibility (Silverstone, 2007, p. 33).

When denominating the mediapolis 'a mediated space of appearance' Silverstone draws on the political philosopher Hannah Arendt in determining the character of this space: 'The polis, properly speaking, is not the city-state in its physical location: it is the organization of the people as it arises out of acting and speaking together, and its true space lies between people living together for this purpose, no matter where they happen to be…'(Arendt, 1958, p. 198).

The mediapolis, conceived as a public sphere contains both totalitarian and liberating possibilities. Silverstone moves on to unfold some of the criteria of media hospitality, media justice and media ethics as morally based reference points which can contribute to achieve a fully effective communication in mediapolis. It is a communication practice which he further argues is based on: (a) a mutuality of responsibility between producer and receiver; (b) a degree of reflexivity by all participants in the communication; and, (c) a recognition of cultural difference. In other words, Silverstone regards this mediated public sphere, 'mediapolis', ideally as a dialogic space which 'is both an encompassing global possibility and an expression of the world's empirical diversity.'

The way he views people in mediapolis, is as participants. He rhetorically asks the question of what to call the person exposed to media: users, consumers, prosumers, citizens, players, etc. And although acknowledging some element of relevance in all the concepts, he opts for audiences and users being participants. Consequently, any form of participation involves agency.

Connecting back to empirical studies, the interesting challenge is to explore what reality is created in the mediated space of appearance, what kind of publicness, and, in that context, to deconstruct the role and relations of online and offline media and communication practices. How is it that ordinary citizens engage as participants in the mediated public sphere of mediapolis?

HUMAN (IN)SECURITY

While Silverstone's 'mediapolis' helps us understand the mediated public sphere which both influences and conditions citizens' everyday practices, the concept of human security helps us understand the social reality they live in, and the socio-psychic situation this reality produces. By exploring the issue of human security we can uncover the condition of existence which is the fundamental context of agency and communication. It helps us to understand the subjective position from which people speak and act.

Understanding human security provides us with yet another argument as to why media ethnographies are important to our research. Human security helps produce a direction or a parameter for the quality and scope of civil society–driven media and communication initiatives, for example, in my current research in Tanzania. The big communicative challenge is enhancing empathy and establishing a relationship of trust. A lot of our contemporary media and communication flows articulate the opposite of empathy and trust. However, living in a risk society, with prevalent cultures of fear and insecurity, what a theory of communication for social change has as its fundamental challenge is to help establish the bases for what we might call emphatic communication.

The late British professor of global politics, Caroline Thomas, defines human security as follows:

> Human security as freedom from fear describes a condition of existence in which human dignity is realized, embracing not only physical safety but going beyond that to include meaningful participation in the life of the community, control over one's life and so forth. This suggests a radical account of politics as freedom from domination/exploitation, not simply the freedom to choose as advocated by the liberal tradition. Thus, while material sufficiency lies at the core of human security, in addition the concept encompasses non-material dimensions to form a qualitative whole. In other words, human security embraces the whole gamut of rights, civil and political, economic and social, and cultural. (Thomas, 2007, pp. 108–109)

This definition is broad. It connects both the material and non-material dimensions that help establish conditions of human security. Human security speaks to the fundamental challenge of having control over one's life and being free from domination and exploitation. While our world of today is influenced by cultures of fear (Furedi, 2002), man-made threats of all kinds in the risk society (Beck, 1992), pandemics and global warming, human security comes to symbolize the ideal scenario of what humanity should pursue.

However, the Spanish social scientist Jaume Curbet makes an interesting point in stating that 'more than deciphering the reality of the insecurity, we dedicate our energies to the search for security' (Curbet, 2006, p. 8). When you think about it, the focus of many policies is on proactively ensuring human security rather than deconstructing the realities sparking insecurity. Rather than seeking the deeper causes of the insecurity people experience and feel in their everyday life, we pursue making security solutions. The surprising and contradictory fact of the matter, Frank Furedi argues, is that we currently are experiencing unprecedented levels of personal security (Furedi, 2002). For our purposes, it is what Curbet calls for that is interesting and relevant: to deconstruct the realities that spark insecurity.

Curbet's point relates well to the field of HIV and AIDS communication and prevention which I have worked with, mostly in Africa, over the past decade. In HIV and AIDS communication many media and communication initiatives have focused on delivering rather simple solutions (messages) thereby hoping the HIV infection did not spread. There has, for example, been solutions like the well-established ABC of HIV/AIDS communication: Abstain, Be faithful or use a Condom. However, as my research in South Africa has shown, there was tremendous discrepancy between the ability of young South Africans to reproduce these messages and the tremendous degree of insecurity expressed by them as to how they were coping with HIV and AIDS in their own lives (Tufte, 2009). While HIV and AIDS *can* be defined narrowly as a health issue, it is also so much more. HIV and AIDS is so entangled with poverty, culture, gender roles, power and spirituality that you cannot prevent the epidemic from spreading by only providing solutions to the immediate physical mode of transmission of the virus.

A similar example can be drawn from examining the sites where the majority of the world population today lives; the cities. Many cities have developed into sites of insecurity, non-places as the French sociologist Marc Augé (1995) denominates urban spaces with potentially many people but limited or no social interaction. A growing body of literature—not least in the Latin American cities—explores the often times chaotic sites (Monziváis work on Mexico, 1995), industrial accidents in the midst of the city (Reguillo, 1996), cities where fear is prevalent due to crime, violence or narco-trafficking (Martínez et al.'s work on Medellin, Colombia, 2003) and where people struggle with deteriorating living conditions in increasingly segregated cities (Holston on Sao Paulo, Brazil, 2008). Jesus Martin-Barbero has also explored the urban modernization and what he calls 'a change in sensibility' in the Latin American cities (Martín-Barbero, 2002, p.277).

The point to be made here is that city studies of an intensified process of urbanization, seen more recently in Africa than in Latin America, is producing some of the 'conditions of existence' Caroline Thomas speaks about, where human *in*security is prevalent. In concrete terms, public spaces are often times compromised, or if they are there, people simply retrieve physically from all but the necessary presence in the public space, due to lack of confidence and trust. Social realities as the ones outlined here provide an important context when exploring the rise, proliferation and uses of new social media in times of human insecurity. This is an emerging research agenda which requires attention to produce better insight into the social and political uses of new social media.

Curbet (2006) further distinguishes between ***objective insecurity*** and ***citizen insecurity***. The first deals with the established material facts that generate insecurity, which in the first case above would be fear of being infected with HIV in a concrete relationship. Other examples of objective insecurity are fear of crime or violence in high-prevalence areas. An interesting question to pose here could be: what material facts are found, or mediated through online territories and influencing objective insecurity? The other concept, citizen insecurity, speaks of a feeling which is less detectable. It can be the insecurity of how to cope with the existence of HIV in your community—you cannot see it, but you know it is there. Or it can be citizen insecurity vis-à-vis unemployment, natural disasters or lack of health insurance. These feelings are not less real, but usually more difficult to establish the actual source of and find a clear solution to. Again, exploring the relation between social uses of new social media and the articulation of citizen insecurity constitutes an interesting emerging research agenda in developing countries like Tanzania.

A human security approach forces us to think about how global, national and local structures/forces interrelate and their cumulative impacts on individuals and communities, be it traveling pandemics, inhumane mega-cities or global warming.

A human security approach results in a strong call to understand in depth the conditions of existence which civil society–driven media and communication initiatives wish to communicate around. It also places new social media in a particular context of usage. And, more fundamentally, it provides a strong argument to focus on the need for deep social change and, not least, the need for governments to address the identified insecurities.

However, human security, in addition to framing a policy agenda for communication research, also reveals the difficulty and complexity of the task of communicating for social change. It helps us focus our attention on the lived experience of globalization, especially the lived experience of the most marginalized—the majority of citizens that, as Caroline Thomas (2007) argues, are 'excluded from international production.'

Communication for Social Change in the Digital Era—Key Concepts

Summing up, I see three conceptual clusters, emerging as central in the exploration of communication and development in the digital era. These are:

Human security and how it relates to both material and immaterial conditions of existence, and thereby is deeply connected to questions of identity, community and subjectivity

Mediapolis, conceived as a mediated public sphere, a space which hosts both possibilities and limitations for the cultivation of civic action and participation. It includes the flows of media and communication practices.

Citizenship, which I conceive not just as a set of rights and responsibilities bestowed by the state, but as a multi-dimensional concept which includes the agencies, identities and actions of people themselves. Holston's concept of insurgent citizenships is a way to conceptualize peoples' actions.

Together, these conceptual clusters constitute building blocks for an analytical framework with which to study online and offline appropriations and practices of media and communication. Particular focus is on how these appropriations and practices articulate empower—or not—citizens to engage in civic action.

Concluding Remarks

Cutting across the above core conceptual entry points, a series of other questions are important to raise by way of conclusion. Firstly, the question of whether the digital era is qualitatively different from former times. I have here argued that old and new media have co-evolved, yet also recognized the potential of the network society whose uniqueness Castells and so many others have long pointed at Castells recognizes and explains in detail 'the potential synergy between the rise of mass self-communication and the autonomous capacity of civil societies around the world' (Castells, 2009, p. 303). Despite this possibly reinforced potential for civil society–driven social and political change, Zygmunt Bauman reminds us of a characteristic of the net. Bauman states that: "Paradoxically, the widening of the range of opportunities to promptly find ready-made 'like minds' for every and any interest pursued narrows and impoverishes, instead of augmenting and enriching, the social skills of the seekers after the 'virtual community of minds" (Bauman, 2010, p. 166). In other words, while some new social and political dynamics have arisen with the coming of the new social media, they don't necessarily represent a communication practice that Roger Silverstone argues is a possibility of mediapolis: that of recognizing cultural (and social) difference. If we buy Bauman's argument, the new social media is not articulating new social relations, rather just reinforcing the old.

This leads me to the second cross-cutting question which this chapter must not forget to address, and with this I will end. The question is: to what degree do these reflections around human security, mediapolis and citizenship have a universal character and value? Are they only relevant for the citizens and realities in the developed world where almost everybody has access to the Internet and can truly said to be living in a mediatized society? Or, to put it more bluntly: are the above reflections about mediatized societies and a digital era just ethnocentric impositions of themes, concepts and theoretical approaches with no relevance to poor countries and to the practice therein of communication for development and social change? At first glance, it could seem so, considering, for example, the fact that only about 3% of the Tanzanians have access to the Internet (TAMPS, in Femina HIP 2009). However, the technological development is *de facto* booming, more than half of the Tanzanians today have mobile phones, and a third of those Tanzanians who actually access the Internet do it via their mobile phones. So, in a not so distant future, many Tanzanians will probably have overcome the digital divide.

In an age of networked society, with development processes interconnected worldwide and global human security issues being prevalent across borders, the first- and third-world distinctions are rendered obsolete, as are many other conceptual dichotomies. I believe, we can use the three-dimensional dynamic interrelationship between human security, mediapolis and citizenship to explore how communication for social change—both offline and online—is enacted in the real world.

Notes

1. This chapter is an expanded and substantially revised version of the keynote address I gave, in Spanish, at the conference 'Comunicación y Desarrollo en la Era Digital' in Malaga, 3–5 February 2010. A slightly modified version was published in Spanish in Revista ALAIC, Sao Paulo, Brazil.
2. "People Speaking Back? Media, Empowerment and Democracy in East Africa' (MEDIeA) is the title of the research project I am coordinating. It runs from 2009 to2013 and deals with the relation between media and communication, processes of empowerment and democratic development in Kenya and Tanzania. For more detailed information see: http://mediea.ruc.dk

References

Arendt, H. (1958). *The Human Condition.* Chicago: University of Chicago Press.
Augé, M. (1995). *Non-Places: Introduction to an Anthropology of Supermodernity*. London: Verso.
Bauman, Z. (2000). *Globalization. The Human Consequences*. Cambridge: Polity .
Bauman, Z. (2003). *Liquid Love*. Cambridge: Polity .
Bauman, Z. (2010). *Living on Borrowed Time.* Cambridge: Polity.
Beck, U. (1992) *Risk Society. Towards a New Modernity*. London: Sage.
Castells, M. (2009). *Communication Power.* Oxford: Oxford University Press.
Curbet, J. (2006). *La glocalización de la (in)seguridad.* La Paz: Plural Editores.

Femina HIP. *(2009)*. Dar es Salaam, Tanzania.
Furedi, F. (2002). *Culture of Fear. Risk-taking and the morality of low expectation*. London: Continuum.
Gaventa, J. (2005). Foreword, In Kabeer, N. (ed). *Inclusive Citizenship. Meanings and Expressions*. London: Zed Books.
Gumucio-Dagron, A. & T. Tufte (eds). 2006. *The Communication for Social Change Anthology. Historical and Contemporary Readings*. New Jersey: Communication for Social Change Consortium.
HIVInSite, (2007). Retrieved 17 July, 2010, from http://hivinsite.ucsf.edu/global?page=cr09-tz-00
Holston, J. (2008). *Insurgent Citizenship. Disjunctions of Democracy and Modernity in Brazil*. Princeton: Princeton University Press.
Hylland Eriksen, T. (2005). How Can the Glocal Be Local? Islam, the West and the Globalisation of Identity Politics. In: Hemer, O. & T. Tufte (eds). *Media and Glocal Change. Rethinking Communication for Development*. Göteborg & Buenos Aires: NORDICOM & CLACSO.
Nederveen Pieterse, J. (2001). *Development Theory. Reconstruction/ Reconstructions*. London: Sage.
Martín-Barbero, J. (2002*). Oficio de Cartógrafo. Travesías latinoamericanos de la comunicación en la cultura*. Santiago de Chile: Fondo de Cultura Económica.
M. I. V. et al. (2003). *El Rostro del Miedo. Una investigación sobre los miedos sociales urbanos*. Medellín: Corporación Región.
Monsiváis, C. (1995/2001). *Los rituals del caos*. Mexico: Ediciones Era.
Reguillo, R. (1996). *La construcción simbólica. Sociedad, desastre y comunicación*. Guadalajara: ITESO.
Robertson, R. (1995). Glocalization: Time-Space and Heterogeneity-Homogeneity. In: M. Featherstone, S. Lash & R. Robertson (eds)., *Global Modernities* (pp. 25–44). London: Sage.
Silverstone, R. (2007). *Media and Morality. On the Rise of the Mediapolis.'* Cambridge: Polity.
Thomas, C. (2007). Globalization and Human Security. In: A. McGrew & N. K. Poku (eds). *Globalization, Development and Human Security*. Cambridge: Polity.
Tufte, T. et al (2009). From Voice to Participation? Analysing Youth Agency in Letter Writing in Tanzania. In: T. Tufte & F. Enghel (eds). *Youth Engaging with the World. Media, Communication and Social Change*. The International Clearinghouse on Children, Youth and Media. NORDICOM & UNESCO: University of Gothenburg.
Tufte, T. & F. Enghel. (2009). Youth Engaging with Media and Communication. Different, Unequal and Disconnected? In: T. Tufte & F. Enghel (eds). *Youth Engaging with the World. Media, Communication and Social Change*. The International Clearinghouse on Children, Youth and Media. NORDICOM & UNESCO: University of Gothenburg.
World Bank (2007). *World Bank Development Report*. Washington: World Bank.

CHAPTER SEVEN

The Rise and Fall of Online Feminism

LIESBET VAN ZOONEN

Some 40 years ago, a Dutch feminist wrote a song called *There is a country where women want to live.*[1] Halfway through the 1990s it seemed that the Internet would provide such a space. Feminists of all kinds thought that the lack of physical presence and the anonymity of the Internet would make it possible to experiment with diverse gender identities and even non-gendered identities. Such practices would reverberate in the offline world, undermine dichotomous gender discourse and liberate women and men from their fixed genders. Only ten years later, anonymity has acquired a bad reputation due to anxieties about predators and bullies. Utopian prospects have been annihilated by a discourse of risk and fear that makes gender experiments suspect. It is a classic story of rise and fall, in this case of online feminism, as opposed to feminism online.[2] The latter concerns the web presence of offline groups and activism and will be only discussed briefly in the conclusion. The more dramatic defeat of online feminism as a politics of gender subversion forms the core of this chapter. I tell it as a linear story of feminist utopia frustrated by the persistence of everyday gender practice, to be finally annihilated by hegemonic mechanisms of control and discipline. While, admittedly, such a narrative may overlook the exceptions and subtleties that typifies good academic discourse, it identifies the changed ideological climate in which feminism, in its on- and offline varieties, has to operate more sharply.

Utopia

The early and mid-1990s were great years for feminist cyber utopia. Browser applications like Mosaic and Netscape, launched in 1993, had rapidly turned the Internet into a medium that was massively used (Naughton, 1999), and imbued with everybody's hopes, whether they were social, cultural and educational or commercial, financial and organizational. Feminists of all kinds shared the rampant optimism, with authors like Sadie Plant (1997), Dale Spender (1995) and Sherry Turkle (1995) delivering the resources to think about the supposed bright futures of online feminism.[3] Spender, by then an already well-known feminist literary scholar from Australia, published her reflections on the Internet in a book entitled *Nattering on the Net: Women, Power and Cyberspace* (1995). "Who we are, what we know, and how we think," she wrote, "are all being changed as we move from a print-based society to a computer-based world. We are becoming different people; we are creating a new community" (p. xiii). Such language of total transformation was ubiquitous at the time, and Spender articulated it with a clear feminist agenda: "Women have to get in there. For the wealth, the power—and the pleasure of it all." (p.xxiv). While Spender extensively acknowledged that existing power structures and abuse would be hard to counter, her book nevertheless reflected a deep confidence about what women could do on the net: "Where women have made the technology accommodate their needs, their success knows no limitations" (p. xxiv). Sherry Turkle's influential book, *Life on the Screen: Identity in the Age of the Internet*, appeared in the same year (1995) and emanated a similar combination of critique and irresistible sanguinity about what the Internet could mean for gender. In her analysis of experiments with gender swapping enabled by the anonymity of the Internet, Turkle wrote that such experiments offered "an opportunity to explore conflicts raised by one's biological gender" (p. 213), "a form of consciousness-raising about gender issues" (p. 214), "a vehicle for self-reflection" (p. 219) and the possibility "to experiment safely with sexual orientation" (p. 223). In addition, Turkle was quoted as saying that community, collaboration and consensus formation are the skills you need to successfully navigate the Internet, and that those are "the skills that women bring to the table" (quoted in Jenkins, 1999, p. 332).[4] Sadie Plant elaborated that particular argument in her 1997 book *Zeroes and Ones: Digital Women and the New Techno Culture*. Plant argued that the network architecture of the Internet bore many similarities to the female technology of weaving. She too emphasized the affordances of anonymity:

> In spite or perhaps even because of the impersonality of the screen, the digital zone facilitates unprecedented levels of spontaneous affection, intimacy and informality, exposing the extent to which older media, especially what continues to be called 'real life,' come complete

> with a welter of inhibitions, barriers and obstacles sidestepped by the packet-switching systems of the net (p. 144).

While the three authors did include critical and cautionary arguments in their analyses, the academic and mainstream reception of their work mostly focused on the potential benefits of the Internet, celebrating the vistas of better futures that digital technology offered. Thus, the British press praised Plant as the most radical "techno-theorist" of her time, and Motorola, the American communications multinational, hired her—as "one of the most advanced thinkers in the field of human relationships with technology"—to examine the effects of mobile phones on social and individual lives (Plant, 2001). Sherry Turkle's work about the postmodern identity laboratory that computers and the Internet constitute was an immediate hit, and has been quoted thousands of times by academics and journalists ever since. In fact, such was the mainstream approval of this kind of work that by the end of the decade, less popular feminist analysts of technology exclaimed: "Where have all the feminist technology critics gone?" Canadian scholar Ellen Balka (1999) wrote, for instance, about Plant, Spender and Turkle, that they had fallen prey to a naïve celebration of cyberspace: "The evident advantages of the Web now lull us into an… uncritical stance."

What exactly did Turkle and Plant and a range of others celebrate? Two things. First, there was the promise of breaking the binaries of traditional dichotomous gender discourse, by experimenting with opposite, 'third-gender' or complete non-gendered identities in the anonymous spaces of chat rooms and online multi-user games. It was a promise that was especially eloquent in Turkle's work, and many others subsequently took it up in their writing (e.g. Danet, 1998; Schaap, 2002). Second, there was the promise of Internet being essentially a 'feminine' space, especially advocated by Plant, due to its networked, cooperative and communicative set up. Some authors added a psychoanalytic twist to this claim, contending that full immersion in cyberspace evoked the experience of the foetus in the womb, protected by the maternal body and hence unconstrained by its own body limits (Ogden, 1986; Springer, 1991, quoted in Robins, 1995). Invariably, all of these authors would conclude their arguments with calls for more research (e.g. Spender, 1995, p. 241), not foreseeing that this research would expose the Internet as offering an ordinary rather than transgressive environment.

Back on Earth

Pretty soon after the happy future was proclaimed, a large scale survey among users of computer-mediated communication demonstrated that about half of them never switched gender online (Roberts and Parks, 1999). Moreover, the gender bending respondents said that they did so only occasionally and for a short period, often for instrumental reasons rather than to interrogate gender and sexuality. This was a first, empirically based modification of both the size and the cultural meanings of gender bending on the net. Other studies followed that gradually undermined utopian discourse about the Internet enabling the collapse of the gender binary. They focused on the Internet use of women and men as both social and symbolic practices through which gender is performed. While there are many good studies in this area, I take the opportunity in this chapter to take stock of our own work about the social and virtual articulations of Internet and gender, as developed in the Dutch Centre for Popular Culture[5] in the past decade,

Internet as Social Practice

What much early discourse about cyber utopia forgot is that access to cyberspace takes place in real social settings of work, the home, the café, the airport, the train; basically from everywhere with the increasing options for mobile Internet. Particularly the domestication of the Internet as a home entertainment technology raised the question as to how gendered family relations and Internet use mutually shaped each other. In early 2000 therefore, we conducted family interviews with 24 young Dutch couples between 20 and 30 years of age, living together without children. We asked them about their daily routines, their media consumption, their usage of the PC and of Internet (Van Zoonen, 2002; Van Zoonen and Aalberts, 2002). Our findings suggested the existence of four kinds of gendered domestic Internet cultures among young couples. We found some couples to have fairly straightforward *traditional* relations with the men in the household having control over the PC and the Internet. They would be the most frequent users, they would know most about it and would be highly interested as well. Their female partners would leave the field to them, claiming that they didn't know much about it and did not see much use of it. While the couples did recognize their arrangements as traditional, they often mentioned perfectly good practical reasons for this, and did not seem to mind very much. Other couples had developed a practice of *deliberation* around their PC and Internet access. These partners would talk to each other about their uses and pleasures of the Internet and made it part of their collective identity as a couple by reminiscing about shared tastes and sites. Potential conflict of interest about PC use and Internet access would be solved by prioritizing use for the purpose of work and study. This is where traditional gender relations crept in again because among most of our respondents, as in the rest of the Dutch society, the male partner was the main provider who thus would always come first. Couples with more equal social positions solved this problem by buying an extra PC or bringing a lap top from work. In these completely *individualized* practices the relevance of gender for the access to Internet at home disappeared (although it still had consequences for use, as we will show shortly). Traditional relations were rarely *reversed* but among two couples we found the women to be in control of PC and Internet use. In both cases, however, the men had jobs linking them to PC-screens all day and did not want to spend their home time behind that screen as well. We concluded that male usage offered the main explanation for three out of the four Internet cultures we found. Only the individualized forms of Internet use, depending on the presence of an extra PC or lap top, enabled the female partner to have access and use on her own terms. Yet, our findings also showed how these particular Internet practices, in turn, contributed to the gender identities of the partners and their relationship; the active abstinence of female partners, for instance, was part of an accepted and appreciated division in the household of male and female tasks and pleasures. In more deliberative use practice, equal rights to the PC were constructed as part of being a modern emancipated woman who would know when to prioritize the 'objective' needs of her male partner. In retrospect, the outcomes of the study may seem dated; not because gender relations have spectacularly changed but because of the rapid individualization of digital technologies that our data only hinted at: in 2008, almost 50% of Dutch households had three PCs or more in the home.[6]

Such individualized use patterns, however, do not mean the disappearance of gender as a factor in social Internet practice. It means that the articulation of gender and Internet moves from access and use patterns to meanings and experience. The research in our group about how video and online games appeal to women makes that clear. Contrary to media stereotypes, female gamers are not a rare species but they do tend to play different kinds of games. Casual and simple games are quite popular among them and among the more elaborate games *The Sims* dominates women's playing field. Research about these gamers is rare and mostly concerned with the lack of interesting games for girls and female adolescents, and the male-dominated game culture. In such research, an incorrect impression of women's social and symbolic absence is constructed which denies the active game play of women. Researcher Mirjam Vosmeer (2010) aimed to counter that image by zooming in on the social and cultural meanings that playing *The Sims* had for women, especially those who had to balance their game play with work and family responsibilities. She talked to some 50 women across the Netherlands who considered themselves to be *The Sims* fans, spending considerable time and money, and keeping up with the newest releases. One of the main pleasures the women experienced was their control over their *The Sims* characters and their environment: they would create characters with physical and psychological features that resembled their own, and—especially—they would design and decorate their house and garden in the optimal way that wasn't financially possible in their own lives. Most gamers therefore used the cheats in the game to get more money without their Sims working more and longer. In fact, Vosmeer's analysis suggests that many women used *The Sims* to create a parallel world to their own everyday life; one that could be easier controlled, but that would also function as a ground for experimentation with different kinds of behavior and ideas. Some gamers even had made different scenarios for their various *The Sims* families. Surprisingly, while these gamers did participate in virtual communities around *The Sims*, they hardly engaged in online *The Sims* play, contributing to the overall failure of *The Sims* online for that matter[7]. Furthermore, Vosmeer found a definitive reluctance of her *The Sims* respondents to play the game with other members of the family, the children in particular. Vosmeer explains both the pleasure and the desire of the women to play on their own from their social position as the main carer in the household. Both married and single mothers in her group of respondents obviously appreciated the total control that *The Sims* offer to the players, a control that is by definition impossible to achieve in everyone's ordinary life, but that is even less possible for parents who need a flexible attitude to the ever-surfacing challenges and surprises of having children. This domestic stress of never-ending availability, and the complete convergence of work and leisure in the domestic space, has been identified, by feminists and conservatives alike, as debilitating to women's individual and social development. Vosmeer's research shows, much like Radway's (1984) research about romance novels, how playing *The Sims* helps to carve out a space in the household, literally secluded behind the screen, for women's individual pleasures without any demands from their environment. No wonder they don't want to play online, or with their children.

Both studies, about the domestication of the Internet, and about playing *The Sims* in gendered family relations, demonstrate that online practices are by definition social practices, and therefore thoroughly embedded and confined by macro and micro discourses of gender, most of them rather traditional. We found similar results in our research about Internet use as a symbolic practice.

Internet as Symbolic Practice.

When it comes to the Internet as a cultural practice, it is especially the Web 2.0 affordances that reveal whether and how users articulate their gender identities and experiments online. While Web 1.0 bulletin boards and chat rooms already were dependent on user participations and thus enabled an analysis of gender performance, the possibility to create one's own blogs, profiles and videos exacerbated—the so-called—user-generated content, making it into a widespread online practice in itself. Web 2.0 has once again been subject to optimistic discourse about is revolutionary potential, to "transform the way we socialize, gather information, and do business," as *Time* magazine wrote when they identified 'YOU' as the person of the year 2006. In a series of case studies, researcher Niels van Doorn examined whether and how content generators , both on the old and the new web, perform gender and sexual identities in new and subversive ways. Focusing on chat rooms, weblogs, social network sites and YouPorn, he asked about the relation between the specific architectures of these applications and the space they offered for performing gender. In all these cases, Van Doorn's conclusion was that the offline binary sexed body remained crucial for the disembodied online performance of gender. Even in the text-based chat rooms, Van Doorn found that conversations contained different types of embodiment. To begin with, many chatters would evoke images of their real-life bodies, by talking about where they were on the globe, what their PC and room looked like, what they were wearing and doing, and so on. Within these sets of comments, phallic references were a typical subgenre, as a form of male bravado in heterosexual exchanges, and as homo-eroticism in gay male chats. In addition to concretely mentioning the body, chatters would also in their language metaphorically call the body to mind, especially through metaphors of movement and material space. This was facilitated by the fact that the chat room software offered separate, so-called 'rooms,' if chatters wanted to have a private conversation. Especially through these forms of embodiment, chat rooms tend to reinforce 'the norms of a binary gender system through the reiteration of a "natural" connection between gender and sexed bodies,' (Van Doorn, Wyatt and Van Zoonen, 2008, p. 371). In blog software, the articulation of online with embodied offline identities seems even easier, especially because of the possibility to use pictures and other visual representations of the offline self and interests. In a subsequent case study, Van Doorn therefore examined the construction of gender identity on blogs of about 100 randomly selected Flemish and Dutch women and men. The overall demographic information on their blogs suggests that the typical blogger is about 30, well educated, living in an urban environment and s/he blogs in Dutch. From the description of their hobbies, a rather traditional gender division appears: women tend to mention domestic hobbies and pets, men list information and communication tecehnology hobbies and some of the bloggers also worked in the ICT sector. Through the choice of pictures, hyperlinks and the activities and experiences they mention, the bloggers in the sample present themselves as 'man' and 'woman' in relation to their everyday lives: 'they do not seem to 'play' with their gender identity, instead choosing to present their identity through rather mundane domains such as hobbies or educational status' (Van Doorn, Van Zoonen and Wyatt, 2008, p. 155). Yet, within the representational categories of 'man' and 'woman,' Van Doorn found a wide variety of ways to perform masculinity and femininity. Van Doorn also argues that blogging opens up spaces for feminine discourse on the net, by bringing the traditional female activity of diary writing into the technological domain: 'the act of diary writing on weblogs can be seen as blurring the gendered connotations of the weblogs as an ICT, thereby showing that the use of technology is pivotal in shaping its gender' (ibid, p. 156). In another study, Van Doorn focused explicitly on embodied presence on the Internet by analyzing the videos that people uploaded to YouPorn, a dedicated channel for amateur porn. There too, the videos mainly replicated a traditional porn script 'adhering to a male-centered, conventional "porno norm" [that] perpetuate an essentialist (and sometimes sexist) gender ideology that ties gender to the male and female anatomy and the heterosexual pleasures derived therefrom' (Van Doorn, forthcoming).

Thus, all three studies suggest that active participation on the Internet, in chat rooms, on one's own blog or by uploading one's own porn, is articulated with the offline material world and physical embodiment. As a result, offline traditional gender patterns are replicated in these activities, be it with a much more visible and prominent variety than the dichotomy of gender usually suggests. Only one of Van Doorn's studies—about the articulation of gender and sexuality on MySpace, a popular social network site—showed online practices that transgressed the heteronormative gender binary as a matter of course. He randomly selected one network of male and female friends of about 20-years-old, who also regularly met offline. He analyzed exchanges and communications for a period of five consecutive weeks and found four main topics of exchange: popular culture, night life, alcohol and drugs, and sex. In all these topics, the communicative tradition built up among the friends—through messages, shared links, visual gifts and other possibilities that MySpace offered—required the continuous expression of desire, for drugs, for partying, for sex. Such public articulations are in itself already a specific challenge to the dominant moral order that is common in other sites of youth culture. In addition, the flows of intimacy through the networks extended indiscriminatingly to male or female friends, 'creating a multidirectional flow of polymorphous desire: everyone loves everyone else. Heterosexual masculinity and femininity are temporarily destabilized or rendered ambiguous' (Van Doorn, 2009, p. 17). For Van Doorn, the micro- network offers a situated example of—what Judith Butler has called—queer performativity that challenges traditional dichotomous gender. It is especially the seemingly ephemeral nature of the exchanges that makes such transgressions possible but also relatively inconsequential.

The use of the Internet, both as a social and as a symbolic practice, did not bring the utopia that was expected; on the contrary, it basically seemed to offer another space for traditional gender discourse, with some exceptions, just like in real life. For some of the early utopians that was hard to stomach. As usual, after we submitted it to the journal our blog article was anonymously reviewed and assessed. Interestingly, the reviewer asked whether we could not 'read more resistance' into our data, because s/he found it hard to believe that online practices would be so traditional. Unfortunately for the utopian project, they were. And even worse, the mere idea of the Internet as utopian space became controversial, naïve and dangerous when public attention moved to the downsides of anonymity: online predators, pedophiles and bullies were said to be the main beneficiaries.

Control and Discipline

Most talk show hosts in the US and the UK, and some in Europe have paid attention to 'stranger danger' on the Internet, very often in sensationalist and scary ways. A typical example comes from the website of Dr. Phil, an American popular psychologist whose TV show revolves around identifying people's personal and family problems and solving them. On his website, the section on Internet pedophiles begins with: "One in four children in chat rooms will be solicited by a child predator. These pedophiles seek a target-rich environment for finding their prey, and the Internet has become their flocking ground."[8] The program that possibly did the most to stir up widespread anxiety about Internet predators was the US reality show *To Catch a Predator* which ran from 2006 till 2008. The program was built around undercover operations of program team members who would set up an under-age profile in a chat room or social networking site to lure possible sex-offenders. If those took the bait, the decoy 'children' would develop an online relation with them and eventually invite the predator to come to their home. There, hidden cameras and the presenter of the show would wait for him, first inquiring about his intentions and then slowly moving towards exposing him as a pedophile. That usually was the moment when the predator would flee out of the house, onto the street, where most of the time the police was waiting for him.

The fear about Internet predators has taken on the features of a full-blown 'media panic,' defined by Drotner (1999) as an exaggerated fear about a new communication technology that focuses mainly on children and young people. They are also recurring features of media and cultural history, with, for instance, famous panics about the comic books in the 1950s, the dance halls in the 1920s and the dime novels or penny dreadful in the late 19th century. The Internet predator panic is widespread; on the Internet itself parents can find an endless series of websites containing warnings and advice about how to protect their children. Concerned pressure groups, commercial entrepreneurs, government agencies and tabloid media, in concert, keep the fear alive by repeating and revealing predator stories. Yet, despite the real presence of predators and pedophiles on the Internet, the question is whether the size of their threat justifies all the panic and precaution. Researchers of the Crime against Children Research Center of the University of New Hampshire in the US have answered this question with a firm 'no.' Janis Wolak and her colleagues (2008) claim that the typical media stories about Internet predators have produced a number of unhelpful and even dangerous myths: (1) online molesters 'lurk' in sites popular among children; (2) they use these children's public profile to target them; (3) they contact these children under false pretense; (4) they entice them into meetings or stalk and abduct them; (5) that there is an epidemic of this 'new' type of crime. To asses whether these myths bear any truth, they investigated real court cases of Internet sexual offenses against under-18 youth. They found that all five myths were in stark contrast with the nature of actually tried crimes. To begin with, Wolak and her colleagues found that the majority of children knew they were in contact with an adult, and they were also aware of his intentions. Only in five percent of the cases that came to court, the offender had actually deceived his victim with false information about his age and intent. 'When deception does occur, it often involves promises of love and romance by offenders whose intentions are primarily sexual' (Wolak et al., 2008, p. 118). The victims mostly had agreed to meet offline, knowing that sexual contact would be on the agenda. The age of the victims, in addition, was usually between 13 and 15, which makes it legally impossible for them to consent to sex. Therefore, Wolak and her team conclude that this kind of crime does not represent a whole new phenomenon, but is much more similar to common offline violations of age of consent laws. In a number of the cases brought to court, the victims did not consider themselves as victims at all, and did not want the offender to be convicted. Strikingly, the research carried out in the Crime against Children Research Center also showed that while 'arrests of online predators increased between 2000 and 2006, most arrests and the majority of the increase involved offenders who solicited undercover investigators, not actual youth.[9]

Is there no risk at all then? Don't the statistics of Dr. Phil (see earlier), like the 'one in four' children having been sexually approached online, tell of a real danger? On the Dr. Phil website, it is unclear where the 'one in four' figure comes from. Often such figures are not accredited to a particular source, or they are selectively interpreted. Benjamin Radford (2006), author of the book *Media Mythmakers: How Journalists, Activists, and Advertisers Mislead Us*, convincingly dismantled a 1-in-5 statistic that dominated US politics and the media around 2005. He traced the figure to a report by National Center for Missing and Exploited Children that asked 1,501 American teens between 10 and 17 about their online experiences. One of the questions was whether they had had a 'request to engage in sexual activities or sexual talk or give personal sexual information that were unwanted or, whether wanted or not, made by an adult.' That definition is so vague, Radford (2006) argues, that questions of friends about a date could be considered as 'sexual solicitation.' The type of danger that the media myths assume was much rarer and occurred only in three percent of the cases. Radford concludes that 'the reality is far less grave than the ubiquitous "one in five" statistic suggests' (ibid).

The media panics about stranger danger on the Internet conceal that other risks may be more significant, and that certain groups of youth are more at risk than others. Radford also reviewed that online sexual harassment more often comes from children's peers than from strangers (ibid); likewise, it has been shown that online bullying is very much related to offline bullying (Li, 2005). Both findings demonstrate that the risks of Internet use are produced by one's known social circle rather than by conniving strangers. In addition, Wolak and her team (2008) have identified that particular groups of children are more at risk than others: girls, especially those with an offline history of victimization, and boys who identify as gay or are questioning their sexuality. These are findings that do not make it into the media with the same frequency and intensity as stranger danger stories. They are less easily framed in individualized villain and victim roles and suggest strongly that online sexual abuse is embedded in the inequalities and oppressions that women and homosexual men face in everyday life. It doesn't need a feminist conspiracy thinker to recognize that such a suggestion is not easily articulated in neoliberal gender discourse of individual choice and responsibility. However, to understand the paradox of previous hopes and current fears that is articulated in relation to anonymity and the gender experiments on the Internet a further feminist deconstruction is necessary, focusing on questions like why this panic has gotten such widespread adherence?

Shayla Thiel-Stern, researcher at the University of Minnesota, calls the current Internet panics "overreactions to a perceived threat to society," that resemble the anxieties of the early 20th century about girls going to dance halls. The dominant cultural discourses surrounding MySpace and Facebook, Thiel-Stern (2008, p.3) writes, 'are reminiscent of the discourses surrounding the rise of dance halls in the late 19th and early 20th century, and specifically, girls' and young women's use of them.' After analyzing the newspaper coverage of the dance hall panic and the social network site anxieties, she brings the two historically different phenomena together by interpreting them as discursive battles about women's use of public space. The modern division of public and private space as the domain of, respectively, men and women, has made women's access and use of public space problematic and threatening to the social order: 'Girls and women who place themselves in public purview have been punished with harassment, depicted as members of a lower class, or worse, viewed as sexually promiscuous' (ibid, p. 13). The dance hall 'problem' of the early 20th century was about young women who would go to dance halls to meet friends, dance and have fun. They were unescorted and—according to the press of the time—often went to the dance halls without knowledge of their parents. Liquor and men, in concert, would lure the girls into indecent behavior leaving their reputation in ruins. Social workers and other moral guardians of the time considered dancing as a 'sex-act' and proposed all kinds of measures to protect girls against the presumed onslaught of men apparently waiting to take advantage of them. Thiel-Stern identifies many parallels with current anxieties about girls' uses of social network sites, and shows how current media limitedly focus on online sexually provocative behavior of girls much in the same way as they curbed their attention when writing about the dance halls. She concludes that both panics are expressions of the same type of patriarchal discourse that punishes women and girls for their active presence in public space. By focusing on the excesses, that discourse constructs all girls and young women as both the 'victims and the blamed,' 'meaning that when they assert themselves, articulate their gender or sexuality (especially in non-traditional ways), they will be punished for it, and they shouldn't expect otherwise' (ibid, p. 35).

Thiel-Stern's analysis of the parallels between the dance hall panics and the anxieties of social networks is helpful to understand the continuous reproduction of distorted statistics about stranger danger on the Internet. They too function as a means to control and discipline girls and young women, especially when it comes to their sexuality. While the instruments of discipline come in forms ranging from the production of outright fear to the moderate warning for possible risks, the discourse of restraint, modesty and sexual discretion as desirable traits of women underlies them all. From the broader perspective of the maintenance of traditional gender and sexual regimes, it is no wonder that the utopian vistas of Internet as the space par excellence to experiment with gender and sexuality had to be contained; by identifying, exaggerating and reiterating danger, and by reminding women and girls, constantly, that they are at risk and vulnerable. Such a discourse of threat and the attribution of vulnerability clearly assigns to the individual a passive and dependent role (Furedi, 2007, p. 7).

A Place Like All Others

The mistake of early feminist utopian thinkers, of course, was to accredit a technology with the potential to bring about revolutionary practice. Such technological determinism is typical for the introduction of new technologies and is usually quite quickly substituted by perspectives of doom, also caused by the new technology (De Wilde, 2000). We see this pattern clearly in the perceptions of the Internet and the paradox of hope and fear surrounding online identity work. While evidently there are sites that foster gender subversion and creativity, the discourse of transgression no longer dominates writing and thinking about the Internet. Empirical analyses like the ones we discussed have shown how the Internet has been taken up as a means to confirm and reiterate traditional gender discourse and practice. Predator panics create an atmosphere of danger that contains possible subversive and playful experimentation. As a result, some 15 years after the Internet has turned into a massively used medium, online feminism has become marginal and gender practices online seem to perfectly mirror their offline counterparts. Professionally produced online gender identities have seamlessly flown into the commercial consumer world of women's portals providing access to women's magazines, online shopping and diet sites (Consalvo, 1997). User-generated content is also by and large traditional, constructed within the binary opposition of women and men, be it with impressive diversity and short-lived transgressions and excess. At the margins of this mainstream, we find feminism and male violence. Yet, whereas media panics exaggerate the violent threat coming from predators, feminism is regularly declared dead, despite all the examples to the contrary that can be found on the Internet. Online subversive gender practice (online feminism) may have become suspect, but 'feminism online,' understood as the presence of feminist groups and activists on the Internet is strong and vibrant. America's oldest organization for the equality of women, NOW, has a large website that lists actions, possibilities to donate and join, support groups and information about feminist issues. In addition to organization websites, the feminist online magazine is common in many languages; *The F-Word*, for instance, is an almost ten-year-old e-zine for young British women that examines women's issues, provides links to other groups and a blog of its editor. The list of examples is possibly endless and likely to cover the whole world, although to date there are few systematic analyses of these online extensions of activism (but see e.g. O'Donnell, 2001 for Ireland; Edwards, 2004 for the Netherlands; Svenning, 2008 for Ecuador). It is questionable whether such a separate analysis of feminism online would be very relevant, given the fact that research about other kinds of activist and civic websites has shown that they are firmly connected to offline practice, and that users perceive their on- and offline participation as one coherent set of converged actions (e.g. Hirzalla and Van Zoonen, 2009), that is, nevertheless, greatly facilitated and sped up by the use of the Internet.

All in all, when it comes to gender relations and feminism, the conclusion must be that cyberspace is pretty much a standard gendered universe that offers options and instruments for feminist identity politics and activism, provided somebody wants to use them for that purpose. On its own, by itself, the Internet is not sufficient to produce change.

NOTES

1. Joke Smit (1984). *Er is een land waar vrouwen willen wonen. Teksten 1967–1981.* Amsterdam: Feministische Uitgeverij Sara.
2. The distinction parallels the one made by Helland (2000), between online religion and religion online.
3. The following section is partly based on a segment from Van Zoonen (2001).
4. Turkle expressed this view in an ABC News Nightline feature about girls and gaming.
5. From 1999 until 2009, the Centre functioned as a network for Dutch scholars and PhD students researching popular culture, hosted by the University of Amsterdam but connecting peo-

ple from all universities. At the moment of writing, the network's base is being moved to Erasmus University, Rotterdam.

6. Research by Packard Bell, reported on http://whizpr-press.blogspot.com/2008/07/onderzoek-packard-bell-bijna-helft.html, accessed January 20, 2010.
7. See, for instance, http://www.techcrunch.com/2008/04/29/ea-land-the-sims-online-joins-the-deadpool/, accessed January 20, 2010.
8. http://www.drphil.com/articles/article/166, accessed on January 26, 2010.
9. http://www.unh.edu/ccrc/pdf/CV194.pdf, accessed January 26, 2010.

REFERENCES

Balka, E. (1999). Where Have all the feminist technology critics gone? *Loka Alert 6*(6). Retrieved December 15, 2009, from http://www.loka.org/alerts/loka.6.6.txt

Consalvo, M. (1997). Cash cows hit the web: Gender and communications technologies. *Journal of Communication Inquiry, 21*(1).

Danet, B. (1998). Text as mask: Gender, play and performance on the Internet. In Jones, S. (eds). *Cybersociety 2.0: Revisiting Computer-Mediated-Communication and Community*. London: Sage.

De Wilde, R. (2000). *De voorspellers: een kritiek op de toekomstindustrie* [The Forecasters: Criticizing the Future Industry]. Amsterdam: De Balie.

Edwards, A. (2004). The Dutch women's movement online: Internet and the organizational structure of a movement. In: W. van de Donk, B. D. Loader, P. G. Nixon, & D. Rucht(eds). *Cyberprotest: New Media, Citizens and Social Movements* (pp. 183–206). London, Routledge.

Furedi, F. (2007). The only thing we have to fear, is the 'culture of fear' itself. *Spiked*, 4 April. Retrieved January 31, 2010 from http://www.frankfuredi.com/pdf/fearessay-20070404.pdf.

Helland, C. (2000). Online -religion/religion online and virtual communities. In J.K. Hadden & D E. Cowan (eds). *Religion on the Internet: Research Prospects and Promise* (pp. 205–223). New York: JAI Press.

Hirzalla, F. & Van Zoonen, L. (2009). De online/offline-deling voorbij. Convergenties van online en offline participatievormen onder jongeren. *Tijdschrift voor Communicatiewetenschap 37*(3): 215–237.

Li, Q. (2005). New bottle but old wine: A research of cyberbullying in schools. *Computers in Human Behavior, 23*(4), 1777–1791.

Naughton, J. (1999). *A Brief History of the Future. The Origins of the Internet*. London: Weidenfeld and Nicolson.

O'Donnell, S. (2001). Analysing the Internet and the public sphere: the case of women's link. *Javnost/The Public 8*(1), 39–58.

Plant, S. (1997). *Zeroes and Ones. Digital Women and the New Techno Culture*. London: Fourth Estate.

Plant, S. (2001). On the mobile—the effects of mobile telephones on social and individual life. Motorola. Retrieved December 15, 2009, from http://www.motorola.com/mot/doc/0/234_MotDoc.pdf

Radford, B. (2006). Predator panic: Reality check on sex offenders. *LiveScience*, May 16, 2006. Retrieved January 26, 2010, fromhttp://www.livescience.com/strangenews/060516_predator_panic.html.

Roberts, L. D. & Parks, M. (1999). The Social Geography of Gender Switching in Virtual Environments on the Internet, *Information, Communication & Society, 2*(4), 521-540.

Robins, K. (1995). Cyberspace and the World We Live In. In M. Featherstone R. Burrows (eds) *Cyberspace, Cyberbodies, Cyberpunks* (pp. 135–155). London: Sage.

Schaap, F. (2002). *The Words That Took Us There. Ethnography in a Virtual Reality.* Amsterdam: Aksant Academic Publishers.

Spender, D. (1995). *Nattering on the Net.* Toronto: Garamond Press.

Svenning, M. (2008). How and why do women's/feminist movements in Ecuador use the Internet. Paper presented to the conference Learning Democracy by Doing Alternative Practices in Citizenship Learning and Participatory Democracy, Toronto, October 16-18.

Swartout, K. (ed) (2007) *Encyclopedia of Associations, Vol 1: National Organizations of the US, 44th Edition.* Farmington Hills, MI: Thomson Gale.

Thiel-Stern, S. (2008) *From the Dance Hall to Facebook: Analyzing Constructions of Gendered Moral Panic in Girls and Young Women in Public Spaces.* A paper presented to the Cultural and Critical Studies Division of the AEJMC (Association for Education in Journalism and Mass Communication, Chicago, IL, August 6-9.

Turkle, S. (1995). *Life on the Screen. Identity in the Age of the Internet*. New York: Simon and Schuster.

Van Doorn, N. (2009). The ties that bind: The networked performance of gender, sexuality and friendship on MySpace. *New Media and Society, 11*(8), 1–22.

Van Doorn, N. (forthcoming). Keeping it real. User generated pornography, gender reification and visual pleasure. *Convergence,* accepted for publication.

Van Doorn, N., Van Zoonen, L. & S. Wyatt (2008). Writing from experience. Presentations of gender identity on weblogs. *European Journal of Women's Studies, 14*(2), 143–159.

Van Doorn, N., Wyatt, S., & Van Zoonen, L. (2008). A body of text. Revisiting textual performances of gender and sexuality on the Internet . *Feminist Media Studies,8*(4), 357–374.

Van Zoonen, L. (2001). Feminist Internet Studies. *Feminist Media Studies,1*(1),. 67–72.

Van Zoonen, L. (2002). Gendering the Internet . Claims, controversies and cultures. *European Journal of Communication, 17*(1), 5–23.

Van Zoonen, L. & C. Aalberts (2002). Interactive television in the everyday lives of young couples. In: M. Consalvo & S. Paasonen (eds). *Women and Everyday Uses of the Internet: Agency and Identity.* New York: Peter Lang.

Vosmeer, M. (2010). Videogames en gender. Over spelende meiden, sexy avatars en huiselijkheid op het scherm [Videogames and gender. Girl games, sexy avatars and screen domesticity]. PhD dissertation University of Amsterdam. ISBN/EAN: 978–90–9025018–2.

Wolak, J., Finkelhor, D., Mitchell, K. & M. Yberra (2008). Online "predators" and their victims: Myths, realities and implications for prevention and treatment. *American Psychologist 63* 111–128.

CHAPTER EIGHT

Social Movement Web Use in Theory and Practice

A Content Analysis[1]

LAURA STEIN

INTRODUCTION

Social movements are important actors in democratic societies. They are key spaces for formulating, advancing and leveraging the interests of civil society against elites and authorities.[2] As such, they serve as sites of public advocacy around social and political issues, which markets and states are more reticent to address (Mueller et al., 2004, p. 5). In essence, social movements are social networks that engage in sustained collective actions, have a common purpose and challenge the interests and beliefs of those with power (Tarrow, 2005, p. 4). While communication processes are integral to their success (Atton, 2003, p. 58; Downing, 2001, p. 26), research shows that movement actors experience several difficulties communicating through mainstream media. Mainstream media often systematically distort, negatively cast, or ignore social movement viewpoints. They may deny social movements access or representation at critical moments in their development (Raboy, 1981, p. 8), employ message frames that undermine or weaken public perceptions of a movement's legitimacy (Gitlin, 1980, pp. 271–4; McLeod and Detenber, 1999; Shoemaker, 1982, 1984), or tacitly encourage movement actors who seek coverage to cater to the questionable values of mainstream reportage on social activism, including a heightened interest in violence, emotionality, and slogans (DeLuca and Peebles, 2002, p. 133; Gamson, 1990, p. 166; Kielbowicz and Scherer, 1986).

Although mainstream media coverage remains a crucial resource for many social movements (Carroll and Hackett, 2006, p. 87), the problematic relations between mainstream media and movements underscore the need for movements to employ alternative modes of communication. On this score, communication scholars have suggested that the Internet can serve as an important resource for social movement communication, providing movements with communication opportunities not available in mainstream media or in alternative forms of movement media. Social movements can use the Internet to bypass mainstream media gatekeepers or repressive governments and communicate directly with their constituencies and the broader public (Boyd, 2003, p. 13). Moreover, the Internet could help level the playing field between social movement organizations (SMOs) and the more resource-rich institutions of business and government through its combination of greater speed, lesser expense, wider geographical reach, and relatively unlimited content capacity compared to older forms of print and electronic media (Downing, 1989, p. 157; Gross, 2003, p. 259; Kidd, 2003, p. 62; Van Aeist and Walgrave, 2002, p. 466). Of course, Internet access is not universal, and these benefits accrue only to those with the requisite skills and resources.

Although many scholars view the Internet as a potentially useful tool for social movement communication, there is a dearth of scholarship examining whether, how, and to what extent most SMOs use the Internet. Most communication scholarship on the Internet and social movements has followed two threads. Scholars have undertaken case studies of social movements, such as the Zapatistas in Chiapas, the anti-corporate globalization movement, and the antiwar movement against the second US War in Iraq, to examine vanguard uses of the Internet designed to facilitate identity formation, mobilization and networking (Ford and Gil, 2001, p. 222; Kahn and Kellner, 2004, p. 94; Van Aeist and Walgrave, 2002, p. 467). Others have focused on the role of the Internet in enabling or shaping transnational movement activism (Bennett, 2003a, 2003b; Cammaerts and Van Audenhove, 2005; Downing, 2003; Kidd, 2003). While current scholarship highlights emerging vanguard and transnational uses of the Internet, it does not investigate how the majority of SMOs utilize this communication resource or employ methods that produce generalizable results (Garrett, 2006, p. 216). Yet, whether and how most SMOs actually use the Internet is of interest to those who theorize the political value of this communication resource.

While communication scholars have studied some aspects of social movement Internet use, social movement scholars have all but ignored this area of study. Social movement literature has failed by and large to consider the role of media and communication in the life of social movements and at best offers thin descriptions of the actual communication practices of movements (Carroll and Hackett, 2006, p. 87; Downing, 2001, p. 26). In contrast, a growing body of communication research on alternative media offers a theoretical foundation for investigating movement media. This research suggests that movement media are best conceptualized as alternative media. Most scholars view social change as the raison d'être of alternative media, whether these media are associated with contesting and negotiating social and cultural codes, identities and relations (Rodríguez, 2001, p. 20), forming an oppositional culture to a dominant order (Raboy, 1981, p. 9), publicly communicating dissident ideas (Kessler, 1984, p. 14), raising awareness of oppressed and marginalized groups (Gillett, 2003, p. 621), or articulating and defending alternative identities and interests (Steiner, 1992, p. 121). Downing's understanding of radical media, though not limited to social movement communication, best describes the overriding purposes of social movement media. Downing defines radical media as media that express the opposition of subordinate groups towards a power structure, engage in lateral communication against policies and power structures, and tend to be more democratically organized than conventional media (Downing, 2001, p. xi). In his view, alternative media and social movements have a dialectical and interdependent relationship; each can mutually constitute and influence the other (Downing, 2001, p. 23). As this view suggests, the study of alternative media is largely concerned with how social movements use media to debate, formulate, articulate, disseminate, and sustain an oppositional culture and politics.

This chapter examines social movement Internet use by surveying how SMOs utilize this communication resource at one of its most visible point of access, the World Wide Web. As established social groups with shared activities and goals, organizations are more likely to be engaged in the sustained collective actions central to our definition of social movements. While previous empirical studies have focused on what the Internet contributes to transnational social movements or on exceptional cases, this study randomly surveys the websites of SMOs operating in the United States. SMO websites are critical spaces for organizational representation online and increasingly are part of the everyday communication repertoire of movement groups. The national context remains of prime importance to movements whose collective actions are intimately tied to the political opportunities provided by reigning governance structures (Tarrow, 2005, pp. 18–19). While national SMOs may not be the most innovative movement actors online, they remain key players with substantial resources who are well positioned to take on long-term social change projects, as well as the discrete campaigns and protests that characterize much of the cutting-edge movement activity online. Drawing on alternative media studies, the chapter develops a typology that identifies the communication functions most central to social movements and identifies a set of features or attributes within each type. Websites are a highly malleable medium, capable of serving any number of communication goals. This typology is meant to illuminate the relevant range of goals that movement actors may or may not choose to enact in practice.

The chapter then randomly surveys the websites of US movement groups to determine the degree to which their websites exhibit various features and functions pertaining to different communicative goals. More specifically, the chapter poses the following research questions: (1) What features or attributes do US movement groups include on their websites, and how prevalent are these features, (2) to what level or degree do US SMOs use the web to engage in primary communication functions, (3) Do US SMOs with greater financial resources or memberships utilize a greater range of features than groups with fewer resources and members? This chapter aims to describe how and to what extent SMO websites enact the range of communicative goals of social movements. Given the dearth of empirical research on SMO communication practices online, this study serves as an exploratory step in charting the uses that US SMOs make of their websites. The study also considers how additional research methods might serve to further answer these questions and to address how offline factors, such as their individual orientations and resource constraints, help shape their everyday website practices.

A Typology of Social Movement Communication

Alternative media scholarship suggests a variety of communication functions in which SMOs might engage, particularly on the web, which is thought to pose low barriers to movement communication and offer unmediated opportunities to reach potential and existing movement members. Drawing on this scholarship, I offer a typology that highlights the functions most salient to social movement communication. The typology classifies communication according to whether it provides information; assists action and mobilization; promotes interaction and dialog; makes lateral linkages; serves as an outlet for creative expression; and promotes fundraising and resource generation. Though these functions are often interrelated in practice, I categorize them separately for heuristic purposes. In this section, I describe each of these categories and identify the features or attributes within web-based communication that contribute to each.

Information

Alternative media allow for the dissemination of information regarding movement identity, views and issues to interested recipients both inside and outside the movement. Groups may provide information in order counter what is seen as misinformation or disinformation in mainstream media and to disrupt prevalent ideologies while offering more radical alternatives (Downing, 2001, p. 16; Rodríguez, 2001, p. 63). They may also communicate movement identities and issues to interested parties, including the public, movement constituencies, the press, academics, and others (Costanza-Chock, 2003, p. 175; Kessler, 1984, p. 155). Alternative information is essential for sustaining social movements and can serve as a record for posterity of alternative views and possibilities (Downing, 2001, p. 31). Because the Internet is relatively free of centralized gatekeepers, it provides a ready site for direct and uncompromised information dissemination by SMOs.

Action and Mobilization

Alternative media also serve as instruments of mobilization, or the organizing of collective actions and initiatives aimed at producing specific outcomes. Alternative media can coordinate initiatives and actions (Barlow, 1988, p. 103; Kessler, 1984, p. 74) and spread viewpoints designed to galvanize actions, a process referred to as 'consensus mobilization' (Tarrow, 2005, p. 113). Alternative media scholars have documented the ability of the Internet to effectively supplement traditional forms of action and mobilization. Social movements have used the Internet to coordinate real-world events and actions, petition political representatives, disseminate action alerts and campaign materials, and engage in virtual civil disobedience (Costanza-Chock, 2003, p. 175; Ford and Gil, 2001, p. 224; Kahn and Kellner, 2004, p. 88; Van Aeist and Walgrave, 2002, p. 481). The ongoing presence of websites allows potential supporters to learn about and participate in SMO campaigns over time and at their convenience, overcoming the temporally sensitive nature of other forms of contact (Boyd, 2003, p. 17; Van Aeist and Walgrave, 2002, p. 469).

Interaction and Dialog

Alternative media function as relatively autonomous sites of interaction and dialog. Fraser (1993, p. 83) argues that such spaces are necessary to further participatory parity between dominant and subordinate groups within larger spheres of discourse. Alternative media function as counter-public spheres, a term Fraser (1993, p. 84) uses to denote alternative discursive arenas where subordinated groups can deliberate, articulate and circulate oppositional interpretations of their own identities, interests, needs, strategies, and objectives. Downing (2001, pp. 34–46) adds that the internal dialog these spaces permit, which often involve shared processes of meaning construction between activist producers and particularly 'active audiences,' help movement participants arrive at common understandings of their problems and strategies. Scholars have begun to examine the role of the Internet in facilitating political discussion and debate among social movement and other actors. Many argue that participatory forums, such as chat rooms and bulletin boards, provide ongoing opportunities for dialog and discussion and facilitate the formation of discursive networks that offer alternative perspectives on both national and international issues (Downing, 2003, p. 252; Ford and Gil, 2001, p. 224; Kahn and Kellner, 2004, p. 91).

Lateral Linkages

Social movement actors also use alternative media to communicate laterally and build networks among movement members. Alternative media can link social change activists by making them aware of one another's views and interests and by uniting communities of interest across national and transnational space (Barlow, 1988, p. 100; Kessler, 1984, p. 158; Steiner, 1992, p. 131). One way the Internet can facilitate lateral linkages is by connecting one organization's site to another through hyperlinks. External links to other sites are a strategic choice that acknowledges the presence of other actors, establishes an interconnected sphere of online sites, and may reflect an organization's desire to offer information provided by others (Foot and Schneider, 2006, pp. 39, 59; Rogers and Marres, 2000). Hyperlinks may also carry movement supporters to sites of news and research and to national or international SMOs affiliated with their primary movement or with other social movements.

Creative Expression

Alternative media can function as a site for creative expression. As Downing (2001) points out, political communication does not always take the form of rational argumentation. Emotion, imagination, and aesthetics are central aspects of much political expression, taking such forms as satire, irony, cartoon, caricature, slander, and pornography (Downing, 2001, p. 28). Costanza-Chock (2003, p. 175) notes that SMOs have used the Internet to display art, including poetry, visual art, video, music, and parody that support social movements. Given the web's ability to display a variety of media forms, we might expect SMO websites to host a variety of creative expression.

Fundraising and Resource Generation

Finally, social movements can use alternative media to engage in fundraising and resource generation. Movement groups attempt to raise financial support and resources through a variety of means, including requests for donations, the sale of merchandise, building member databases, and recruiting new members, personnel and volunteers (Costanza-Chock, 2003, p. 175; Van Aeist and Walgrave, 2002, p. 481).

A Comparison of Social Movement and Political Communication Typologies

The above typology is similar to others designed to analyze the communication functions of political party and candidate websites. However, it differs in accord with its theoretical grounding in the communication aims of social movements. Political communication typologies tie communication functions to goals associated with persuasion or winning elections. For example, Foot and Schneider (2006, p. 22) construct a typology of political communication during elections that is driven by the need to persuade. This typology includes presenting information, promoting interaction between supporters and organizations in order to garner resources, connecting with others through hyperlinks or cognitive links, and mobilizing supporters to promote a candidate to others (Foot and Schneider, 2006, pp. 22, 46, 67, 70, 103, 132). Gibson and Ward (2000, p. 306) utilize a typology that includes information provision, campaigning to recruit voters, resource generation, building links between organizations, and promoting participation in political processes. The typology used here employs some of the same concepts, but does so from the vantage point of social movement communication, which is theoretically motivated by a broader and more far reaching set of goals. While many social movements engage in campaigns and aim to persuade, they may also emphasize participatory democratic communication forums, the need to mobilize on multiple fronts, and long-term efforts toward society-wide change. These goals merit special attention in a communication typology of social movements. Consequently, in addition to a focus on information provision, lateral links and resource generation, this typology foregrounds the concept of mobilization broadly conceived, forums for creative works, and interaction in association with dialog (rather than interaction aimed at extracting resources from supporters). Several other studies focused on democratic participation online, while not concerned with mobilization or creative works, similarly highlight the role of dialog and discussion in their communication typologies (Gibson and Ward, 2003, p. 143; Jankowski and Van Selm, 2000, pp. 151–152; Tsagarousianou, 1999, p. 195). Thus, while the typology presented here draws support from other political communication typologies, it is uniquely tailored to the purposes of social movements.

The Study

The sampling frame consisted of a master list of US-based groups identified with six social movements: the environmental movement, the gay, lesbian, bisexual and transgendered (GLBT) movement, the anti-corporate globalization movement, the human rights movement, the media reform movement, and the women's movement.[3] These movements are notable for their contemporary prominence and for their efforts to harness media for public communication.[4] Ten researchers compiled a list of 749 organizations, from the *Encyclopedia of Associations: National*, the most comprehensive source of information on US nonprofit membership organizations with national scope (Hunt, 2005).[5] The *Encyclopedia's* publishers estimate that the volume includes ninety to ninety-five percent of all US-based organizations at any given time (Swartout, 2006). A search of *WorldCat* for other handbooks and directories related to these movements yielded no results. So while a cross check of our principal source with others was not possible, this master list likely represents the vast majority of established and active organizations from our selected social movements. Researchers reviewed the *Encyclopedia of Associations* keywords index to identify keywords that could be broadly associated with one of the movements cited above. Multiple keywords were found for each movement. For example, keywords associated with environmentalism included 'environment,' 'environmental education, 'conservation,' 'forestry,' 'pollution control,' and 'rain forests,' among others. Researchers then identified the organizations associated with each keyword and reviewed the self-reported descriptions of each in order to determine whether they qualified as a SMO. Organizations that self-identified with a movement and its goals, engaged in some form of collective action, and advocated social change met our criteria for SMOs. Our criteria followed Castells (2001, pp. 69–70) in presuming that we should identify SMOs largely on the basis of whether they consider themselves to be part of a broader social movement. However, researchers did scrutinize organization's self descriptions, and later their websites, to eliminate those that utilized some movement rhetoric, but espoused purposes contrary to commonly recognized movement goals.[6] Finally, researchers recorded the annual budgets and membership numbers of each organization whenever that information was available.

Once I had a master list, I assigned researchers organizations for further examination using a stratified random sample. Researchers surveyed more than ten percent of the websites of each type of SMO on the list. Nearly all of the eighty-six SMOs selected had their own websites. In the few instances where this was not the case, additional SMOs of the same movement type were randomly selected for inclusion. In total, researchers surveyed the websites of thirty-three environmental, ten GLBT, six anti-corporate globalization, sixteen human rights, ten media reform, and eleven women's organizations. While sixty-eight percent of the SMOs did not give their annual budget, the remaining thirty-two percent had annual budgets that ranged from $3,000 to over $3,000,000. Sixty percent of the organizations did not list the number of members affiliated with their organization. Of the remainder, thirteen organizations had less than a thousand individual or group members, fourteen had between one and ten thousand members, four had between fifteen and one hundred and fifty thousand members, and one listed one million and eight hundred thousand members. Researchers examined the SMO websites over a four-month period between February and May of 2006.

The typology constructed here was the basis for a survey instrument that enumerated the features or attributes of each communication function, categorizing them according to whether they (1) provided information, (2.) engaged in action and mobilization, (3) promoted interaction and dialog, (4) made lateral linkages, (5) hosted creative and cultural works, and (6) attempted fundraising and resource generation. While the survey instrument did not include an exhaustive list of all possible features, it did represent activities theorized as central to the purposes of social movements. Researchers were asked to examine all web pages on each organization's site and to code each feature or attribute on the survey form as present or absent. An open coding process allowed researchers to record additional features found on the websites through an 'other' category, which subsequently led to the inclusion of some additional items. Researchers coded a total of sixty-three items (reported in Tables 1–6).

I determined intercoder reliability by asking three researchers to code a reliability sample consisting of ten percent of the total number of websites selected for analysis. I measured reliability according to Krippendorff's alpha, an appropriate measure for studies that include more than two coders and multiple variables involving both ratio and nominal levels of measurement. Each coder coded all websites included in the randomly selected reliability sample and completed all sections of the survey instrument. Utilizing three coders for analysis enhanced the inherent conservativeness of Krippendorff's alpha, already one of the most conservative and reliable measures of intercoder reliability (Lombard et al., 2005). Several variables had an alpha of less than .60 and were eliminated from the study. Five of the remaining sixty-three variables had an alpha of between .60 and .69; twelve had an alpha of between .70 and .79; twenty-six had an alpha between .8 and .89; and the remaining twenty had an alpha of .90–1.0.[7] Given the exploratory nature of this study, the enhanced reliability gained through the use of three coders rather than two, and the inherent conservativeness of Krippendorff's alpha, I included in this study all the variables that had an alpha of .6 or greater.

After obtaining the completed surveys, I compiled data on the frequency of activities in each category and then aggregated the data to describe the percentage of organizations that made no, low, moderate or high uses of the web to engage in each of these communication functions. An SMO's overall engagement in each function was coded as low, medium or high according to whether the observed activities fell into the bottom, middle or top third of all recorded activities in that function. I report both sets of data next. The survey's results provide some insight into whether and how national SMOs are using web-based communication to further the central communicative functions of social movements.

RESULTS

This study first examined what features were present on US SMO websites and how prevalent various features were within each communication type. One set of features involved the provision of information related to an organization's views, issues and identity. As Table 1 shows, two-thirds or more of the organizations offered information about their identity, providing organizational descriptions and histories. More than half of the organizations also offered information about movement perspectives and issues through self-published articles and reports, newsletters, and law and policy analysis. A quarter to under a half offered alternative and mainstream news articles, suggestions for further research, speeches or articles by movement leaders, and press releases. Less than a quarter offered video or audio reports, critiques of mainstream media coverage, frequent updates (really simple syndication), and read only listservs.

Table 1. Providing Information

Features of Providing Information	Frequency	Percent	Alpha
Podcast Video or Audio Reports	5	5.8%	.76
Stream Audio Reports	9	10.5%	.85
Stream Video Reports	12	14%	.85
Mainstream Media Critiques	12	14%	.82
RSS/Frequent Updates	15	17.4%	.83
Read Only Listservs	19	22.1%	.80
Alternative News Articles	23	26.7%	.75
Mainstream News Articles	26	30.2%	.84
Further Research Suggestions	34	39.5%	.98
Leadership Speeches & Articles	37	43%	.83
Media/Press Releases	39	45.3%	.67
Self-published Articles/Reports	49	57%	1.00
Newsletter	51	59.3%	.94
Law & Policy Analysis	54	62.8%	.83
Organization History	57	66.3%	.90
Organization Description	80	93%	.91

Another set of features, categorized as action and mobilization, involves the organization of actions and initiatives aimed at specified outcomes or the propagation of viewpoints that can spur action. I was also interested in whether SMOs employed the web to mobilize locally, nationally or internationally. Although more than four-fifths of the websites featured project descriptions or news and three-fifths coordinated offline actions of some kind, other action and mobilization features were far less frequent, as Table 2 indicates. A little over two-fifths of the SMOs posted calendars of events or planned national actions online, and only about a quarter issued urgent action alerts, or planned local or international actions. Fewer than one-fifth solicited participation in online petitions, email campaigns, online actions, or surface letter-writing campaigns. Only one group engaged in virtual denial of service attacks.

Table 2. Action and Mobilization

Features of Action & Mobilization	Frequency	Percent	Alpha
Denial of Service Attacks	1	1.2%	.70
Online Petitions	11	12.8%	.98
Email Campaigns	12	14%	.82
Coordinates Online Actions	16	18.6%	.69
Surface Letter Writing Campaigns	16	18.6%	.87
Plans International Actions	20	23.3%	.86
Urgent Action Alerts	23	26.7%	.87
Plans Local Actions	27	31.4%	1.00
Plans National Actions	36	41.9%	1.00
Calendar of Events	37	43%	.77
Coordinates offline actions	51	59.3%	.90
Project or Campaign Descriptions or News	71	82.6%	.73

A third set of features facilitates two-way communication between and among movement supporters and social movement organizations. Alternative media theorists consider this category, labeled interaction and dialog, essential to the formation of a collective meaning and identity among members of social movements. While nearly all organizations provided their contact information, other activities related to interaction and dialog did not feature heavily on the websites. Table 3 shows that little over a third of the organizations provided support services or advice of some kind, but fewer offered first-person accounts of movement activities, online discussions of organizational strategies, or participatory forums. About a tenth of the sites made member profiles or member contact information available, and only six or fewer groups discussed organizational issues, held meetings, made web logs available to site visitors, or conducted polls and surveys online.

Table 3. Interaction and Dialog

Features of Interaction and Dialog	Frequency	Percent	Alpha
Discusses Organizational Issues Online	1	1.2%	.85
Holds Meetings Online	2	2.3%	1.00
Web Logs	5	5.8%	.71
Online Polls and Surveys	6	7%	.78
Member or Supporter Profiles	7	8.1%	.64
Member or Supporter Contact Information	10	11.6%	.83
First Person Accounts of Movement Activities	17	19.8%	.81
Discusses Strategies Online	17	19.8%	.82
Participatory Forums	18	20.9%	.85
Support Services or Advice	29	33.7%	.79
Organization Contact Information	81	94.2%	.97

Theory suggests that SMOs should foster lateral linkages in order to build networks among social movement actors and to lead supporters to sources of news and information related to their cause. This study operationalized lateral linkages as hyperlinks to other SMO sites and to alternative news, mainstream news, and independent research sites. While about a fifth of the SMOs linked to mainstream news sites, over a quarter offered links to alternative news sites, as shown in Table 4. National SMOs linked to other national organizations more often than to international SMOs, with over half linking to the former and a little more than a quarter to the latter. More than three-fifths of SMOs maintained links to other SMOs affiliated with their movement, and over a third linked to sites affiliated with other social movements, as well as to research sites.

Table 4. Lateral Linkages

Features of Lateral Linkages	Frequency	Percent	Alpha
Links to Mainstream News	17	19.8%	.82
Links to Alternative News	22	25.6%	.73
Links to International SMOs	24	27.9%	.78
Links to SMO Sites of Other Movements	30	34.9%	.88
Links to Research Sites	33	38.4%	.61
Links to National SMOs	44	51.2%	.95
Links to SMO sites of Primary Movement	53	61.6%	.95

The category of creative expression encompassed features involving imaginative cultural works that communicated political ideas without relying on rational argumentation or reportage. Less than a third of the organizations acted as a platform for any type of creative expression, as Table 5 shows. About a third offered visual art on their sites, and half that number offered creative videos. Only a handful of SMOs offered cartoons or poetry, and only one or two offered parody or music.

Table 5. Creative Expression

Features of Creative Expression	Frequency	Percent	Alpha
Parodies	1	1.2%	1.00
Music	2	2.3%	.94
Poetry	4	4.7%	1.00
Cartoons and Comics	6	7%	.94
Video	14	16.3%	.76
Visual Art	28	32.6%	.78

The last set of features related to fundraising and resource generation, or attempts to solicit financial and human resources in order to support the organization or its activities. Nearly three-fourths of the organizations attempted to recruit volunteers or solicit donations online, and over half enabled new members to join online. Table 6 summarizes the results. A little over a third of the SMOs collected information from members or supporters, sold merchandise online, or posted job listings. Over a quarter of the sites enabled members to renew their membership or pay dues online.

Less common activities included facilitating membership recruitment by offering a form that supporters could print and mail to the organization or allowing members to subscribe to products or services. Only two SMOs carried advertising from third parties on their sites.

In addition to examining the frequency of features found online, the study also asked to what level or degree do SMOs engage in different types of communication considered essential to their purposes. Table 7 summarizes the levels of activity found across each communication type.

While very few organizations engaged in high levels of communication across any type, some types of communication were more prevalent than others. The most common type of communication was information provision. All the organizations engaged in information provision, with over half engaging in informing activities at medium or high levels. The next most common communication categories were action and mobilization and fundraising or resource generation. More than ninety percent of the organizations engaged in these activities online, with around thirty-five percent undertaking them at medium to high levels. While comparatively fewer organizations (eighty-one percent) made lateral linkages, a greater number (nearly forty-nine percent) performed these activities at medium or high levels, although more than eighteen percent made no lateral linkages. The least commonly exhibited communication functions included interaction and dialog and creative expression. More than ninety-five percent of the SMOs engaged in interaction and dialog, but in this case only about thirteen percent did so at medium or high levels. Only about thirty-nine percent of the organizations acted as a platform for creative expression, with less than five percent doing so at medium or high levels.

Table 6. Fundraising and Resources

Features of Fundraisings and Resource Generation	Frequency	Percent	Alpha
Advertising	2	2.3%	1.00
Allows New Members to Join Offline (Print Form)	9	10.5%	1.00
Sells Subscriptions to Products or Services	18	20.9%	.82
Membership Renewal	23	26.7%	.82
Pay Dues	26	30.2%	.88
Collects Member or Supporter Information	29	33.7%	.89
Sells Merchandise	30	34.9%	.68
Job Listings	31	36%	.82
Allows New Members to Join Online	46	53.5%	.81
Volunteer Sign Up	63	73.3%	.94
Solicits Donations	66	76.7%	.87

Finally, the study asked whether SMOs with greater budgets or memberships use the web more than those with fewer financial resources and members. I hypothesized that organizations with greater amounts of resources and supporters would be more likely to offer more elaborate websites that utilized more features. However, my analyses revealed no significant correlations between levels of web-based communication types and the estimated yearly budget or membership size of the SMOs.

Table 7. Levels of Communication Activity

Type of Communication	No Activity	Low Activity	Medium Activity	High Activity	Total Activity	Alpha Range
Providing Information	0 0%	39 45.3%	42 48.8%	5 5.8%	86 100%	.67-1.00
Action and Mobilization	5 5.8%	50 58.1%	26 30.2%	5 5.8%	86 100%	.69-1.00
Fundraising and Resources	6 7%	51 59.3%	26 30.2%	3 3.5%	86 100%	.68-1.00
Lateral Linkages	16 18.6%	28 32.6%	35 40.7%	7 8.1%	86 100%	.61-.95
Interaction and Dialog	3 3.5%	72 83.7%	8 9.3%	3 3.5%	86 100%	.64-1.00
Creative Expression	52 60.5%	30 34.9%	2 2.3%	2 2.3%	86 100%	.76-.94

Discussion

Communication theory suggests that social movements have multiple incentives for utilizing new communication resources, like the web, to bypass traditional media gatekeepers and get their messages out to supporters and the public at large. Alternative media scholarship further suggests a typology of social movement communication that encompasses the range of communication functions vital to social movements. However, this survey indicates that by and large some of the most prominent, contemporary social movements are not engaging heavily in such communication on the web.[8] The SMOs examined here show moderate or high levels of activity in four areas deemed central to social movements: providing information, coordinating action and mobilization, engaging in fundraising and resource generation, and making lateral linkages. However, with the exception of information provision, the majority of SMOs exhibit no or low activity in these areas. Furthermore, substantial activity across the remaining two categories, interaction and dialog and creative expression, is uniformly lacking. These findings are somewhat comparable to those of Foot and Schneider (2006, pp. 159–160) in their study of political campaign sites during elections. They found that while most campaign organizations engaged in informing activities online, very few engaged in connecting (similar to lateral linkages) or mobilizing. Lateral linkages appear to be more common among SMOs, though, as with political campaign sites, mobilization fared more poorly. These findings are also consistent with a study of anti-globalization websites that found few opportunities for interaction and dialog (Van Aeist and Walgrave, 2002, p. 478).

Given the gap between the theory and practice of social movement communication online, these findings beg the question, why are national SMOs not utilizing the full potential of the web to engage in the types of communication that are thought to be central to their very existence? In this section, I draw on scholarship examining social movements and the Internet to consider three factors that might account for SMOs' relatively low levels of web use and to consider how additional research methods could help explain or refine my findings.

Underutilization of the web as a communication resource may relate to three factors having to do with the social and economic conditions surrounding Internet use: organizational orientations (including goals, strategies and beliefs); organizational resources; and resource sharing among organizations. In their accounts of social movement communication, alternative media scholars have yet to differentiate among types of SMOs and their varying communication needs. Diani (2001, p. 122) suggests that different types of SMOs have different needs and goals when it comes to Internet use. For example, those focused on organizing professional resources may be motivated to use the Internet to diffuse information online, but not to increase member interaction with the organization, while those focused on broader participatory resources may have more incentive to cultivate grassroots participation in organizational activities and actions (Diani, 2001, p. 123). In addition, organizations may have different ideas or beliefs about the efficacy of the web as a communication tool, preferring to focus their efforts on more traditional or direct communication forms, such as face-to-face contact, print materials, or more targeted forms of Internet communication, like email. Finally, scholars have identified potential drawbacks of using computer-mediated communication for traditional social movement activities, such as building trust, community or commitment on the part of movement members, when not reinforced by offline relationships (Calhoun, 1998; Diani, 2001, p. 121). Multi-method studies could combine surveys or interviews of key personnel responsible for website content with comparative quantitative analysis of websites. Researchers could utilize surveys or interviews to gather data about the experiences, attitudes and goals of website producers and assess whether differential orientations correlate with different types of SMOs or whether differences among these variables explain varying levels of web use. For example, Foot and Schneider's (2006, p. 168) work on political websites found that different levels of website features correlated with the aims and orientations of webmasters and with the target audiences and goals of campaign organizations. Taking account of different organizational orientations, including different types of SMOs, different strategies and goals, and different perceptions about the utility of communications technology, may reveal calculated, strategic, and efficient uses of the Internet on the part of SMOs, rather than broad-based uses.

Alternately, established, national SMOs may simply be less likely than other types of social movement formations to make full use of web-based communication. Some scholars suggest that new media are precipitating the development of online movement formations that are decentralized, flexible, and issue-oriented, and these formations may rely more heavily on Internet communication than traditional SMOs who may be more inclined to simply integrate Internet use into their pre-existing routines (Bennett, 2003a, p. 145; Ward et al., 2003, p. 656). One study of political party websites found that smaller parties experimented more with participatory and dialogic features than larger, more entrenched parties (Cunha et al., 2003). Moreover, some scholars and activists have criticized mainstream national SMOs for their professionalized, hierarchical structures and lack of commitment to discursive and participatory processes (Mitchell et al., 1992; Schlosberg, 1999). The present study surveyed relatively established, national SMOs and did not include local SMOs or social movement formations that exist primarily or exclusively online. Consequently, it did not capture information about these other important fields of social movement actors. Comparative analyses of the strategies, goals and online activities of local or primarily web-based social movement actors with more established, national organizations could reveal significant differences in how they think about and use the web.

Another reason why SMOs may not take advantage of the full range of web functions is that they may lack the necessary resources, including time, money and knowledge. Kirschenbaum and Kunamneni (2001, p. 7) have identified an 'organizational divide,' particularly among the community-based nonprofit sector. Many such organizations lack the technological capacity to generate relevant online content, develop new applications, and use computer-mediated communication to address their needs and goals. Lacking the time, skills, or funds to integrate computer technology into their work, the majority of community based nonprofit organizations make only limited use of computers and the Internet as communication tools (Kirschenbaum and Kunamneni, 2001, p. 7). Scholars have begun to study the actual resources, including time and money, required to design and maintain websites (See De Cheveigné, 2007; Foot and Schneider, 2006),[9] and one study suggests that resource limitations may be a major factor inhibiting SMO Internet use (León et al., 2009). Ethnographic observations of website producers within selected SMOs, or interviews or surveys with respectively larger numbers of website producers, could be used to assess the human, financial, temporal and technical resources devoted to different levels of web use and whether the availability or allocation of resources affect web use among differently resourced organizations.

A third explanation for relatively low levels of activity among the majority of national SMO websites may be found within larger patterns of web use within social movements themselves. Scholars have documented the existence of Internet sites designed to pool the resources of social movement actors and act as hubs allowing movement members to exchange information and analysis, coordinate action, and formulate discourse and dialog around movement agendas (Downing, 1989, p. 156: León et al., 2009). Such sites exist on both sides of the national political spectrum, as exemplified by the conservative movement's Townhall.com or the sites managed by the progressive Institute for Global Communications, including AntiRacismNet, EcoNet, PeaceNet, and WomensNet. The existence of central sites devoted to information dissemination, networking, and mobilization may render unnecessary the duplication of these functions by others. Even if such central sites do not exist, organizations may offer links to direct visitors to information or features available elsewhere in order to avoid resource duplication. Studying the anarchist community's use of the web, Owens and Palmer (2003) found a core-periphery structure among the movement's websites. Moreover, an informal division of labor existed between the peripheral sites that addressed movement insiders and the small number of core sites that served to present anarchism to movement outsiders. Core sites offered the public an introduction to anarchist philosophy and identity that peripheral sites could reference, rather than reiterate (Owens and Palmer, 2003, p. 347). Similarly, Foot and Schneider (2006, p. 59) found that many political organizations link to other sites in order to offer information without having to collect, maintain or display it themselves. Further research could map the linkages between social movement organizations, determine whether core-periphery structures exist among websites affiliated with particular movements, and perform qualitative and quantitative content analysis of the function of links between actors.

Finally, research on how users engage with the functions and features of SMO websites would complement our understanding of the form these sites take. Web-based surveys can illuminate how and why visitors use SMO websites, how frequently they use them, and who these users are.

Conclusion

Communication scholarship suggests that the Internet can serve as an important resource for social movements. This study draws on alternative media scholarship to construct a typology of communication functions that encompasses the potential usages movement actors can make of the Internet. The study designs a survey instrument based on this typology that enumerates the features and activities linked to each communication function, investigates which features actually appear on the websites of a random sample of national SMOs, and highlights the discrepancy between the potential and actual uses of the web by the SMOs. The survey results suggest that the majority of national SMOs are not utilizing the web to its full capacity and posits a number of reasons why this might be the case, including organizational goals, strategies and objectives, organizational resources, and organizational efforts to share or pool resources made available to movement supporters. This chapter also suggests additional methods researchers might use to investigate these issues and to hone theory addressing social movement web use in future empirical research.

While alternative media scholarship elaborates the potential purposes and functions of social movement communication, scholars can advance the study of movement communication and new media by broadening their empirical and theoretical horizons. Empirically, scholars can undertake generalizable and comparable studies of movement web use, paying particular attention to: the attitudes, strategies and goals of website producers; the different orientations and types of SMOs; the disparate resources, opportunities and constraints available to movement actors; the presence and nature of online linkages between movement groups; and how actual users experience movement websites. Theoretically, communication scholars can draw on social movement studies to enhance their understanding of the broader dynamics that shape movement structures, strategies and practices, and of how these dynamics condition the production and use of movement media. Scholars should also draw on the intersection of social movement and organizational studies to delineate organizational forms, functions, structures, goals, and repertoires of action and mobilization, employing these categories to differentially investigate and analyze the range of actors and activities operating within social movement communication. These theoretical and empirical avenues of research can help scholars gain a better understanding of how and why social movement actors use the Internet.

Notes

1. Originally published in *New Media & Society* (Stein, L. (2009). Social Movement Web Use in Theory and Practice: A Content Analysis of US Movement Websites. *New Media & Society*, 11(5), 749–71
2. Civil society denotes that segment of collective social life that is separate from markets and states.
3. I selected a number of movements in order to gain a more general picture of SMO web use, but limited the number to six due to resource and time constraints. While comparative data of web use between movements might prove valuable, the small sample size of some of these groups effectively ruled out valid comparisons.
4. In his book analyzing the role of technology in modern social change, Castells (2001) focuses on environmentalism, anti-corporate globalization, and the women's movement, and highlights, though to a lesser extent, the GLBT movement. In their book examining networks and strategic communication, Keck and Sikkink (1998) focus on the human rights, women's and envi-

ronmental movement. Scholars have also identified the media reform movement as an emerging social movement (Carroll and Hackett, 2006).

5. While all of the organizations included operated in the US, thirty of them also operated in other countries.
6. Most of these organizations appeared to be lobbying groups for industries targeted by social movement actors.
7. I used KALPHA, an SPSS macro, to generate Krippendorff's alpha (see Krippendorff, 2003 and Hayes and Krippendorff, 2007 for more information). Lombard, Snyder-Duch and Bracken (2005) maintain that 'Coefficients of .90 or greater are nearly always acceptable, .80 or greater is acceptable in most situations, and .70 may be appropriate in some exploratory studies for some indices. Higher criteria should be used for indices known to be liberal (i.e. percent agreement) and lower criteria can be used for indices known to be more conservative (Cohen's kappa, Scott's pi, and Krippendorff's alpha).'
8. However, given the lack of other research on the parameters of the SMO population as a whole, there is no basis by which to compare this sample population with the overall SMO population.
9. Foot and Schneider (2006) found that political campaign sites that engaged in higher levels of involving and mobilizing devoted more staff and time to website production than sites lacking those features.

References

Atton, C. (2003). Infoshops in the Shadow of the State. In N. Couldry & J. Curran (Eds.), *Contesting Media Power: Alternative Media in a Networked World* (pp. 57–69). Lanham, MD: Rowman & Littlefield.

Barlow, W. (1988). Community Radio in the US: The Struggle for a Democratic Medium. *Media, Culture & Society,* 10(1), 81–105.

Bennett, W. L. (2003a). Communicating Global Activism: Strengths and Vulnerabilities of Networked Politics. *Information, Communication & Society,* 6(2), 143–68.

Bennett, W. L. (2003b). New Media Power: The Internet and Global Activism. In N. Couldry & J. Curran (Eds.), *Contesting Media Power: Alternative Media in a Networked World* (pp. 17–37). Lanham, MD: Rowman & Littlefield.

Boyd, A. (2003, Aug. 4/11). The Web Rewires the Movement. *The Nation,* 277(4), 13–18.

Calhoun, C. (1998). Community without Propinquity Revisited: Communication Technology and the Transformation of the Urban Public Sphere. *Sociological Inquiry,* 68(3), 373–97.

Cammaerts, B., & Van Audenhove, L. (2005). Online Political Debate, Unbounded Citizenship, and the Problematic Nature of a Transnational Public Sphere. *Political Communication,* 22(2), 179–96.

Carroll, W. K., & Hackett, R. A. (2006). Democratic Media Activism through the Lens of Social Movement Theory. *Media, Culture & Society,* 28(1), 83–104.

Castells, M. (2001). *The Power of Identity.* Cambridge, MA: Blackwell Publishers.

Costanza-Chock, S. (2003). Mapping the Repertoire of Electronic Contention. In A. Opel & D. Pompper (Eds.), *Representing Resistance: Media, Civil Disobedience, and the Global Justice Movement* (pp. 173–88). Westport, CT: Praeger.

Cunha, C., Martín, I., Newell, J., & Ramiro, L. (2003). Southern European Parties and Party Systems, and the New ICTs. In R. Gibson, P. Nixon & S. Ward (Eds.), *Political Parties and the Internet: Net Gain?* (pp. 70–97). New York: Routledge.

De Cheveigné, S. (2007). Internet Practices around Environmental Questions. Paper presented at 50th IAMCR Conference, Paris, 23–25 July.

DeLuca, K., & Peebles, J. (2002). From Public Sphere to Public Screen: Democracy, Activism, and the Lessons of Seattle. *Critical Studies in Mass Communication,* 19(2), 125–51.

Diani, M. (2001). Social Movement Networks: Virtual and Real. In F. Webster (Ed.), *Culture and Politics in the Information Age* (pp. 117–28). New York: Routledge.

Downing, J. (1989). Computers for Political Change: PeaceNet and Public Data Access. *Journal of Communication,* 39(3), 154–62.

Downing, J. (2001). *Radical Media: Rebellious Communication and Social Movements.* Thousand Oaks, CA: Sage Publications.

Downing, J. (2003). The Independent Media Center Movement and the Anarchist Socialist Tradition. In N. Couldry & J. Curran (Eds.), *Contesting Media Power: Alternative Media in a Networked World* (pp. 243–57). Lanham, MD: Rowman & Littlefield.

Foot, K. A., & Schneider, S. M. (2006). *Web Campaigning.* Cambridge, MA: MIT Press.

Ford, T. V., & Gil, G. (2001). Radical Internet Use. In J. D. H. Downing (Ed.), *Radical Media: Rebellious Communication and Social Movements* (pp. 201–34). Thousand Oaks, CA: Sage Publications.

Fraser, N. (1993). Rethinking the Public Sphere: A Contribution to the Critique of Actually Existing Democracy. In H. Giroux & P. McLaren (Eds.), *Between Borders: Pedagogy and the Politics of Cultural Studies* (pp. 74–100). New York: Routledge.

Gamson, W. (1990). *The Strategy of Social Protest* (2nd ed.). Belmont, CA: Wadsworth Publishing Company.

Garrett, R. K. (2006). Protest in an Information Society: A Review of Literature on Social Movements and New ICTs. *Information, Communication & Society,* 9(2), 202–24.

Gibson, R., & Ward, S. (2000). A Proposed Methodology for Studying the Function and Effectiveness of Party and Candidate Web Sites. *Social Science Computer Review,* 18(3), 301–19.

Gibson, R., & Ward, S. (2003). Letting the Daylight in? Australian Parties' Use of the World Wide Web at the State and Territory Level. In R. Gibson, P. Nixon & S. Ward (Eds.), *Political Parties and the Internet: Net Gain?* (pp. 139–160). New York: Routledge.

Gillett, J. (2003). The Challenges of Institutionalization for AIDS Media Activism. *Media, Culture & Society,* 25(5), 607–24.

Gitlin, T. (1980). *The Whole World Is Watching: Mass Media in the Making and Unmaking of the New Left.* Berkeley, CA: University of California Press.

Gross, L. (2003). The Gay Global Village in Cyberspace. In N. Couldry & J. Curran (Eds.), *Contesting Media Power: Alternative Media in a Networked World* (pp. 259–72). Lanham, MD: Rowman & Littlefield.

Hayes, A. F., & Krippendorff, K. (2007). Answering the Call for a Standard Reliability Measure for Coding Data. *Communication Methods and Measures,* 1(1), 77–89.

Hunt, K. N. (2005). *Encyclopedia of Associations: National Organizations of the US,* (42nd ed.). San Francisco, CA: Thomson Gale.

Jankowski, N., & Van Selm, M. (2000). The Promise and Practice of Public Debate in Cyberspace. In K. L. Hacker & J. Van Dijk (Eds.), *Digital Democracy: Issues of Theory and Practice* (pp. 149–65). Thousand Oaks, CA: Sage Publications.

Kahn, R., & Kellner, D. (2004). New Media and Internet Activism: From the "Battle of Seattle" to Blogging. *New Media & Society,* 6(1), 87–95.

Keck, M. E., & Sikkink, K. (1998). *Activists Beyond Borders: Advocacy Networks in International Politics.* Ithaca, NY: Cornell University Press.

Kessler, L. (1984). *The Dissident Press: Alternative Journalism in American History*. Newbury Park: Sage Publications.

Kidd, D. (2003). Indymedia.org: A New Communications Commons. In M. McGaughey & M. Ayers (Eds.), *Cyberactivism: Critical Theories and Practices of On-Line Activism* (pp. 47–69). New York: Routledge.

Kielbowicz, R. B., & Scherer, C. (1986). The Role of the Press in the Dynamics of Social Movements. In L. Kriesberg (Ed.), *Research in Social Movements, Conflicts and Change* (pp. 71–96). Greenwich, JAI.

Kirschenbaum, J., & Kunamneni, R. (2001). *Bridging the Organizational Divide: Toward a Comprehensive Approach to the Digital Divide: A PolicyLink Report*. Retrieved from http://www.policylink.org/pdfs/Bridging_the_Org_Divide.pdf#search=%22Organizational%20divide%22.

Krippendorff, K. (2003). *Content Analysis: An Introduction to Its Methodology*. Thousand Oaks, CA: Sage.

León, O., Burch, S., & Tamayo, E. (2009). Societies-in-Movement: The Latin American Minga Informativa. In L. Stein, C. Rodriguez, & D. Kidd (Eds.), *Making Our Media: Global Initiatives Towards a Democratic Public Sphere, v. 2*. Cresskill, NJ: Hampton Press.

Lombard, M., Snyder-Duch, J., & Bracken C. C. (2005). *Practical Resources for Assessing and Reporting Intercoder Reliability in Content Analysis Research Projects*. Retrieved from http://www.temple.edu/mmc/reliability.

McLeod, D. M., & Detenber, B. H. (1999). Framing Effects of Television News Coverage of Social Protest. *Journal of Communication*, 49(3), 3–23.

Mitchell, R. C., Mertig, A. G., & Dunlap, R. E. (1992). Twenty Years of Environmental Mobilization: Trends Among National Environmental Organizations. In R. E. Dunlap & A. G. Mertig (Eds.), *American Environmentalism: The US Environmental Movement, 1970–1990* (pp. 11–26). Washington, DC: Taylor and Francis.

Mueller, M., Kuerbis, B., & Pagé, C. (2004). *Reinventing Media Activism: Public Interest Advocacy in the Making of US Communication-Information Policy, 1960–2002: Ford Foundation Report*. Retrieved from http://dcc.syr.edu/ford/rma/reinventing.pdf.

Owens, L., & Palmer, L. K. (2003). Making the News: Anarchist Counter-Public Relations on the World Wide Web. *Critical Studies in Media Communication*, 20(4), 335–61.

Raboy, M. (1981). Media alternatives and social movements: Quebec 1960–1980. *Graduate Communications Program: Working Paper Series*. Montreal, Canada: McGill University.

Rodríguez, C. (2001). *Fissures in the Mediascape*. Cresskill, NJ: Hampton Press.

Rogers, R., & Marres, N. (2000). Landscaping Climate Change: A Mapping Technique for Understanding Science and Technology Debates on the World Wide Web. *Public Understanding of Science*, 9(2), 141–63.

Schlosberg, D. (1999). *Environmental Justice and the New Pluralism*. New York: Oxford University Press.

Shoemaker, P. J. (1982). The Perceived Legitimacy of Deviant Political Groups: Two Experiments on Media Effects. *Communication Research*, 9(2), 249–86.

Steiner, L. (1992). The History and Structure of Women's Alternative Media. In L.F. Rakow (Ed.), *Women Making Meaning: New Feminist Directions in Communication* (pp. 121–43). New York: Routledge.

Tarrow, S. (2005). *Power in Movement: Social Movements and Contentious Politics*, 2nd ed. Cambridge, MA: Cambridge University Press.

Tsagarousianou, R. (1999). Electronic Democracy: Phetoric and Reality. *Communications*, 24(2), 189-208.

Shaping the "Anti-Globalization" Movement. *Information, Communication & Society,* 5(4), 465–93.

Tsagarousianou, R. (1999). Electronic Democracy: Phetoric and Reality. *Communications*, 24(2), 189-208.Shaping the "Anti-Globalization" Movement. *Information, Communication & Society,* 5(4), 465–93.

Van Aeist, P., & Walgrave, S. (2002). New Media, New Movements? The Role of the Internet in Tsagarousianou, R. (1999). Electronic Democracy: Rhetoric and Reality. *Communications*, 24(2), 189–208.

Ward, S., Gibson, R., & Lusoli, W. (2003). Online Participation and Mobilisation in Britain: Hype, Hope and Reality. *Parliamentary Affairs,* 56(4), 652–68.

CHAPTER NINE

Identity and Surveillance Play in Hybrid Space

DAVID PHILLIPS

This chapter approaches identity as the sharing, creating, and performing of socially meaningful relationships. Identity negotiations occur in space, but spatial settings themselves are negotiated and created along with the performances, the roles, and the identities that they support.[1]

Two historical shifts are causing deep structural changes to the negotiations of space and identity. The first is the increasing prevalence of surveillance as a mode of knowledge production. Surveillance may be visual or actuarial, or a hybrid of these. The second shift is the development of mobile and "cloud" ubiquitous computing—sensors, computation devices, terminals, and responders embedded into mobile devices and stationary objects, and networked to mesh physical and data environments.

Depending on how and to whom resources of surveillance and mobile computing are made available, they may facilitate the generation of spaces that are open and public, in that they sustain and are sustained through more or less egalitarian, emergent, collective activity. Or they may become mechanisms of normativity and the re-entrenchment of politically and economically useful identity categories.

This chapter explores these issues, using location-aware social media as the site of structured but messy interplay among modes of surveillance, interactions between data space and physical space, and performative play.

Introduction

This chapter explores the interrelations among public space, surveillance practice, and identity play. It is especially concerned with the possibility of counter-normative praxis within that matrix. Briefly, the first section argues that identity and space are both performed and performative. They are mutually sustained by the actions they mediate. The following section describes how the availability of resources for this co-construction is shifting with two historical trends. The first is the increasing institutionalization of visual and actuarial surveillance practices. The second is the overlaying and intermeshing of data space and physical space through mobile cloud computing. The next section describes counter-normative practices in new spaces of visual surveillance. This is followed by a discussion of the possibility of extending those practices to spaces of actuarial surveillance. Finally, I look at three location-aware social media applications to try to understand how particular infrastructures mediate the possibilities of space, surveillance, and identity.

Identity and Spatiality

Following Goffman, I approach identity as the sharing, creating, and performing of socially meaningful relationships. Identity is socially meaningful in that it carries with it a set of rights and expectations; an identity claim is a claim to be treated in a particular way. Whether it is the two-year-old insisting that she is no longer to be treated as a baby, or the adult citizen demanding access to the voting booth, we deploy our identities in a quest for social position. What is being negotiated is not only the identity label ("baby," "citizen") but the structure of rights appertaining thereunto (Goffman, 1959).

Identity mediates between the individual and the social; it is a product of exterior relations and interior subjectivity. It is concerned both with "positionality in the social-structural system of social category relations based on power, exchange, distribution of resources, [etc.] and the habitus of embodied dispositions to action of particular meaningful kinds…" (Lemke, 2008, p. 21). Identity is performed and performative; it is not fore-given but is continually brought into being, straining or reinforcing the structures which are both medium and outcome of the performance (Goffman, 1959; Giddens, 1984; Butler, 1998).

Spatiality is one of the social resources called upon in identity negotiation. Spatiality is implicated in identity negotiation in at least two ways. First, and most obviously, spatial arrangements structure resources of visibility and co-presence. Goffman's notions of front and back stages are inherently spatial and architectural. People retire behind closed doors before they drop one role, one set of social demands, and take on another. Curtains, one-way mirrors, street lighting, ha-ha's—all shape the interactions between individuals and their audience.

Secondly, though, spatial organization permits of different possibilities for the "mutual coordination of actions" and so for collective meaning-making (Dourish and Bell, 2007, p. 419). Certain settings carry expectations of appropriate actions and exchanges; certain roles and relations are easier to sustain in certain places. As Dourish and Bell put it, "the organization of space [is] an infrastructure for the collective production and enactment of cultural meaning." (Dourish and Bell, 2007, p. 415). Similarly, McGrath holds that "space is the fundamental subjective condition of perception, of knowing and understanding the external world" (McGrath, 2004, p. 10).

Yet the logic of space does not precede the logic of interaction. Settings themselves are negotiated and created along with the performances, the roles, and the identities that they support. Space is produced by the actions it mediates. Standards of appropriate activity, and hence the meaning of spaces, are under constant renegotiation. Dodge and Kitchin refer to this as "transduction"—"the constant making anew of a domain in reiterative and transformative practices" (Dodge and Kitchin, 2005, p. 162). McGrath (2004) refers to this as the performativity of space—actions suggest meanings and those meanings are acknowledged or "taken up" to create spatial consciousness.

This performativity, this transduction, this continual recursive construction of habitus and space is itself structured by laws, economics, and cultural norms. Berlant and Warner (1998), for example, talk of how zoning laws, especially those that relegate erotic bookstores and clubs to unpopulated areas, deeply impact the street life of urban neighborhoods and hence the opportunities for queer interactions, identities and visibilities. The boundaries of propriety and the potential for practical ambiguity are imbricated with economic interests and physical structures. The strategic ambiguity of department store windows, for example, provides opportunities legitimately to dawdle and tarry, and perhaps, covertly, to flirt and cruise (Chauncey, 1994).

Visual and Actuarial Surveillance

Increasingly, the techniques and practices of surveillance are reshaping the resources available for this mutual negotiation of space, interaction, and identity. Surveillance has many forms. For now, we will consider two of those forms, and refer to them as the visual and the actuarial.

Visual surveillance is a differential unbounding of space, through enhanced or extended visuality or cognitive awareness. Perhaps the most mundane form of this is found in closed-circuit TV (CCTV) cameras in public places, which relay local images to remote watchers (Norris and Armstrong, 1999). But we also use "visual surveillance" to refer to the troubling reconfigurations of presence and place mediated through cellphone cameras, webcams, and online social media like Facebook. All of these practices both trouble and reinforce norms. They trouble in that distinctions between public and private, work personae and play personae are redrawn (Grimmelmann, 2009). Yet CCTV operators are, from a distance, alert to people and activities that are "out of place," and are ready to call in police to reinforce those norms of place (Norris and Armstrong, 1999).

But perhaps a more profound and fundamental shift is the move toward actuarial surveillance—the systematic, analytic, methodical creation of normativity. In its idealized form, actuarial surveillance individualizes each member of the population, and permits the observation and recording of each individual's activities, then collates these individual observations across the population. From these conglomerated observations, statistical norms are produced. These norms are then applied back to the subjected individuals, who are categorized and perhaps acted upon according to their relation to the produced norm. Thus actuarial surveillance produces both discipline (that is, conformity to the norm), and the disciplines (regulated fields of knowledge and expertise) (Foucault, 1977). It alters both the structures of visibility and the structures of meaning-making. It renders us visible, it identifies us, in relation to the norms it produces.

Actuarial surveillance as a technique of knowledge production and population management is becoming a central organizing principle of modern institutions. It is being adopted more and more in institutional settings, from insurance companies to marketers to police agencies.

These two forms of surveillance cannot be equated. Visual surveillance does not necessarily imply (and usually does not imply) the statistical understanding at the heart of actuarial surveillance. Nevertheless they are related. CCTV is more and more directly linked to actuarial surveillance as cameras record license plate numbers and refer them to data processing facilities, thus discovering and creating norms of traffic flow (and variances by individuals from those norms) through the monitoring of vehicles. The grail of CCTV system developers is to extend this identification capability to individual human bodies through facial recognition or iris scans (Phillips, 2008).

The structure of lived places, which we encounter visually and with our bodies, (neighborhoods, malls, streets) is informed by actuarial surveillance through geodemographic analysis, or the "codification and spatial mapping of habitus" (Parker et al., 2007, p. 905). Geodemographic systems correlate residential data with other personal data, including credit card purchases, subscription data, and public records, to produce statistical identity categories such as "Blue Blood Estates" ("The nation's second-wealthiest lifestyle,...characterized by married couples with children, college degrees, a significant percentage of Asian Americans and six-figure incomes ...") and "Shotguns and Pickups" ("... young, working-class couples with large families...living in small homes and manufactured housing"). These lifestyle clusters are associated with particular neighborhoods. Subscribers to geodemographic services can then choose a neighborhood and discover the prevalence of particular lifestyles within that neighborhood. As one of these services puts it, "You Are Where You Live" (Claritas, Inc., 2008).

These categorization techniques work. They seem to correlate both with the observations of ethnographers and with the subjective perceptions of inhabitants. Judging by their success in the market, the places they help produce accord with the needs of organized capital. Thus, they can be seen as one mechanism of normativity, entrenchment, and stabilization in the co-creation of identities and places (Parker et al., 2007).

The same dialectic of visual and actuarial surveillance is at work in the organization of online spaces, such as chat rooms and other social media. Operators of these spaces are often "Janus-faced," offering extended embodied presence to their users, while making actuarial sense of those users' activities, and selling that knowledge to advertisers. (Campbell, 2005; Steeves, 2006)

Thus, in practice, both online and sense-making occurs both visually and statistically, differently placed actors making a different kind of sense. And just as visual and actuarial surveillance are analytically distinct but practically entwined, so are data spaces and physical spaces enmeshed in complicated ways.

It is at this messy intersection that this piece tries to dwell. It asks about the structuring of performative possibilities in a world mediated by both images and data flows and understood through both visual and actuarial surveillance. In particular, it looks at location-aware social media, since they facilitate both modes of surveillance practice, while integrating data space and physical space in a variety of configurations.

We are dwelling in this mess to ask how best to engage in counter-normative world-making. Scholars of surveillance have suggested that people are adept at negotiating and embracing visual surveillance to generate new senses of self, community, space, and place, often distinctly at odds with prevalent norms of placed behavior. Yet this insight leaves us with two questions. The first is whether and how these tactics can be useful in actuarial surveillance. The second is how those negotiations are constrained and enabled by the structural properties of the media environment in which they occur. The object of this inquiry is not to halt or oppose surveillance practice. Rather, it is to suggest how we might "engage productively, creatively, libidinally, with systems of surveillance," and further, how we can structure media environments to facilitate that engagement (McGrath, 2004, p. 209).

To answer these questions, we will first look at some of the counter-normative ways in which artists and activists have engaged with visual surveillance. We will then explore whether we can imagine analogous engagements with actuarial surveillance. Then we'll look at three forms of location-aware social media, to see how each structures the potential for counter-normative practice.

Counter-Normative Performativity in Visual Surveillance

Recently, scholars have chronicled the actions of agents negotiating visual surveillance as they create new identities, new social relations, new subject positions, and a new kind of space. This new space is not merely more of the same; it is not an extension of past experience. It is instead a new experiential topography ordered by a novel geometry. Koskela (2009) refers to the practices that "challenge [...] the conventional ethics of seeing and being seen, of presenting and circulating images" as "hijacking surveillance" (p. 148). Albrechtslund (2008) calls it "participatory surveillance." McGrath's *Loving Big Brother* (2004) offers perhaps the most extensive exploration of the ways in which people are finding creative agency in surveillance space.

People negotiate surveillance space for many reasons, from the overtly political, to the instrumental, to the just plain fun. In all of these negotiations, people are forming and reforming social relations and, so, identities. As examples of political action, consider the Surveillance Camera Players or the Institute for Applied Autonomy. The former stage literal performances for street cameras, explicitly addressing their operators as the audience (http://www.notbored.org/the-scp.html). The latter produce maps of the locations street cameras, suggesting routes of least surveillance through urban space (http://www.appliedautonomy.com/isee.html). The band "Get Out the Clause" produced a music video by performing in front of CCTV cameras, then requesting the footage under Britain's Freedom of Information Act (Chivers, 2008). All of these are at some level engaged consciously in the political project of using surveillance to make surveillance visible, "hijacking" the system to turn it upon itself.

Steve Mann's (1997) work on "sousveillance" likewise adopts and adapts surveillance technique. The term "sousveillance" is a neologism intended to connote watching from below and to contrast that with "surveillance" or watching from above. Mann wears cameras and microphones on his body, both recording and transmitting his interactions. Often those interactions are intentional—purposely undertaken to expose surveillance systems, to "watch the watchers." Mann will confront workers in surveilled space, such as department stores, and ask about the surveillance cameras in the ceiling, recording the ensuing conflict and discomfort. While his work suffers from a naïve understanding of "above" and "below" in the social order (the people he confronts are usually low-level, low-wage employees, and so among the world's most surveilled populations), nevertheless it is an example of a possible redistribution and co-opting of surveillance technique.

Visual surveillance technique is also engaged to ontological, rather than political, effect. McGrath cites examples of activists using cameras and text messages not only to coordinate street demonstrations, but to foster a sense of self, history, and community. People gather to watch and exchange images, to remember and reconstruct the event and its participants. Thus the 'camera's gaze [becomes] a kind of ontological guarantee of [one's] being" (Žižek, 1992, p. 203).

The construction of novel ontological sense, of new forms of embodied self-awareness, is almost necessarily erotic. Webcams, blogs, and erotic video exchange sites like xtube.com trouble notions of presence and absence, of public and private. Bringing a camera to an event, sexual or otherwise, generates disruption and dislocation in space and time. When or where will that event recur? Before whom? In the midst of this uncertainly, an erotic space opens for the subjective negotiation of propriety, of risk, of boundaries and their transgression. We wonder whether we are ourselves or our image, or some charged hybrid. We are displaced to a future where we are watching ourselves. We are no longer certain where we are and where the other is. We are in the midst of "multiple narcissisms,...a deliciously homosexual position to be in." (McGrath, 2004, p. 68). We might feel, sense, or imagine the absent viewer as a Peeping Tom, or a benevolent Big Brother. Or instead that absent presence might serve as a blank slate on which to project a personal moral code. In any case, the erotic interplay of bodies in space is reconfigured and reconstructed.

If one agrees with McGrath that "space...is the result of social relations," and that "'abstract space'...is a particular historical construct,...an ideologically naturalized screen hiding the dynamics of power," then the co-construction of new subject positions in new spatial topographies is inherently ideological and political (2004, p. 114). Through camera play and sousveillance the normative hierarchies of visibility are thrown into question. Through transgressive erotic presence, the ideology of unitary bodies in Newtonian space and time is de-naturalized. This is semiotic activism. It requires not only the production and circulation of images, but also of the codes with which to make sense of those images. As McGrath says:

> The challenge of communication under surveillance is to develop a continual proliferation of codes, beyond any one authority's translation skills...It will be in this encodedness that any state or corporate claim to *know* through surveillance will be revealed as false. Encodedness...will be the key to counter-hegemonic practice...subverting any tyranny of meaning. (McGrath, 2004, pp. 218–219)

It is clear from the above examples that the space of visual surveillance is one where "knowledge and images of unexpected intensity...emerg[e] in unpredictable configurations and combinations" (Haggerty and Ericson, 2006, p. 29). Through new forms in the production of images, new patterns in their circulation, through transgressive interpretation, through an active and productive ambivalence toward visibility and presence, actors are straining hegemonic concepts of space, place, power and identity.

This leads us to the two questions we set out to explore: First, whether these activities in visual surveillance space have analogues in the space of actuarial surveillance, and second, what structural properties of mediated interactions facilitate or constrain these subversive practices.

Constructing the Space of Actuarial Surveillance

Just as visual surveillance practice troubles an already existing hegemonic ideology of Newtonian space, so non-normative actuarial surveillance must trouble the ideology of data space. This received ideology holds that identifying, monitoring, and "informatizing" entities in physical space, and manipulating that information to better understand the physical entities is a straightforward organizational tool. While there may be issues of power and knowledge, these are problems not of knowledge production, but of access to already existing knowledge. They are issues of privacy. The records in databases are assumed to refer unambiguously and unproblematically to pre-existing "real" bodies in Newtonian space. When they do not, it is understood as an aberration, an error, an unfortunate artifact to be ignored or corrected. Activist intervention must challenge and strip away this normalizing ideology to reveal the production of selves, identities, and relations in actuarial surveillance. Activism must help us "rethink these records of ourselves as doubled bodies, allowing for the possibility of agency, and replacing a fearful relationship to embodiment...with the idea that these other data bodies are our products, our performances" (McGrath, 2004, pp. 162–163). It must help us understand and employ data as self, as performative utterance, as a presentation rather than as a representation.

Remember that one of the tactics in destabilizing and denormalizing visual surveillance is to draw attention to its operation. Scholars and journalists are certainly doing this, but there are very difficult, perhaps intractable, problems in making surveillance space visible and palpable. This is at least in part due to the different ways in which these kinds of surveillance are institutionalized.

Consider the issue of the proliferation of data images, of disturbing their flow and problemetizing their relation to the physical world. We can perhaps see in the intractable problems of "data theft," in the intrinsic impossibility of controlling the flow of data images, just how powerful and productive dataveillance is, and to whom. The institutions that have adopted dataveillance are among the most powerful in capitalist society. They are the banks, the credit agencies, the marketers, the tax authorities. Actuarial surveillance is tied to the global flow of money in ways that visual surveillance is not. Lives are altered when credit reports are hijacked. While hijacking visual surveillance certainly disrupts power, it doesn't disturb the operation of capital in the way that hijacking dataveillance does. Therefore we can expect powerful reactionary responses to that disruption. The borders of data space are well-policed.

Consider, too, that visual sense-making is second nature. Give people a camera, an Internet connection, and some kind of publishing software, and there you have the possibility for new patterns of production and distribution. As these new images proliferate, so do the codes for their interpretation. That is not so with data. Some media must be developed for the facilitation of sense-making. In this way, too, actuarial surveillance is institutionalized differently than visual surveillance. The tools and techniques of engagement with visual surveillance (cameras, web browsers, familiar ways of seeing) are fairly well-distributed and democratically available. Yet the analogous techniques in dataveillance (sensors, displays, statistical software) are nascent, restricted, and specialized. Thus, both in its institutionalization and its cognitive operation, actuarial surveillance is more difficult to hijack than visual surveillance.

Infrastructures of Surveillance: Location-Aware Social Media

While McGrath's examples elucidate Koskela's point that "authorities cannot control how and where surveillance is used," nevertheless all action in surveillance space is constrained and enabled by the structural properties of the media environment in which it occurs (Koskela, 2009, p. 148).

This section looks at three location-aware social media applications—Dodgeball (and its successors Foursquare and Latitude), Area/Code, and Blast Theory. Each of these is questioned in terms of how they structure opportunities for public engagement, public sense-making, and identity formation. In particular, we will look at the interplay of actuarial and visual surveillance, and the interplay of the system operators and system users. We will try to discern how each presents opportunities to circulate images, and to decode, interpret, or translate images. Likewise we will explore questions of watchers and presence, and of the ideology of space.

Dodgeball/ Latitude/ Foursquare

Dodgeball is a smartphone app developed by Dennis Crowley and Alex Rainert in 2000. In 2005, Dodgeball was bought by Google, who continued its development as Latitude. In 2007, Crowley and Rainert left Google and joined with Naveen Selvadurai to further develop the Dodgeball concept as Foursquare.

These three applications share a basic paradigm of operation, in that they permit participants to do two things. First, participants can "tag" physical locations; that is, participants can make notes about locations and publish those notes on a map. Second, participants can publish their own current location on the map. The tags and the locations are stored on a server by the application operators. Participants decide which other participants are able to see how they've tagged locations and where they are.

These apps share a paradigm of linking sociality and urban space. Foursquare's developers claim that their intent is to encourage people to do new things, to explore places they generally would not, to take them out of their usual scene or comfort zone. They do this by turning social life into a goal-based video game. Participants are awarded points for "going out more times than [their] friends, and hanging out with new people and going to new restaurants and going to new bars—just experiencing things that [they] wouldn't normally do" (Reagan, 2009). Yet they also encourage a sort of competitive sedentary egocentrism: those who are most often in a location become the "Mayor" of that location. The bar or restaurant owner may offer discounts to the mayor. Thus Foursquare mediates the exchange of social for economic capital. One's social capital may increase if one becomes mayor of a cool place by hanging out there often and spending money. One's economic capital may increase if you get free drinks for enticing your friends in to buy.

Despite the stated intent of encouraging novelty and exploration, research has suggested that in fact Dodgeball (Foursquare's predecessor) re-inforced existing networks of friends and acquaintances. It did not bring its users into contact with strangers. Instead, it acted to "tame" urban space by promoting "social molecularization, whereby [participants] both experience and move through the city in a collective manner" (Humphreys, 2008, p. 353). It was another vehicle for urban gentrification.

So, in current practice at least, Foursquare integrates existing social identities with existing physical places. It perhaps facilitates novel uses of those spaces by social groups previously unfamiliar with them, but it does nothing to problematize space, place, or identities as organizing principles in themselves. Through its incentive structure (points are not awarded between 8:00 a.m. and 4:00 p.m., for example, and the places it makes most visible are places of consumption), it re-entrenches the hegemonic relation of work and leisure, production and consumption.

Participants experience an extended visual surveillance—they see where their friends are; they can read what a particular friend thought of a particular place. System operators, meanwhile, record the participants' locational histories and their networks of friends, producing through actuarial surveillance an enhanced understanding not just of those individual participants, but of the leisure patterns of an emergent urban class.

Area/Code

Area/Code (www.areacodeinc.com) produces multi-media games to order. According to their "manifesto," they intend these games to "build [...] a future in which socially aware networks, smart objects, location sensing and mobile computing open up new ways for people to play, [to] create a conscious confusion between...information and space, [and to facilitate] collaborative hallucinations, and serious fun." The clients for these games are significant media and marketing corporations, including Disney, CBS, Nokia, MTV, and Nike. Often these games are intended both to create new kinds of interactions in hybrid space, and to extend the markets and visibility of entertainment products within those spaces. (http://areacodeinc.com/about/).

For example, in the game "Chain Factor," Area/Code and ABC worked to together to extend and transmute the space inhabited by the TV series "Numb3rs." ABC wrote the character "Spectre" into the show; Area/Code incorporated messages from Specter into an online game. Those messages referred participants to other media. Eventually, the game's landscape extended across all CBS media, including billboards, mobile and online news services, and prime-time television. Players worked collaboratively through a wiki to discover and make sense of clues embedded in this hybrid landscape, and eventually to discover the truth behind Spectre's character and motivation (http://areacodeinc.com/work/numb3rs/).

Thus Chain Factor significantly reconfigured space and troubled the relations between offline and online. Images and data flowed through this media space in bizarre and novel ways. Identity and presence were troubled. But this novel space was entirely owned and operated by a global entertainment conglomerate, and the altered senses of embodiment were directed toward that corporation's profit. By entertaining and intriguing individual participants, CBS hoped to cohere them into an audience, and deliver them to its advertisers.

"Code of Everand" is another example of a potentially transgressive hybridity of "real-world" information and online space used to hegemonic ends. This online game is intended for an audience of 9–13-year-olds. In many ways a standard "Dungeon and Dragons" type game, players attain and use "powers" to navigate a treacherous landscape of "monster-filled spirit channels" and uncover a secret code. But the behavior of the monsters of Everand is affected by real-world traffic data, and the young players can only succeed if they "learn to stop and think, [and] look both ways" before crossing the spirit channels. Commissioned by the United Kingdoms's Department of Transport, the game explicitly intends to reinforce proper road-crossing habits (http://areacodeinc.com/projects/code-of-everand/).

Blast Theory

Like Area/Code, Blast Theory produces games that meld online and offline spaces. However, they populate that space, not with personae created by the game organizers, but with the intimate memories and feelings of the participants. For example, in one game, a chase occurs involving players both online and in physical space. Players in physical space are equipped with GPS-enhanced PDAs through which they can access data space. Likewise, online players track the physical location of the offline players through those PDAs. Before the chase begins, players are asked for the name of a friend whom they haven't seen for a long time, but often think of. That would be, for me, Sue Ballou (Hi Sue!). At the end of the chase, when I was finally caught, my captor would inform me of that fact by saying, from my elbow or through my PDA, "I've just seen Sue Ballou."

Another game begins by asking participants if they would consent to providing their contact information to a stranger, and promise to be available to that stranger if she suffers a personal crisis in the following year. Then the game continues in online and offline space, players chasing each other, calling to each other with intimate knowledge, asking for intimate connections, asking for trust, and empathy. At the end, players who are still willing to exchange contact information with a stranger, and contract to be available to the stranger should a crisis occur.

Blast Theory games perhaps epitomize the artful, deeply serious play that McGrath celebrates. They are intensely affective and personal, evoking new and transgressive ways of sensing presence, and being present, in a hybrid world of images, physicality, and data.

Blast Theory is a group of seven artists located in the United Kingdom. They have been funded by Arts Council of England, Microsoft Research, and British Telecom (Jardin, 2003).

Conclusion

This chapter set out to explore two questions. The first was whether the practices of "hijacking" visual surveillance could provide us with insight into the potential for hijacking actuarial surveillance. It suggests that in fact, both the institutional and the cognitive structures of these two types of surveillance differ enough that analogies tied to understand how media structured between the two are difficult, if not impossible. The second exploration questioned how the structure of the mediation of surveillance space made certain formations of space, interaction, and identity possible. In this, we note several points. First, there is nothing about the melding of data space and physical space per se that facilitates transgressive identity construction or world-making. Foursquare, for example, is normative to its core. Like many other social media applications, it uses enhanced visuality as a lure to entice participants into the purview of invisible actuarial surveillance. It extends already hegemonic understandings of identity and space. Second, even truly generative hybrids of spaces and presences can be colonized by hegemonic cultural interests. As we have seen, Area/Code's "Numb3rs" game leverages CBS's extensive media holdings to shape this new space according to their corporate needs. Finally, the only structures where participants were truly present as primary occupants and constructors of new space were the Blast Theory games, supported by Arts Council grants. These examples reflect contests, usually invisible, over the structure of the hybrid space of data and embodied action. We are left to wonder if the exigencies of capital necessarily dominate these contests, claiming and shaping these emergent spaces as territory, mystifying the presence of the actuarial watcher, and continually reconstituting actors and their desires as agents of consumption.

Notes

1. This research has been supported by Social Science and Humanities Research Council of Canada grant #72035904. I would also like to acknowledge the assistance of Karen Pollock and Ginger Coons.

References

Albrechtslund, A. (2008). Online Social Networking as Participatory Surveillance. *First Monday, 13.* Retrieved 22 February 2010, from http://firstmonday.org/htbin/cgiwrap/bin/ojs/index.php/fm/article/viewArticle/2142/1949

Berlant, L. & Warner, M. (1998). Sex in Public. *Critical Inquiry, 24,* 547–566.

Butler, J. (1998). Imitation and Gender Insubordination. In D. H. Richter (Ed.) *The Critical Tradition: Classic Texts and Contemporary Trends,* 2nd ed. (pp. 1514–1525). New York: Bedford/St. Martin's.

Campbell, J. (2005). Outing PlanetOut: Surveillance, Gay Marketing and Internet Affinity Portals. *New Media and Society,* 7(5), 663–683.

Chauncey, G. (1994). *Gay New York: Gender, Urban Culture, and the Making of the Gay Male World, 1890–1940.* New York: Basic Books.

Chivers, T. (2008). The Get Out Clause, Manchester Stars of CCTV. *The Telegraph,* 8 May. Retrieved 22 February 2010, from http://www.telegraph.co.uk/news/newstopics/howaboutthat/1938076/The-Get-Out-Clause-Manchesters-stars-of-CCTV-cameras.html

Claritas, Inc. (2008). Customer Segmentation > 66 PRIZM Marketing Segments, Claritas Customer Segmentation. Retrieved 22 February 2010, from http://www.claritas.com/claritas/Default.jsp?ci=3&si=4&pn=prizmne_segments

Dodge, M. & Kitchin, R. (2005). Code and the Transduction of Space. *Annals of the Association of American Geographers, 95*(1), 162–180.

Dourish P. & Bell, G. (2007). The Infrastructure of Experience and the Experience of Infrastructure: Meaning and Structure in Everyday Encounters with Space. *Environment and Planning B, 43*, 414–430.

Foucault, M. (1977). *Discipline and Punish : The Birth of the Prison.* Translated by Alan Sheridan. New York: Pantheon Books.

Giddens, A. (1984). *The Constitution of Society: Outline of the Theory of Structuration.* Berkeley, CA: University of California Press.

Goffman, E. (1959). *The Presentation of Self in Everyday Life.* New York: Doubleday.

Grimmelmann, J. (2009). Saving Facebook. *Iowa Law Review, 94*, 1137–1206 .

Haggerty, K. & Ericson, R. (2006). The New Politics of Surveillance and Visibility. In K. Haggerty & R. Ericson (Eds.) *The New Politics of Surveillance and Visibility* (pp. 3–25). Toronto: University of Toronto Press.

Humphreys, L. (2008). Mobile Social Networks and Social Practice: A Case Study of Dodgeball. *Journal of Computer-Mediated Communication, 13*, 341–360.

Jardin, X. (2003). On Your Mark, Get Set, Unwire! *Wired,* 12.25.03. Retrieved 22 February 2010, from http://www.wired.com/gadgets/wireless/news/2003/12/61721?currentPage=1

Koskela, H. (2009). Hijacking Surveillance? The New Moral Landscapes of Amateur Photographing. In K. F. Aas, H. O. Gundhus & H. M. Lomell (Eds.), *Technologies of InSecurity: The Surveillance of Everyday Life* (pp. 147–167). New York: Routledge-Cavendish.

Lemke, J. (2008) Identity, Development, and Desire: Critical Questions. In C. Calda-Coulthard & R. Iedema (Eds.). *Identity Trouble* (pp. 17–42). New York: Palgrave Macmillan.

Mann, S. (1997). Wearable Computing: A First Step Toward Personal Imaging. *Computer*, 30(2), 25-32

McGrath, J. (2004). *Loving Big Brother: Performance, Privacy, and Surveillance Space.* London: Routledge.

Norris, C. & Armstrong, G. (1999). *The Maximum Surveillance Society: The Rise of CCTV.* Berg: Oxford.

Parker, S., Uprichard, E. & Burrows, R. (2007). Class Places and Place Classes: Geodemographics and the Spatialization of Class. *Information, Communication, & Society, 10*, 902–921.

Phillips, D. (2008). Locational Surveillance: Embracing the Patterns of Our Lives. In P. Howard & A. Chadwick (Eds.) *Handbook of Internet Politics* (pp. 337–348). London: Routledge.

Reagan, G. (2009). Foursquare, Hot New Phone App, Is Dodgeball on Steroids. *New York Observer*, March 10. Retrieved 22 February 2010, from http://www.observer.com/2009/media/foursquare-hot-new-phone-app-dodgeball-steroids

Steeves, V. (2006). It's Not Child's Play: The Online Invasion of Children's Privacy. *University of Ottawa Law & Technology Journal, 3,* 169–188.

Žižek, S. (1992). *Enjoy Your Symptom/Jacques Lacan in Hollywood and Out.* New York: Routledge.

CHAPTER TEN

Hacking, Jamming, Boycotting, and Out-Foxing the Commercial Music Market-Makers[1]

PATRICK BURKART

New social movement politics is a variety of civil disobedience directed at the modern technical interdependencies of system and lifeworld, especially those that seem unfairly or arbitrarily imposed by remote and unknown powers. "Colonization" of the lifeworld by the system is a useful and relevant way to explore music and cyberliberties activism, media activism, and NSMs together in a critical theoretical project for media and communication studies. This chapter addresses the revolts of music fans and their allies in the music production and performance business who actively oppose the imposition of the new consumption norms for digital distribution of commercial music.

Colonization is an observable process of media rationalization in late capitalism. The "colonization of the lifeworld" is the theory of communicative action's interpretation of the reification thesis as introduced by Marx. Reification was later adapted by the Frankfurt School of critical theorists and by Georg Lukács. Colonization is detectable in visible processes of bureaucratization, and commercialization of media cultures, and the cutting off of flows of music and other media. From my perspective, the colonization thesis describes processes that contribute to the conversion of the Internet into a mass-media distribution platform for the entertainment industry (or a "Celestial Jukebox") and regulated by the needs of e-commerce, police and military, and corporate network security administrators.

Principally, the strategic use of law and technology by the major music labels, and the responses of typically passive consumers and regulators to juridification and reification, have led to (1) the desiccation of music scenes organized around practices of collecting and sharing recorded music in physical places; (2) the emergence of music as a commercial service provided online in private and atomized streams; (3) the insinuation of DRM[2] into music-listening experiences; (4) the erosion of fair use as a public-interest exception to copyright; and (5) the invasion of contract law and lawsuits into the lives of music fans. These effects are all symptoms of the colonization of the "music lifeworld" (Burkart, 2010). The music lifeworld is comprised of all the "places of music" where people meet to share and discuss music, and generally, of all the socializing interactions of music scenes and other social spaces mediated by discourses. As these places of music go online in commercial music portals, they die off as a collective resource. They become atomized and sealed off from the full range of robust social interactions as situations negotiated in person give way to programmed activities executed in isolation from others. The music lifeworld is not a romantic concept or even a Utopian ideal, but the empirically observable domains of private and social life that are continually undergoing processes of social rationalization. When rationalization through colonization erodes into rituals that support cultural reproduction, the consequences can become pathological. Software enclosures exclude non-payers, and thereby remove music's public good characteristics, and rely on non-linguistic steering media to enforce their policies, removing any possibility of deliberation and dissent about the terms of access to music culture online. The reactions against these transformations of the music lifeworld raise objections to the replacement of free music cultures with locked-down software "enclosures" (Andrejevic, 2002).

In music scenes, and within virtual music communities, socialization, cultural reproduction, and social integration are symbolically reproduced communications that constantly come under pressure to "monetize" (to borrow a word from e-business that means turning non-payers into payers in an online community) music and fandom more efficiently. Willing participation by fans in the industry and complicity in the process can render colonization effects invisible, which in turn serves to support the growth and expansion of the system. Updating the colonization thesis, Mosco (1996) refers to "cybernetic commodification," or market making for information-based products and services. In the context of digital music distribution online, cybernetic commodification becomes colonizing when it converts communications dependent on the lifeworld institutions and processes (such as personality/identity formation, socialization, acculturation) into non-linguistic and functionalist operations of an e-commerce enterprise.

Insofar as audiences are both constituted and marketed online, the users of media software and services become cybernetic commodities, partly through their "work of being watched" (Andrejevic, 2002, p. 230). The pathological aspect of digital enclosures is seen in the fact that the laboring relationship of the user for the provider is actually acquired and paid for by the user. Paying to labor in a portal is the juridical basis for the negative privacy consequences to the user, because it represents a total capitulation and acquiescence to the Celestial Jukebox model of media consumption. Popular music portals, like iTunes, that distribute DRMed music also commodify users' profiles and personal information through contractual relationships that give the portals nearly complete control over users' access, activity logs, personally identifying information, and terms of service. A music fan's extra value to a music-service provider consists in his or her "digital shadow" or "second self," in Andrejevic's terms (2002, p. 137). The second self appears as a surplus value to the provider and benefits the firm in addition to the first self, or user, who is the purchaser or renter of music, watcher of advertisements, and a generator of site content.

The value of the work of being watched online increases as more people switch from CDs to digital downloads. Contract law has expanded into digital music by forcing users to enter into legal contracts for each transaction required to access music; these juridifications occur in online work. Confinement and control of the user is their principal objective. Logins, authentications, digital signatures, clickwrap agreements, terms of service, end user license agreements (EULAs), and purchase agreements bury the music fan with the real work of demonstrating, over and over, who he or she really is, and swearing, repeatedly, to not having criminal intents with respect to IP law. The juridification process requires consumers to make allegiances and overcommitments of their own rights and money to corporate entities. The new regime of legal-economic transactions for basic access to culture exposes the lifeworld to system pressures by converting undisturbed lifeworld experiences into contractually obligating, technologically mediated legal agreements. The coercive power of the service-client relationship, forged in contract law, resembles the wage relationship insofar as it is individually arranged, unavoidable, and non-negotiable, with all-or-nothing terms. In the process of digitally "signing" all those agreements (typically without reading any of them), music fans turn over old and new rights and freedoms to the entertainment industry while participating in a broader privatization of the music lifeworld.

Try as they might, however, the culture industries have not yet been successful in getting people to shed basic expectations about music as a part of the cultural commons, even if they submit to the confinement of the digital enclosure. When the Celestial Jukebox eventually becomes the consumption norm for media delivery, it will for most music fans become a natural and ordinary way of doing things; as with the other structures of the lifeworld, it will become "the result of the emergence of a 'second nature,' in which we habitually orient ourselves in a changing 'space of reasons'" (Honneth, 2002, p. 500). In the mean time, there has been a public showing of agents who refuse to participate in the new control systems.

Conflicts over the Juridification of Music

Music studies' interest in DRM grew from legal studies with a juridical approach to IP law, and has made a property-rights–based case against legal DRM by focusing on the loss of use and possession of recorded music, rather than on the communicative impacts of this loss. But because no theory of social action stands behind a normative theory of cyberliberties law, the injunctions and warnings of music and media studies do not yet connect with a framework for explaining or understanding how the work of change and reform in the culture industries can begin or end. Critical social systems theory provides a robust framework for describing and critiquing the social structures supporting the technological lockdown of culture, and for understanding the social sources of political and cultural resistance.[3] A variety of this, the theory of communicative action, allows for the critique of DRM to be expanded to question the moral authority of the regulationist, legal realist, and "copyright maximalist" (Samuelson, 1996) views of copyright, telecommunications, and information law and policy. These dominant views have brought us to a basic political conflict between keeping or rejecting civil libertarian values for the Internet. The nominal victories at the European Parliament by the Swedish Pirate Party express this conflict (Keating, 2008; "Pirates ahoy!", 2009).

Power and money are the "steering mechanisms" that coordinate colonization. They transfer socialization, cultural reproduction, and social integration to system-integration processes that typically handle material reproduction: "Pathologies result only at the battle front between 'system' and 'lifeworld,' and even there only in one direction, namely the switch of the 'system' over into the 'lifeworld'" (Habermas, 1987, p. 175). McCarthy (1985) and others who have critiqued the rigid system and lifeworld dualism have offered possibilities for thinking of it in different ways, including thinking of it as different standpoints with respect to the phenomenon of complex communication processes between and among formal and informal organizations (Koivisto and Väliverronen, 1996; Calhoun, 1995). "System integration and social integration...seem to be extremes rather than alternatives that exhaust the field of possibilities: The denial of one does not entail the other" (McCarthy, 1985, p. 41). These standpoints facilitate for observation the "uncoupling" of system and lifeworld processes which Habermas describes. Uncoupling is more visible in the Celestial Jukebox than in previous modes of music production, distribution, and consumption. Users are, indeed, integrated into formalized systems logic when they join the Celestial Jukebox: they join "a strongly hierarchical, formally organized setting in which the actors have no clear idea of why they are ordered to do what they do" (McCarthy, 1985, p. 41).

Today, online portals can create airtight, hierarchical, formalized, organizational enclosures enforcing norm-free compliance through contract law and technology. New music distribution systems create pressures in society to "clientelize" and juridify private life and cultural life, while technocratic controls substitute for interpersonal and negotiated transactions in acquiring music. Fans are already to a large extent beholden to the consumerism, clientelization, and juridification of social rationalization that occurred as the consumer role became subject to management and protection by the welfare state. Yet "clients" of music companies must accept rights that are increasingly juridically restricted. Juridification "seals the deal" for the system. It formally legitimates online legal relationships, but without using normatively sanctioned communication. Instead, juridification allows money and power to help replace meaningful social practices with cybernetic routines. Music delivered as a service is neither purchased nor sold, but licensed for temporary and highly conditional access. Acquiescence in these technical means of coordination is not grounded in communicative action.

The nonnegotiable terms imposed by DRM on behalf of system technologists are protected by law, but have not been grounded in "moral consciousness" (Habermas, 1987, p. 174), which is willful, deliberative, and discursive. The impositions of the technocratic controls on music fans in exchange for access to culture cannot be a relationship that develops into "socially recognized norms of action" (Habermas, 1987, p. 175), because they detach music acquisition and distribution from the legitimating lifeworld. In other words, the legalistic terms of the Celestial Jukebox are not legitimate from the standpoint of record collectors. Perhaps similarly, because their communications are not socially recognized, "spamculture" and "functional trash" appear to be varieties of disposable waste (Gansing, in this volume).

After nearly a decade of delays, and Apple's cajoling, the music industry is now experimenting with DRM-free online distributions of digital music. But the Big Four still nurture ecstatic visions of "ubiquitous" DRM that locks down all users into reliably leakproof digital enclosures. Ubiquitous DRM may seem undesirable and implausible to most music fans, but it remains the current objective of the culture industries. Ubiquitous DRM would use legal and software controls to fully automate the Celestial Jukebox. Cory Doctorow presents the dystopian endpoint of ubiquitous DRM (Doctorow, 2005):

- Every copy of the song circulated, from the recording studio to the record store, has strong DRM on it;

- No analog-to-digital converters are available to anyone, anywhere in the world, who might have an interest in breaking the DRM;
- Peer-to-peer networks cease to exist;
- Search engines cease to index file-sharing sites; and
- No "small worlds" file-sharing tools are in circulation.

What seems dystopian to a cyberlibertarian is Xanadu for the Internet regulationists. Ubiquitous DRM would make it possible for the culture industries to bypass any remaining checks and balances on IP law completely by automating the coordination of money and access power across all music players.

Assuming for a moment that it were technically feasible, ubiquitous DRM would introduce what Habermas calls a "second nature" in a music lifeworld that is currently sustained in cultural reproduction and social integration. A technocratically constructed second nature would provide a "normfree sociality that can appear...as an objectified context of life" (Habermas, 1987, p. 173). Andrejevic describes similar conditions that currently obtain in digital enclosures. As technically encoded power and money transactions increasingly transform negotiated agreements into automated exchanges between buyers and sellers of music and cultural objects, and people are habituated to participate in clientelistic transactions, expressions of communicative rationality become "secularly threatened" (Berger, 1991, p. 168).

In other words, even though ubiquitous DRM depends upon institutions of the musical lifeworld for demand and for legitimacy as a new technology practice, it can erode the lifeworld enough to threaten it as a future resource for juridification and legitimation. Samuelson (2003) expresses the concern that the "main purpose of DRM is not to prevent copyright infringement but to change consumer expectations about what they are entitled to do with digital content" (p. 41). Gillespie (2009) illustrates the pro-DMCA and anti-piracy propaganda campaigns run by the RIAA and MPAA in US public schools. Touraine (1988) describes changes to the lived experiences of individuals in the "programmed society" as creating the conditions for future participation in an oppositional new social movement (p. 106).

Ideal Types of Cyberliberties Activism

The theory of communicative action can render instances of the system-lifeworld dialectic observable in processes of rationalization and colonization. The clash between the norm-free capitalist economy and the "independent communication structures" of the lifeworld that support political participation, identity formation, and cultural integration (Habermas, 1987, p. 391) is clearly visible in the operation of the Darknet, CopyLeft musicians, hacker communities, and indie labels outside the Celestial Jukebox. So too, the "rebellious communication and social movements" (Downing, 2001, p. 10) of alternative and radical media activists render the margin of the system and the lifeworld visible. Habermasian critical theory considers these examples of opposition to be evidence of the survival and struggle of "communicative rationality" under conditions of colonization. They are not, however, evidence of a flourishing of revolutionary political agency. The radical roots of communication-oriented media activists are present, but they are not expressions of anything like class consciousness or movement consciousness. Music and cyberliberties activists are engaged in a process of identifying risks to existing communication practices of the technoculture and the online music lifeworld that are worth keeping, and singling out corporate strategies or trends in industry and policy that are worth challenging or rejecting. They see colonizing encroachments into the music lifeworld and counter them with symbolic politics and "tools for de-colonization" (Hadl, 2004), often in solidarity with affinity groups.

In the United States, public-access cable TV, children's programming, public television, and public radio were polic-making accomplishments of alternative-media activists before becoming targets, in turn, of media deregulators in the 1980s under the Fowler-led FCC in the Reagan administration. Mosco calls the rehabilitated "public interest media reform movement" (1996, p. 241) a social movement that has helped influence "the development of the means and content of communication" (p. 240) by promoting "universal access to the information infrastructure, freedom to communicate, democratic policy-making, and privacy protection" (p. 241). These reformers drew considerable participation from media professionals and civil society institutions that worked within state agencies to win policy reforms. Ironically, alternative media often work to attain objectives that may already be formally enshrined in federal communication policy, but are dead or at risk of extinction as existing practices—dead letters of communication law.

Radical media, on the other hand, exhibit counterhegemonic attributes insofar as they resist or challenge existing power structures, empower diverse communities and classes, and enable communities of interest to speak to one another (Downing, 2001). Radical media have captured the attention of media scholars with a Gramscian perspective on non-state–based forms of resistance that begin in rebellious countercultures and social movements in civil society.

Popular case studies of alternative media include Paper Tiger TV, carried via satellite to public-access cable channels; the Prometheus community radio project; microbroadcasters on FM radio; and the IndyMedia news portal. The Future of Music Coalition, Downhill Battle, and online "eLabels." (Edgecliffe-Johnson, 2005) or "Netlabels" (Gumiela, 2007) are alternative media efforts by independent musicians and fans to decolonize the music lifeworld on the model of alternative-

and radical-media activism. Decidedly capitalistic in nature, these alternative distribution networks have an entrepreneurial spirit and innovativeness that have gone lacking in the music business for decades. Lessig (1999, 2001, 2004) and Benkler (2007) both acclaim the interplays between entrepreneurs and social movement constituencies as generators of media diversity and innovative business models and technologies. In many cases, innovation among new music labels offers artists alternatives to major label contracts. The creator of Elektra records, Jac Holzman, launched Cordless Recordings to publish "clusters" of three or more songs online from new artists. Artists with Cordless Recordings sign simple contracts that exchange exclusive distribution rights for twenty-one months for the retention of all masters and copyrights by the artists (Edgecliffe-Johnson, 2005).

The alternative-radical distinction bedevils media studies. Although there is a basic conceptual problem with the notion of "alternative media" that Downing (2001) points out—namely, "Everything, at some point, is alternative to something else" (p. ix)—I hold that the concept is sufficiently parsimonious to qualify alternative-media activism as a set of social and technical demands promoted as challenging or contrary to the Celestial Jukebox and business as usual. These demands are consistent with Hesmondhalgh's (2007) criteria for access and participation, Lessig's (2004) free culture, and Vaidhyanathan's (2005, 2006) and Benkler's (2007) radically democratic political philosophies. They all support the preservation and expansion of civil liberties and civil society in cyberspace. Whereas Paper Tiger TV, Prometheus Radio, pirate radio, and microradio programmers labored to eke out access and carriage for a small number of analog channels in regulated media markets, cyberliberties activists labor to exploit locked-up music and make music more broadly available on the Internet. Conflicts in the music and cyberliberties domain are principally related to IP claims, and disclose a politics of symbolic action that resembles social movement politics in its potentials for "resistance and transformation" (Carroll and Hackett, 2006, p. 83) in the industry, in local music scenes, and in other places for music.

Alternative-media activism loosens the hold of the culture industries on the music lifeworld by promoting alternatives to the existing copyright regime. Examples of formal organizations devoted to copyright activism include the Creative Commons and the Free Software Society (FSF). As alternatives to industry IP practices, the Creative Commons and FSF's CopyLeft scheme promote creativity (Demers, 2006) and serve to establish a creative commons that pools volunteered resources. Alternative-media theorists locate these alternative practices in civil societies of countries and identify global constituencies that are forming around them and nearby them.

Although the alternative-media label risks becoming a grab bag of leftovers unsuitable for other categories, continuing empirical research and theoretical refinements to the "alternative" category are yielding valuable studies of public-sphere institution building and practices of global civil society (Downing, 2001; Benkler, 2007; Kidd, Rodriguez, and Stein, 2009; Stein, Rodriguez, and Kidd, 2009). In an update of the project undertaken in *Radical Media*, the contributors to the *Making Our Media* volumes provide international examples of cyberliberties activism, including opposition to copyright reform, censorship, and privacy violations (especially Lee, 2009; Schweidler and Costanza-Chock, 2009). Public-sphere research in communication, sociology, and political science finds examples of social institutions and media systems that can gather a plurality of groups and interests into a common forum for debating and deliberating the weightiest social problems of the day. Alternative communication channels can challenge or undermine those of dominant, legitimated, and "official" media systems. Alternative-media groups represent their own work and missions to one another at conferences and conventions. The National Conference for Media Reform (NCMR) is a hybrid organization of "old line" media reform activists and cyberlibertarians working with music and new media. Besides policy-oriented activism, the NCMR also supports promoters of community WiFi networking, community radio, cable-access TV, and indie music labels.

An interaction of alternative non-profit organizations working for cyberliberties includes the ACLU, EFF, Software Freedom Law Center, Consumers Union, Public Knowledge, the Ford Foundation, the Social Science Research Council, the Center for Digital Democracy, the American Library Association, the Electronic Privacy Information Clearinghouse (EPIC), Computer Professionals for Social Responsibility, the Nieman Foundation for Journalism (Harvard), the Brennan Center for Justice (New York University School of Law), the American Library Association (ALA), and the Glushko- Samuelson Intellectual Property Law Clinic (American University's Washington College of Law). This list of peers in the network of alternative-media and cyberliberties activists is hardly exhaustive. Each group contributes to a broader project of preserving and expanding a place for music, digital culture, and knowledge in their many formats and expressions.

Globally, media activists were drawn to the World Summit on the Information Society (WSIS) meetings (2003–2005), which reenacted longstanding disputes over ideological purity and pragmatic compromise as divergent ways of democratizing media institutions. The diverse nature of the grievances and agendas reflected broad differences in identity among the participants (Cunningham, 2009; Hamelink, 2004).

Entrepreneurial and "idealistic business ventures" (Schmidt, in Rappeport, 2006) are firms operating alongside alternative media, commercial businesses, and noncommercial civil-society organizations promoting access to cyberculture. For example, Rappeport (2006) describes Internet service providers (ISPs) who operate proxy servers that permit Web surfers on restricted networks to break through most restrictions: in a "cat-and-mouse-war with government censors the world over...scores of small companies...have made it part of their mission to defy censors, unlike Google, Yahoo and Microsoft, which recently succumbed to demands to filter the Internet in China." Unipeak, Peacefire, Your Freedom, Cloak, and Proxify are other for-profit examples. By providing anonymizing services and anticensorship platforms, these ISPs meet needs in the markets for Internet access services that otherwise go unmet. The OpenNet Initiative, a joint project of, among others, the University of Toronto, Harvard Law School, and Cambridge University, is an entrepreneurial non-profit example of a company reducing the impacts of state-sanctioned Internet censorship. BitTorrent's innovative software for delivering media content through P2P has won awards from technology industry groups and enjoys a wide user base. These firms provide access to cultural and political information to individuals and groups otherwise denied of it, even though they are enmeshed in the new economy of "digital capitalism" (Schiller, 1999).

Whereas alternative-media activists are frequently content to adapt to existing power structures, and may even personally or collectively benefit from them, radical-media activists want to subvert or overturn the conventional system. They "have very often experienced state repression—execution, jailing, torture, fascist assaults, the bombing of radical radio stations, threats, police surveillance, and intimidation tactics" (Downing, 2001, p. 19).

A significant contribution of *Radical Media* (Downing, 2001) is its search for the distinctive characteristics of radical (versus merely alternative) media activists worldwide, particularly those struggling to propagate messages under conditions of extreme social repression, political and social revolution, and other social conditions marked by crisis, rapid change, and suffering. Radical media are identifiable in historical studies of international communication that focus on periods of violent political conflict and suppression. They are more likely than alternative media to try to mobilize resources that are not otherwise under their direct ownership or control and to participate in direct action, sometimes using counterhegemonic force.

There are fewer examples of radical-media activism among the music and cyberliberties projects than there are in the larger universe of audiovisual media producers, but pirate radio, Internet Zapatismo, and cable-access TV programming by labor and gay, lesbian, bisexual, and transgender (GLBT) groups exemplify radical media that encounter continual bias and resistance from nearly all corners of society. Radical media take many forms: "Dance, street theater, cartoons, posters, parody, satire, performance art, graffiti, murals, and popular songs or instrumental music are...only some of the most obvious forms of radical media whose communicative thrust depends not on clearly argued logic but on their aesthetically conceived and concentrated force" (Downing, 2001, p. 52).

We move on to the third and fourth ideal types of opposition to cultural colonization of music. Part political thespians, part cultural studies mavens, culture jammers appropriate symbolic—and primarily pop-culture—resources from mainstream institutions and turn the resources against those institutions and their operating logics. In the process, culture jammers rupture the suspension of disbelief required to maintain the ongoing commercial advertising spectacles of consumer society (Baudrillard, 1998; Klein, 2000; Kellner, 2005). Culture jammers deflate the proudest corporate images and subvert their branding, hijack established communication channels, and cast seeds of doubt among audiences, where trust was once presumed to operate. In the process of disruption, opportunities for critical thinking among audience members and media consumers arise, and can stimulate reflection about the function of media in everyday thinking as well as about background expectations that are part of the media lifeworld.

The new communalist spirit (Turner, 2006) of anarchistic, spontaneous, unplanned, and experimental "happenings" inhabits culture-jamming endeavors. Culture jammers tend to use humor to disarm establishment targets. Examples from the contemporary audio world include Plunderphonics, Negativland, DJ Dangermouse, and Brother Russell Ministries. Each enterprise is performative, creates musical and discursive "situations," and, in its own way, purposefully and playfully flouts or skirts the fringes of legality of telecommunications law or the legality of copyright. In the case of each artist or band, sampling and remixing is *de rigueur*. In the case of Negativland, in 2002,the band conducted pirate FM performances and a pirate radio teach-in in the close physical proximity of the National Association of Broadcasters meeting in Seattle (Holdorf, 2002). Brother Russell remixes ludicrous prank telephone calls to live call-in conservative talk radio programs.

The legal defense of these artistic activities can become challenging if the jammers are caught and confronted by hostile opponents; Negativland encountered such a situation when it was sued by Island Records for its U2 spoof album. Some, but not all, culture-jamming artifacts are varieties of "illegal art" by dint of their civil disobedience aimed at IP laws. However, popular-music scholar Kembrew McLeod's "Freedom of Expression" trademark was a prank that entailed a legalistic gesture that splendidly gilded the lily. After trademarking "Freedom of Expression," McLeod sued AT&T for trademark violations after AT&T launched its "Freedom of Expression" ad campaign.

Other examples of musical culture-jamming geniuses include the numerous projects loosely associated with the SubGenius Foundation and its long-running radio shows. The SubGenius media machine promoted a faux-fundamentalist Christian church, and produced volumes of humorous, anti-authoritarian, and anarchistic cartoons, books, radio shows, newsletters, and Web epistles. *RE/Search*

Magazine's "Pranks" edition featured case studies of intertwined music and culture-jamming cultures since the Yippies movement (RE/Search, 1986). Ever increasing in sophistication and slickness, some full-blown integrated marketing campaigns have emerged around the culture-jamming activities of the Yes Men and the Barbie Liberation Organization (the BLO). Culture jamming is among the most performative varieties of media activism, incorporating ironic attitudes from situationism and a predilection for *détournement*, hijacking, and diversion (Downing, 2001, p. 140).

Culture jamming needs cyberliberties to flourish. Personal identities are purposefully obscured to cultivate needed anonymity and obscurity. Many culture jammers depend upon viral modes of communicating news and artifacts; in retro fashion, they also send physical recordings, zines, and home-taped videocassettes, audiocassettes, and disks through the postal mail. The Yes Men use established public relations firms and media production and distribution companies to launch radical messages through mainstream channels. The Yes Men also receive financial sponsorship from philanthropic patrons. Legendary musician and record producer Herb Alpert donated funds to the work of the Yes Men (Curiel, 2004). Other groups mobilize resources by hijacking mainstream cultural objects and communication channels and by dropping conceptual "mind bombs" (Downing, 2001, p. 140) that may take awhile to explode into the mediascape.

Antiauthoritarianism and rebellious communications also characterize the efforts of software hackers to empower music fans who rely on software for access to their music. The hacker's allure in critical cultural studies is his or her identity as a code manipulator. Female hackers have played an important, though rare, role in hacker culture (Newitz, 2001). Hacker culture is anarchistic, and oriented to autonomy and independence. It is also skills-based, reflecting the needs of the technoculture for knowledge work. While there are professional hacktivists who have work arrangements that allow them to contribute to free software full-time (for example, Linus Torvalds and Richard Stallman), there are also examples of famous full-time hackers who have contributed their labor to commercial enterprise. These include such diverse luminaries as Mitch Kapor, Steve Wosniak, Kevin Mitnick, Napster's Shawn Fanning, BitTorrent's Bram Cohen, and, arguably, Bill Gates.

Hacktivism may well be but another "weapon of the weak" (Lovink, 2002; Scott, 1986), but its oppositional activities force confrontations with mainstream media systems and politicize unchallenged and "natural" power relationships in the technoculture. Hacktivist projects can resemble radical media activism, which also uses communication technology to chisel away at the influence of the big media distribution channels. For ethical and even aesthetic reasons, hacktivists may prefer a subversive mode of oppositional engagement with media and online monopolies. Yet software hackers have helped create the music lifeworld, from the underlying

Internet protocols to MP3 players and encoders, to P2P, to the myriad music library managers, file taggers, and so forth, with the intention of enjoying greater freedom to hear, share, and enjoy music.

Hacktivism challenges emerging or extant information regimes that are perceived to be antidemocratic or imposed by bureaucrats or police. The TOR Project, supported by EFF, is a non-profit organization that responds to the loss of anonymity on the Internet with a new model of networking software. The juridification and colonization of the Internet, not only by corporate players but also by state bureaucrats and police, have created the need to evade online surveillance and preserve anonymous surfing and online speech. The TOR developers have explained that their software "aims to defend against traffic analysis, a form of network surveillance that threatens personal anonymity and privacy, confidential business activities and relationships, and state security" (TOR, 2007).

Hacktivists make use of anonymous and pseudonymous speech in cyberspace to make or communicate about technology that can disrupt the plans of monopolist gatekeepers, play pranks on the political and legal establishment, and promote tools for sharing culture. Organizationally, they are informally connected to one another. They can employ collaborative software tools, such as the open-source CVS (concurrent versions system) code management system and Internet Relay Chat, to compile code contributed by hackers working across far-flung international networks. Yet in creating new approaches to sharing music and relevant information, either where none exist or where there are unjust prohibitions on tinkering and sharing, hacktivists make themselves targets of the criminal justice systems of most countries. In this respect, they can be said to engage in forms of civil disobedience.

Building the Alternative Jukebox

The variety of music and cyberliberties activism (alternative and radical media, culture jamming, and hacktivism) demonstrates the influence of lifeworld-based politics on music cultures, cybercultures, and democratic culture. Oppositional and rebellious actors' aims oscillate between the redistributive justice characteristic of collective and class-based politics and the right to be recognized in society as citizens with autonomy (fans, musicians, legal persons with rights, and identity-based countercultures). In both modes, the activism targets the client and consumer roles into which music fans are being pushed by the system.

This chapter has presented an expanded view of the predicament of the music lifeworld, which is still second nature for us, but at risk of depletion through relentless colonization. Once damaged, it seeks to repair itself, thereby indicating the field of social spaces remaining for communicative action around music. It identified our varieties of music and cyberliberties activists: alternative-media activists, radical media activists, culture jammers, and hacktivist-cyberwarriors. Privacy, free speech, and access are three norms around which all three types of activism have oriented their tactical speech acts. These norms underpin a diverse set of grievances and claims against business and legal institutions that fail to recognize an interest in protecting fair use in the digital domain, many forms of cultural production and sharing, and privacy threats. Marginality has become a feature of the Celestial Jukebox that could sponsor a "new culture and a social counterproject" (Touraine, 1988, p. 106) under favorable conditions.

Notes

1. This chapter is adapted from Burkart 2010.
2. DRM, or digital rights management, is the software code added to a media file or format for copy protection.
3. A purely systems-theoretical approach cannot account for communicative rationality or access lifeworld processes, and therefore has a limited impact on technology and policy debates affecting culture. See Luhmann (1982). The ability of social systems theory to become critical is widely debated (McCarthy, 1985; Baxter, 1987; Calhoun, 1995).

References

Andrejevic, M. (2002). The work of being watched: Interactive media and the exploitation of self-disclosure. *Critical Studies in Media Communication 19* (2), 230–248.

Baudrillard, J. (1998). *The consumer society: Myths and structures.* Thousand Oaks, CA: Sage.

Baxter, H. (1987). System and life-world in Habermas's *Theory of Communicative Action. Theory and Society 16*, 39–86.

Benkler, Y. (2007). *The wealth of networks: How social production transforms markets and freedom.* New Haven: Yale University Press.

Berger, J. (1991). The linguistification of the sacred and the delinguistification of the economy. In Honneth, A. & Joas, H. (eds.) *Communicative action: Essays on Jurgen Habermas's The Theory of Communicative Action*, pp. 163–180. Cambridge, MA: The MIT Press.

Burkart, P. (2010). *Music and cyberliberties.* Middletown, CT: Wesleyan University Press.

Calhoun, C. (1995). *Critical social theory: Culture, history, and the challenge of difference.* Cambridge, MA: Blackwell.

Carroll, W. K., & Hackett, R. A.. (2006). Democratic media activism through the lens of social movement theory. *Media Culture and Society 28* (1):83–104.

Cunningham, C. (2009). The right to communicate: Democracy and the digital divide. In Kidd, Rodriguez, & Stein, pp. 207–222.

Curiel, J. (2004). The Yes Men. *San Francisco Chronicle.* October 1. Retrieved October 22, 2009, from http://www.sfgate.com/cgi-bin/article.cgi?file=/chronicle/reviews/movies/YESMEN.DTL&type=movies.

Demers, J. (2006). *Steal this music: How intellectual property law affects musical creativity*. Athens: University of Georgia Press.

Doctorow, C. (2005). Why some "piracy" can increase overall revenues. *Boing Boing,* August 24. Retrieved October 22, 2009, from http://www.boingboing.net/2005/08/24/why-some-piracy-can-.html.

Downing, J. (2001). *Radical media: Rebellious communication and social movements*. With Tamara Villarreal Ford, Genève Gil, & Laura Stein. Thousand Oaks, CA: Sage.

Edgecliffe-Johnson, A. (2005). A new musical arrangement for the Internet age. *Financial Times*, November 15, 14.

Gillespie, T. (2009). Characterizing copyright in the classroom: The cultural work of antipiracy campaigns. *Communication, Culture & Critique 2* : 274—318.

Gumiela, J. (2007). Netlabels. Paper presented at Global Fusion 2007, St. Louis, MO.

Habermas, J. (1987). *The theory of communicative action. Vol. 2, Lifeworld and system: A critique of functionalist reason*. Translated by T. McCarthy. Boston: Beacon Press.

Hadl, G. (2004). Civil society media theory: Tools for decolonizing the lifeworld. *Ritsumeikan Social SciencesReview40*(3): 77–96. Retrieved October 22, 2009, from http://www.ritsumei.ac.jp/acd/cg/ss/sansharonshu/403pdf/hadl.pdf.

Hamelink, C. (2004). Did WSIS achieve anything at all? *International Communication Gazette 66* (3–4): 281–290.

Hesmondhalgh, D. (2007). *The Cultural Industries*. 2nd Ed. London: Sage.

Holdorf, A. (2002). Another battle of Seattle. *Real Change News*. September 5. Retrieved October 22, 2009, from http://www.realchangenews.org/2002/2002_09_05/features/another_battle_of_seattle.html.

Honneth, A. (2002). Grounding recognition: A rejoinder to critical questions. *Inquiry 45* (4): 499–520.

Keating, J. (2008). Pirate politics. *Foreign Policy* Jan/Feb, 104.

Kellner, D. (2005). *Media spectacle and the crisis of democracy: Terrorism, war, and election battles*. Boulder, CO: Paradigm.

Kidd, D., Rodriguez, C., & Stein, L. (Eds.). (2009). *Making our media: Global initiatives towards a democratic public sphere, V. 1 (Creating New Communication Spaces)*. Cresskill, NJ: Hampton Press.

Klein, N. (2000). *No space, no choice, no jobs, no logo: Taking aim at the brand bullies*. New York: Picador.

Koivisto, J., & E. Väliverronen. (1996). Resurgence of the critical theories of public sphere. *Journal of Communication Inquiry 20* (2), 18–36.

Lee, K. (2009). The electronic fabric of resistance: A constructive network of online users and activists challenging a rigid copyright regime. In Kidd, Rodriguez, & Stein, Eds., 2009, 189–206.

Lessig, L. (1999). *Code, and other laws of cyberspace*. New York: Basic Books.

Lessig, L. (2001). *The future of ideas: The fate of the commons in a connected world*. New York: Random House.

Lessig, L. (2004). *Free culture: How big media uses technology and the law to lock down culture and control creativity*. New York: Penguin.

Lovink, G. (2002). An insider's guide to tactical media. In *Dark fiber: Tracking critical Internet culture*. Cambridge, MA: MIT Press.

Luhmann, N. (1982). *The differentiation of society*. Translated by Stephen Holmes & Charles Larmore. New York: Columbia University Press.

McCarthy, T. (1985). Complexity and democracy, or the seducements of systems theory. *New German Critique 35* (Spring–Summer): 27–53.

Mosco, V. (1996). *Political economy of communication: Rethinking and renewal*. Thousand Oaks, CA Sage.

Newitz, A. (2001). Not your girlfriend: The next generation of women hackers are doing it for themselves. *SFGate*. October 11. Retrieved October 22, 2009, from http://www.sfgate.com/cgi-bin/article.cgi?file=/gate/archive/2001/10/11/womhackers.DTL.

Pirates ahoy! (2009). *New Scientist*. June 27, p. 5.

Rappeport, A. (2006). Web warriors play cat-and-mouse with censors to free the Internet. *Financial Times*. March 4–5, p.6.

RE/Search. (1986). *RE/Search #11: Pranks!* Edited by V. Vale. San Francisco, CA: RE/Search Publications.

Samuelson, P. (1996). The Copyright grab. Wired 4 (January). Retrieved October 22, 2009, from http://www.wired.com/wired/archive/4.01/white.paper_pr.html.

Samuelson, P. (2003). DRM (and, or, vs.) the law. *Communications of the ACM 46* (4), 41–45.

Schiller, D. (1999). *Digital capitalism: Networking the global market system*. Cambridge, MA: MIT Press.

Schweidler, C., & S. Costanza-Chock. (2009). Common cause: Global resistance to intellectual property. In Kidd, Rodriguez, & Stein, Eds., 2009, 161–188.

Scott, J. C. (1986). *Weapons of the weak: Everyday forms of peasant resistance*. New Haven: Yale University Press.

Stein, L., Rodriguez, C., & Kidd, D. (Eds.). (2009). *Making Our Media: Global Initiatives Towards a Democratic Public Sphere, V. 2 (National and Global Movements for Democratic Communiation)*. Cresskill, NJ: Hampton Press.

TOR [The Onion Router]. (2007). Tor: Anonymity online. September 1. Retrieved October 22, 2009, from http://tor.eff.org/.

Touraine, A. (1988). *Return of the actor: Social theory in postindustrial society*. Minneapolis: University of Minnesota Press.

Turner, F. (2006). *From counterculture to cyberculture: Stewart Brand, the Whole Earth Network, and the rise of digital utopianism*. Chicago: University of Chicago Press.

Vaidhyanathan, S. (2005). *The anarchist in the library: How the clash between freedom and control is hacking the real world and crashing the system*. New York: Basic Books.

Vaidhyanathan, S. (2006). Critical information studies: A bibliographic manifesto. *Cultural Studies* 20 (2–3), 292–315. Retrieved October 22, 2009, from Http://ssrn.com/abstract=788984.

Part III
Transnational/ Translocal Nexuses

Possibilities latent in online mediations on the one hand, and transnational human mobility and cultural and economic globalization on the other, point to both continuities and disjunctures in the historical institutions and practices contained within certain communities (e.g., the nation) and a given territory. In an effort to critically explore transborder and transcultural histories and experiences, this section draws upon theories such as transnationalism/translocality, globalization, cosmopolitanism, space, and social control and surveillance. In media and communication studies research on the transnational condition in the two decades following the so-called information revolution, there has been a tendency toward overplaying the place/territory-annihilating forces of technologically enhanced global mobility based on a binary model of globality and locality. Empirically informed accounts, however, point to the primacy of the micro processes of the experiential realm of "mobility" and their embeddedness in the immediacy of the everyday and materialities of the locale.

New media and communication technologies mediate new forms of absence and presence, thereby simultaneously destabilizing and restabilizing social structures. While local absence or abandonment of place has often been construed as a precondition of global presence, in transnational contexts what remains really intriguing is how spatial and communicative dynamics of within the locale undercuts the banality and presumed sedentarism of place creating pockets of globalization and diverse forms of connectedness. As such, new media domains and online sociality enable, enrich and include as much as they disable, silence and generate new deprivations and new informational/representational impoverishments. The chapters in this section address, from a variety of perspectives and empirical angles, the compelling question of "multiplicity"—of scales, scapes, norms and forms—generated by transnational mobilities beyond the binaries of placeless space and territorial fixity.

In the opening chapter of this section, "Diaspora, Mediated Communication and Space: A Transnational Framework to Study Identity" (Chapter 11), written by Myria Georgiou, the author asks how we can understand culture and identity at present, if not through mobility, immediate and mediated intersections and juxtapositions of difference. And, how can we understand situated identities if not through the practices that interconnect or interrupt human action in and across places? Georgiou argues that a spatial approach to identity and mediation provides the necessary cross-fertilization of these three concepts as none of them can be fully understood if not through the other. Georgiou further contends that we need to record qualities of human mobility, of meetings and human assemblages and their interruptions in order to capture the meanings of diasporic and migrant identities—and as an extension, of cosmopolitan identities which are not exclusively owned by diasporas. Furthermore, we need to record some of the increasingly central qualities of culture: the mediated connections within and across space that provide human subjects information and communication for being and becoming. This chapter offers a theoretical invitation to think through the relation of space, identity and the media.

In Chapter 12, "Online Social Media, Communicative Practice and Complicit Surveillance in Transnational Contexts," Miyase Christensen argues that, after an initial enthusiasm in media and communications literature celebrating the uniqueness of online spaces and mediated transnational experiences, research into the complexities surrounding the dialectical relationship between online and offline milieus and practices has gained significance in recent years. The center spot exclusively occupied by "the medium" and "the technology" was, Christensen argues, abandoned in favor of an understanding that foregrounds the migrant subject, social practice and urban space in relation to a variety of factors such as power relations, spatial dynamics and social/cultural capital that play into media use. The aim of this chapter is to reflect on the complex process of media and ICT use and their integration in the daily social and cultural lives of transnational migrants, by taking people of Turkish origin living in Sweden as a case in point. Christensen argues that there is a multi-layered interplay between the surveillant gaze—intricately embedded in communication technologies (mobiles, email lists, online fora, social network sites, etc.)—and everyday communication routines operating at the level of subjectivity, giving way to novel forms of communicative practice and reflexivity and complicit surveillance.

The question of surveillance is also addressed by André Jansson in Chapter 13 ("Cosmopolitan Capsules: Mediated Networking and Social Control in Expatriate Spaces"). Jansson explores theoretically and empirically the relationship between Ulrich Beck's notion of cosmopolitanization and Lieven De Cauter's capsularization thesis, focusing particularly on the role of new social media within this tension field. The point of departure of the analysis is that these media attains a dual potential, sustaining on the one hand lifestyles and lifeworlds that are partly deterritorialized, while on the other hand establishing new forms of surveillance and social control. Applying empirical data from fieldwork among a Scandinavian group of expatriate professional it is shown how the stratified uses of social media, within the aggregate ensemble of social practices, tend to reproduce the gap between those groups who desire and enjoy deterritorialized lifestyles, and those who do not. One way of conceptualizing this power geometry of cosmopolitanization, Jansson argues, is to scrutinize the status of "cosmopolitan capital," here understood as a combination of cultural capital (Bourdieu) and network capital (Urry). Jansson's analysis demonstrates how mediated networking (in this particular empirical context) interacts with the accumulation of cosmopolitan capital, working simultaneously as a resource of global socio-cultural distinction, and as a counterforce to the logic of capsularization.

The broad repertoire of modes of social and political engagement mediated by the web has allowed ethnic minorities to expand and widen their participation in a civic culture that crosses the border between off and online spaces. In the closing chapter of the book, "Reconfiguring Diasporic-Ethnic Identities: The Web as Technology of Representation and Resistance" (Chapter 14) Olga Bailey discusses how these online practices are fundamentally important for their politics of identity—symbolic representation—as well as to obtain expertise and knowledge for living with the uncertainties and challenges of everyday life. It is in this respect, Bailey argues, that the online space becomes crucial, as a transnational site of contradictions where power, alienation, and possible resistance are experienced and enacted, as well as a resource for the competing reactions and coexistence of different strategies and discourses of ethnicity. The focus of Bailey's chapter is on how ethnic groups represent their identities, negotiate their cultural and political experiences and traditions in the online space which might reproduce their local everyday life, marked by global economic and political anxieties; how these online spaces construct, de-construct, contest, revise and shift ethnic signifiers; how the web might inform, constrain, and enable many of their practices, decisions, and behaviors, and, in the process, how ethnic groups might re-configure this technological space.

CHAPTER ELEVEN

Diaspora, Mediated Communication and Space

A Transnational Framework to Study Identity

MYRIA GEORGIOU

INTRODUCTION

This chapter proposes a spatial approach to identity and mediated communication as, it is argued, this approach provides a framework for grasping and analyzing the complex and changing formation of identity in current times and in an interconnected world. More specifically, it looks at the significance of mobility, immediate and mediated intersections and juxtapositions of difference as necessary elements of the analysis of current formations of identity. The chapter proposes a multi-spatial analysis and looks at the spatial matrix of diasporic belonging. The case of diaspora is chosen as an exemplary case of cultural formation, which is largely dependent on physical and mediated mobility that links places—either in distance or in proximity. The chapter discusses how the case of diaspora provides a key example in understanding how human mobility and (re-)settlement are not opposites, not cause and effect, but rather as co-existing elements of a world connected through flows and networks. Diaspora also presents an exceptional case of intense mediation, as communication networks and information exchange develop across various locations and they follow different directions with consequences for identity and community.

Space is composed of intersections of mobile elements, assemblages and meeting points; space is a practiced place, produced by the operations that orient it, situate it, and temporalize it, writes De Certeau (1984) providing an inspiring starting point for the discussion on identity, everyday life and space. De Certeau's proposal provides a key reference in the present discussion, not only because it links space to lived experience, but also because it emphasizes the significance of mobility in defining the meaning of space. The interconnection between space and mobility is now more important than ever, especially as two intertwined conditions have intensified: human mobility across boundaries and mediated communication that have advanced the role of virtual and transnational networks. These developments come with an equal number of pressing research questions. How can we understand culture and identity at present, if not through mobility, and through the immediate and mediated intersections and juxtapositions of difference? How can we understand situated identities if not through the practices that interconnect or interrupt human action in and across places? A spatial approach to identity that interrogates the meanings of place, space, and mediation provides the necessary cross-fertilization of these key concepts, and highlights the need to understand them through each other.

The reconfiguration of space through increased connections between individuals and groups—especially through digital networks—has given rise to limitless debates on virtual communities and virtual identities. The inspiring works on time and space distantiation (Giddens, 1989) and on time-space compression (Harvey, 1990), as well as the growing interest in transnational online communication, transnational communities and cross-boundary action have challenged the limits of territoriality. For a large body of literature, the *nonplace* and the *distant* have become more attractive compared to the proximate, grounded, placed experience. However, conceptualizations of place and space as opposing forces might undermine and underestimate lived experience. Especially in the study of identity, the continuity of action in places (as geographical territories), in dialogue or in conflict to action in nonplaces (as virtual locations) needs to be further studied and understood. Some work has indeed examined how the online space might not be *other* to the offline space and how territoriality might not be limited to the grounded and to the introvert. Works such as those by Morley (2000), Moores (2004), and Massey (2005) have been inspiring points of reference in thinking beyond the divisions between *the real* place and the mediated/virtual space. Inspired by such works, this chapter looks at a particular area of mediated networked communication, with significant consequences for understanding identity in its complex construction between interpersonal and mediated communication. The discussion focuses on diaspora as a social formation that captures the interconnectedness between online and offline spaces and the links between mediated practice and articulations of place as meaningful space. Diaspora provides a vivid—though by no means unique—case of interconnected and multiple trajectories (Massey, 2005), a case of identity construction in the context of mediated and networked communication within and across space.

The discussion draws directly and indirectly from empirical research in two separate projects: a recent project on Arab speakers and their media use in London, Madrid and Nicosia, and a transatlantic ethnography studying identity construction and media use among the Greek Cypriot diaspora in London and New York City. The discussion is constructed around three sections. The first one develops a conceptual framework that emphasizes the significance of diaspora and space as two key concepts in understanding current mediated articulations of identity. The second section proposes a multi-spatial approach to the study of identity. The discussion there focuses on the spatial matrix for diasporic identity and mediation, which consists of the home, the city, the nation, and the transnational space. In the third section, the interconnections and co-dependence of identity, space and the media are brought together into a final synthesis.

Thinking through Diaspora

In no other case is the close relation between space, identity and the media more apparent than in the case of contemporary diaspora. Diaspora is a concept that captures human mobility and (re-)settlement not as opposite points, not as cause and effect, but rather as co-existing elements of a world connected through flows and networks. In the recent reincarnation of the concept of diaspora, mobility between places and the meanings of diasporic identity is articulated as a condition emerging at the meeting of roots and routes (Gilroy, 1993). Such articulations have questioned assumptions about migration being a linear and single journey between origin (a place left behind) and destination (a new location of settlement). Physical and imagined connections between places have been discussed in their significance in the construction of diasporic identities. In fact, Gilroy's *The Black Atlantic* (1993) has been extremely influential, especially in its articulation of a matrix of geographical, cultural, and historical elements that inform diasporic identity. He has discussed the dialectic interdependence between geography (the territories around the Atlantic and their particular socio-cultural dynamics), politics of migration (the slavery trade and the Black migration between countries surrounding the Atlantic in seeking refuge, work, and freedom), and the flows of mobility and imagination (diasporic links across territories of identification through particular cultural repertoires and a shared history). In works such as those by Gilroy (1993) and Hall (1990), as well as in the numerous studies that they inspired, the links between networks and flows that surpass geographical restrictions are reaffirmed as central in the process of identity construction. In such works, especially as developed in media and communications, the interconnections between locations (of lived present or past) and between places (physical and virtual) has been further developed, especially in relation to television consumption (cf. Georgiou, 2006a; Gillespie, 1995; Aksoy and Robins, 2000) and virtual networks of diaspora (cf. Brinkerhoff, 2009; Georgiou, 2006b).

Such studies on mediated communication and diaspora have discussed the qualities of space as lived and as imagined, as context for identification and struggle, as dependent on memory, experience and ideology of deterritorialization and reterritorialization, of mobility and of contact or interruptions of contact with old and new others. Diasporic populations live within specific locales—urban places especially—and occupy particular positions in transnational networks of family, kin, and community. The social interaction and communication within the diasporic communities, among dispersed sections of the same diaspora and beyond the limits of diasporic communities, all take place in space. Some of these spaces are grounded in very specific places—such as the neighborhood—while others exist virtually and in nonplaces (Urry, 2000). Social interaction and relations are no longer exclusively dependent on simultaneous spatial co-presence. Next to the face-to-face communication, relations develop with *absent others* through-mediated communication, especially in digital communication. It is in the occasions of long-distance communication, that the singularity of place becomes predominantly challenged. When the neat equation between culture, community and geography is challenged (Gillespie, 1995) increased potentials for imaginative geography and history emerge (Said, 1985). When Greek Cypriot diasporic subjects in New York City, for example, engage in debates about the political situation in Cyprus in online fora, at the same time as members of the Cypriot diaspora in London and in Melbourne do, they reinvent the limits of territoriality. For those active participants of online Greek Cypriot fora, an online territoriality of a community emerges against a grounded territoriality that excludes them from participating in what happens at the actual place (Cyprus as a nation-state and as a grounded territory).

Cultural mobility that defines the limits of space and territoriality through mediated communication is recognized by Cohen (1997) as an element of the growing relevance to diaspora. Diaspora refers to people who cross boundaries and who settle in locations different to those of their origins. Diaspora is also a category that implies multiple connections across space and flows of ideas and information beyond a singular nation. Work done within studies of transnationalism has also contributed in thinking of diasporas as networked, transnational formations (cf. Portes, 1995; Vertovec, 2009). Within this literature, diasporas as transnational communities have been discussed for their contribution in thinking beyond the binary opposition global/local. Diasporic cultures and political activity, emerging across geographical boundaries and taking specific forms in local domains have also attracted interest across the social sciences and have been discussed as forms of counter-hegemonic practices and alternative forms of (selective) community participation and re-articulation of identities (Sandercock, 2003; Schiller et al., 1995; Werbner, 2008). As such, diaspora is not only a descriptive category but it can also become a metaphor in understanding identity as a multi-positioned and networked category.

About Space

Space is a product of interrelations and of 'interactions, from the immensity of the global to the intimately tiny,' Massey (2005, p. 9) argues, emphasizing the coexisting heterogeneity and multiplicity in the possibilities and trajectories within space. Benhabib (2008), explains the tension between human and cultural transnationalism on the one hand, and state persistence on territorial boundedness and loyalties attached to it on the other. She challenges the concept of territoriality, as this is understood to be attached to the nation-state: '*[T]erritoriality* has become an anachronistic delimitation of material functions and cultural identities; yet, even in the face of the collapse of traditional concepts of sovereignty, monopoly over territory is exercised through immigration and citizenship policies' (2008, p. 5). What Benhabib argues is that there is an important tension between identity as attached to multiple locations and cultural references against a rigid and territorial definition of formal/political identity. However, conceptualizations of territoriality as rigid and grounded in a singular place—and more specifically as dependent on a political entity, such as the nation-state—are caught between the realities of political determinism of identity (as dependent upon territory) and the realities of identities that are inevitably surpassing the nation-state. How can an ideology of territorial being and belonging capture the connections of migrants with family in various continents, or their multilingual spheres of belonging developed through their (virtual) travels? And to move beyond diaspora, how can a territorial approach reflect the sense of belonging among participants in global social movements, or even among young people who share transnational Facebook or musical fandom cultures? Territoriality—as perceived to be grounded in a singular place—is indeed anachronistic but remains deeply rooted in political conceptualizations of identity; this is why formal citizenship remains (or has become reinvented to be) property of those having long and usually rooted into territory rights. At the same time, the cultural landscapes and the flows of people, technologies, and information suggest otherwise. As Appadurai argues 'people, machinery, money, images and ideas now follow increasingly nonisomorphic paths' (1996, p. 37) with speed, scale and volume that are (re-)defining culture on a global level. The interconnected world relates more to Appadurai's scapes (1996) than to a stable political and bounded geographical zone. Flows and networks, captured in works such as those of Appadurai since the 1990s, have become further articulated in recent literature on cosmopolitanism. Cosmopolitanism literature has revisited territoriality and challenged the definitions of territory as grounded in a singular political territory; it has recognized the importance of human mobility, boundary crossing and close encounters of difference for the production of meaning, identity and political action (Beck, 2006; Benhabib, 2008). As an analytical concept, cosmopolitanism has captured the complexities of multiple forms of belonging and of heterogeneous and fragmented publics by challenging essentialist interpretations of identity and bounded communities, as well as assumptions about stable and ever-present hierarchies that have indiscriminately defined social relations, politics and culture for more than a century (e.g. social and geographical divisions, meanings of citizenship, key elements of identity). As such, it can explain how those 'in charge' of time and space compression might drive change across boundaries, but also how—especially in times of global financial turmoil—they can also lose control; how those in the receiving side of power geometry (Massey, 1993) might drive elements of change in culture and politics, as observed, for example, in the emergence of heterogenous publics and the transformations of cities of the global north into cosmopolitan locations. 'Through a study of the history of spaces and an understanding of their heterogeneity, it became possible to identify spaces in which difference, alterity, and "the other" might flourish,' argues Harvey (2006, p. 537).

Space is reassessed as a concept that can best frame analyses and interpretations of conflict and dialogue, especially as encountered in the intense juxtapositions of difference in (trans-)urban worlds (Beck, 2006; Harvey, 2006). Space can capture the links between the cultural and the political spheres of representation; if 'space is fundamental in any form of communal life,' then space must also be 'fundamental in any exercise of power,' argues Foucault (in Harvey, 2006, p. 538). These debates point towards one key quality of space as an analytical framework for studying identity; space is a framework that allows us to study identity through mobilities, meeting points, settlement/re-settlement, but also meaningful relations—including those of subordination, exclusion, and participation in publics and communities—as they emerge at the meeting of the political (e.g. formal citizenship) and the cultural (e.g. education, health, communication) spheres. In Lefebvre's words: '[t]he concept of space links the mental and the cultural, the social and the historical' (2003, p. 209). According to Lefebvre, space reconstitutes complex processes that include *discovery* (of new or unknown spaces, of continents, or, of the cosmos), *production* (of the spatial organization characteristic of each society) and *creation* (of landscapes, specifically the city (2003). Space is real and/or virtual and imagined; its natural substance and boundedness are increasingly challenged as permanent characteristics.

It is in the meeting of the cosmopolitan articulations of space and the study of transnationalism and diaspora that territoriality might be reconceptualized to capture the juxtapositions of mediated and physical mobility and the meaning of space as practiced place. The limits of place as single and grounded territoriality are challenged within conceptual frameworks and methodological strategies that allow us to record the significance of various spaces in their own standing and in their interconnections, so that we go beyond *either/or* (Beck, 2006) and instead analyze 'this and that' as well as possible *neither/nor*. The cosmopolitan outlook comes together with a (multi-)spatial approach to identity that unfolds a framework for this discussion. This discussion recognizes a number of significant conditions attached to identity and mediated interconnections across space:

- *The multiple spaces of belonging*: All human subjects move between various spaces of belonging—physical and symbolic. Though not all individuals or groups enjoy the same levels of mobility (and in particular restrictions need to be taken into serious account, especially those relating to class, gender, sexuality, and age), increasingly human subjects develop their sense of being and becoming as they move between online, offline, local, and transnational territories.
- *The transnational connections*: Transnational mobility and the intensification of transnational politics challenge the nation-state from below and above and many of the taken-for-granted assumptions around identity; migrant and diasporic cultural and political activities develop across boundaries and across mediated and physical places.
- *The juxtapositions (and not just hierarchies and linear relations) of difference in the cosmopolis*: The increasingly diverse cities and the complex social action among migrants, diasporas and other groups marginalized in the national and supranational formations of citizenship and economic engagement illustrate the significance of the physical—and more often than not, urban—points of contact in understanding identity, its meanings and its limitations.

A Multi-Spatial Approach

Space brings together into meaningful relations and formations places and practices; it carries social meanings—and these are always plural. The home, the public, the city, and the national, in its interconnections with the transnational, form layers of the spheres of belonging; they together form the puzzle of the context where social relations, communication and action take place and shape the meanings of identity and community. Each of these elements of space is an autonomous node and each has its own dynamics, its morality, economy, and its social and cultural meanings, yet they all depend on their interconnection. 'Knowing where we are is as important as knowing who we are, and of course the two are intimately connected' argues Silverstone (1999, p. 86). The link with mediated communication here is clear, obvious, yet increasingly assumed and banal as mediated communication supplements, or even has replaced, face-to-face communication. The *where* in Silverstone's argument is a key element renegotiated in the context of mediated communication. Online territories, vis-à-vis the interrupted and fragmented territorialities of diaspora (as they emerge in the movement between places of origin and destination, between imagined locations of belonging and locations of living) have provided new possibilities for individuals and groups to control the *where we are*. Online newspapers from the country of origin or the Arabic region have been some of the most widely used media among Arab speakers asked in the context of the *Media and Citizenship*[1] project in London and Madrid. Online newspapers, which are easily accessible for most diasporic Arabs in European capitals, expand the sense of belonging in an imagined community that cannot be constrained within the territory of the nation. The consumption of online news does not involve only a process of (re-)connection with a nation, but it also relates to the desire to keep in touch with the mundane nature of news that can then be shared within familial, domestic, and transnational contexts. This is only one simple example of the daily and ordinary processes of redefining place and space in the context of diasporic everyday life.

The domestic, the local, the public, the urban, the national and the transnational form an interconnected spatial matrix that shapes the possibilities for belonging, for choosing not to belong, or for being forced not to belong.

Home

Home is the symbolic and real place that becomes a synonym to familiarity, intimacy, security and identity against the unknown, the distant and the large. 'Home, of course, needs to be understood in both literal and metaphorical senses. The defence of home is a defence of both the private spaces of intimate social relations and domestic security—the household; as well as of the larger symbolic spaces of neighborhood and nation—the collective and the community,' Silverstone (2004, p. 442) argues. Massey (1993) questions the stable and inward-looking nature of the home and argues that a sense of place does not depend on its stability and purity; rather it depends on its unique position as a point of intersection in a wider context of relations. Nostalgia, strangeness and the sense of loss intensify the efforts of making a house a home. The search for ontological security becomes one of the main reasons for reinventing close family relations, relations that often become even more intense in the diaspora than they would ever be in the country of origin. Silverstone and Hirsch (1994) note the significance of the domestic in the modern world as a place enhanced, mediated, contained, even constrained by our ever-increasing range of information and communication technologies and the systems and services that they offer the household. In the case of diasporas—as the exemplary transnational community—home becomes even more complex: it can be the home where people live now, one that they left behind, one that they imagine constructing or returning to. In this case, mediated communication can play a dual role. It can destabilize the role of the home in achieving ontological security. '[B]oth [the domestic and the collective home] are threatened by the media extension of cultural boundaries: both laterally, as it were, through the globalisation of symbolic space, and vertically through the extension of accessible culture into the forbidden and the threatening. In both cases home has to be defended against material breaches of symbolic security' (Silverstone, 2004, p. 442). But, mediated communication can also sustain a sense of homeliness in being at home in many places or through mobility. The continuity of online communication plays a particular role in sustaining simultaneity of communication and thus possibly in redefining the limits and boundaries of home. One of the Arab-speaking participants in the cross-European research project *Media and Citizenship* describes a sense of homeliness emerging in a world connected through the Internet:

> I don't consider myself immigrant, I don't like this feeling, I feel as an international citizen. I don't feel a conflict between being Lebanese Arab and living here ...Thanks to the Internet, you can live anywhere because you can connect with everyone from a distance (Nicosia, male, 46–65).

This commonly observed position reflects the desire (and possibly the achievement) of a sense of homeliness through the renegotiation of physical positions associated with networked and physical mobility. At the same time, and as pointed out by Silverstone earlier, the constant availability of information about places threatens the perception of home in its totality. In the same research project, a number of participants have referred to the information they receive on satellite television and the Internet as 'eye opening' and as advancing their sense of critical distance from places they might have otherwise idealized.

City

The culturally diverse, and at the same time, segmented urban space is of key relevance in the production of culture, especially since the city brings together intense media production and diversified consumption. Robins (2001) suggests that we should think through the city, instead of through the nation. The city, he argues, is a more useful category of analysis, especially since it allows us to reflect on the cultural consequences of globalization from other than a national perspective. 'The nation, we may say, is a space of identification and identity, whilst the city is an existential and experimental space' (2001, p. 87). With reference to London—a reference that can extend to other cosmopoles—Robins argues that in the city people can re-think and re-describe their relation to culture and identity: '(T)he urban arena is about immersion in a world of multiplicity, and implicates us in the dimension of embodied, cultural experience' (2001, p. 87). City, and especially the culturally diverse and global city, is a symbolic and physical landscape of cultural contestations, of conflict, but also of coexistence of difference. As Massey (2005, p. 169) puts it:

> 'Cities' may indeed pose the general "question of our living together" in a manner more intense than many other kinds of places. However, the very fact that cities (like all places) are home to the weavings together, mutual indifferences and outright antagonisms of such a myriad of trajectories, and that this itself has a spatial form which will further mould those differentiations and relations, means that, *within* cities, the nature of that question—of our living together—will be very differentially articulated.

The particular dynamics of the city and of urban life, with its demographic diversity, its cultural differences and its heterogeneity are directly related to the emergence of politics of representation in the city outside the formal political sphere. Cities tend to be destinations of refuge, but also of openness and possibilities of expression. Especially the streets, community centers, libraries, schools, pubs and clubs of culturally diverse working class neighborhoods reveal imaginings and reimagining of belonging which are often distant from the rules and the imagination of the nation. Urban life comes with potentials and restrictions for expression, representation and identity. These often find expression through digital connections. Why does this matter in the discussion about space? Digital and local face-to-face connections renegotiate the meanings of territoriality and create spaces for people to find their own *sense of place*. Mediated connections involve those between the local in the local (e.g. in social networking that links young people between them within neighborhoods or in engaging parents of different backgrounds in their children's school life). A number of participants in the *Media and Citizenship* project have emphasized that London or Madrid would be locations they would refer to in identifying themselves, but not UK or Spain, respectively—their sense of place is attached to the city as they imagine and live it but not in relation to the nation. At the same time, urban-mediated connections within and between cities lead to the emergence of transurban networks, where cultural and financial exchanges and shared experiences of social life might sustain the sense of continuity and community within diaspora. Greek Cypriots in New York City, taking part in an ethnographic study conducted by the author, for example, repeatedly made references to assumed commonalities between their media and leisure habits and those of the members of the Greek Cypriot diaspora in London.

National and Transnational Space

The nation—which in modernity has been formalized as a nation-state—depends on ideologies and practices of clear-cut borders and requires the formation of identities and communities within defined borders and territory (Holton, 1998). There is a key contradiction in the position of the nation in our times. On the one hand, culture and communication become increasingly mediated and transnational, while on the other hand, the national political boundaries become increasingly reinforced. This condition creates a rupture between the politics of the nation and the human condition within nations and even more so for those human subjects who cross national boundaries, especially through migration. As Beck argues, mediated mobility has transformed 'the experiential spaces of the nation-state *from within*' (Beck, 2006, p. 101), (emphasis in original). But the physical contains and grounds the mediated. The mediated 'freedom of flows' is not by definition liberating but contested; it is grounded by the constraining powers of the physical and the national (Bauman, 1998). The nation-state aims at sustaining its power and legitimacy based on ideologies of singularity—of singular loyalties, of the singularity of the national territorial ownership and boundedness. Diaspora challenges national ideologies, but it often finds itself trapped in them. The nation-state of origin requires loyalty and commitment, so does the nation-state of settlement. The nation-state in modernity—and this has not changed in late modernity as radically as is often claimed—forms its own project of progress and harmony based on social, economic and, inevitably, cultural assimilation of its population. It is in this context that cultural difference—as often expressed in diasporic cultural ideologies and practices—is marginalized, excluded and alienated within countries hosting diasporic populations.

In a context of tension between diaspora and the nation, the latter still keeps its important role in the scheme of diasporic belonging. Even as a source of restrictions and ideological polarization (e.g. expressed in discourses of nationalism, demands for singular loyalty, legitimating of national military power), it should be acknowledged as an element of the diasporic space of identity for at least two reasons. On the one hand, in democratic societies, national law (should) protect (diasporic) minorities from discrimination, racism and exclusion and thus (should) broaden participation in multiethnic societies with consequences for diasporic representation. On the other hand, restrictions, polarizations and exclusions initiated in the actual practices and ideologies of the nation play their part in the construction of identities. Identities are not shaped only through positive and creative processes of participation and communion, but also in processes of exclusion, marginalization and regressive ideologies—expressed in the mainstream ideologies of the country of settlement or of the country of origin, but also voiced from within the diasporic communities.

While the national is a space of a constant struggle between processes of diasporic social and cultural inclusion and exclusion, the space that is almost homologous to diaspora is the transnational. Transnational space emerges as a meeting place of lived and mediated connections and travels, and it increasingly depends on virtual and other mediated networks for sustaining within and across it meaningful relations. For this reason, the transnational emerges in the intense interaction and tense co-dependence with the national context and imaginary. Transnational space acknowledges the development and sustaining of connections and networks across geographical, cultural and political borders (Vertovec, 2009) and emphasizes the possibility for development of meaningful relations and social formations through dense networks (Portes, 1997). It is in the context of transnationalism that contemporary theorizations of diaspora become useful in thinking of continuity, community and attachment across space. While there is a temptation to interpret the often observed attachment of dispersed people with a transnational community as a reproduction of the imagined community of the nation beyond boundaries, the diasporic case is significantly different from both the nation and from any primordial bounded community. Diasporic continuity is as much about the imagining of a common origin and a common fate as it is about the transnationalization of possible common imaginings, which are particular and specific to a group but also global in their relevance. As captured in the words of a female participant in the *Media and Citizenship* project, shared repertoires of diaspora across boundaries reaffirm shared identities in transnational contexts:

> For me, when it comes to Arabic, *Bab al-harah* is interesting … no matter where you are, what country you are in Western or Middle East, people all watching it the same time. All my cousins watch it in Iraq. Also, in the USA, Germany. That's what I'm saying, almost everyone around the world watches Bab al-harah (London, females, 18–24).

Space, Identity and the Media

The intensification of mobility and thus of diasporic scattering, the development of communication technologies and networks across the world, and the relocation of diasporic populations in global cities play a key role in the construction of diasporic identities. Massey's discussion (2005) of Graham's thesis on the mutual constitution of technology, space and place is inspiring in arguing about the reconstitution of territoriality through the cultural imaginings of space as juxtaposed with mobility. In the case of diaspora, we see constant mobility, but also changes in directions, quantities and qualities of communication exchange across transnational spaces. These changes, for example, are observed when television broadcasts become replaced by online viewing of television productions, as part of transnational mobility between different locations, only to then be replaced again by television viewing at home and in grounded in place locations. Thus, images and sounds of different origins, destinations and flows, as well as a shifting articulation of technology, all become part of the everyday emotional and communicational experience of the diaspora. As a consequence, the social relations which are located in place are shaped and in turn shape technologies of communication and of connection. One of the consequences of the available diverse cultural repertoires at the push of a button is the contestation of the mainstream, national and top-down ideologies of identity as equated to national boundedness. If nothing else, networked online communication forces us to think outside the national, grounded territoriality as a unique framework for identity construction. It might provide a challenge in considering the possibilities of parallel territorialities—online, offline, national, and transnational—that might co-exist and compete. Diversity in cultural repertoires exchanged, consumed and constructed in virtual spaces can possibly destabilize the authority and hegemony of national media and redefine transnational and local ones. Digital communication forms one of the major challenges to bounded and formally constructed political geographies. Think of the *Al Jazeera effect* in the days and years after 9/11. Western states suddenly noticed a player in transnational communication which could not be controlled by the established policy and corporate players dominating the (inter-)national systems of politics and communication. The Al Jazeera phenomenon, like other satellite and digital media phenomena of lesser or similar effect, is directly linked to the mediated communication users' desire to link and think across cultural territories beyond, next to and vis-à-vis the imagined community of the nation (Anderson, 1983).

In the context of diasporic scattering and in the development of digital, familial, and interpersonal networks across the world, especially across global cities, diasporic identities are lived and imagined. Massey (1993, 2005) talks about the changing geography where social relations, movements and communication are transformed and come together in places, which in turn become unique points of their intersection. Geographical places become meeting places:

> Instead then, of thinking of areas with boundaries around, they can be imagined as articulate movements in networks of social relations and under-standings ... but where a large proportion of those relations and understandings ... are constructed on a far larger scale ... And this, in turn, allows a sense of place which is extroverted, which includes a consciousness of its links with the wider world, which integrates in a positive way the global and the local (Massey, 1993, p. 239).

Massey stretches out a basic element of (inter-)spatial meetings—the hybridization of human relations, of identities and of places. The city especially, becomes a hybrid place, both for contact and for communication. On the one hand, everyday close encounters in the city and struggles over the control of local resources bring together urban dwellers in shared realities. On the other hand, the close encounters reveal how the city as a place is formed through movement (Massey, 2005). So are identities. Positioned between the place, the encounters, the connections. These can be better understood in revisiting the questions of where and with whom, of spatial proximity and distance. Diasporization, as combined with intense mediation, has advanced the sense of proximity in two ways: (i.) in relation to increased (mediated) proximity among members of diaspora located in various places but occupying a common diasporic (symbolic) space; and (ii.) in relation to the mediated and physical proximity shared among people with different backgrounds, especially in cosmopolitan cities—i.e. the intense juxtapositions of difference in the cosmopolis. In relation to mediated diasporic proximity across spaces, the level and continuity of contact is unprecedented in human history and this creates possibilities for reinforcing a common sense of belonging. Increased exchange of images and sounds from the country of origin and other diasporic people becomes a constant reminder of diversity and of the real and present face of the country of origin and the broader diaspora. The second sense of proximity, which brings close people of different backgrounds—the intense juxtaposition of difference in the cosmopolis—also has important consequences for identity and the sense of belonging for diasporic and other cosmopolitan subjects. Possibly more than any other location, the city brings people, technologies, economic relations, and communication practices into unforeseen constellations and intense juxtapositions of difference (Benjamin, 1997). The intense urban juxtapositions of difference in the unglamorous, and often marginalized and deprived, quarters of the global cities are usually invisible in tourist brochures; these are, however, locations where the potential of communication technologies to connect people in the locale and across boundaries, in shared attempts to seek citizenship, to find a location in the city and the world, and to shape identity in a cosmopolis are revealed in intensity rarely observed elsewhere. Networked political movements for the rights in the city directly challenge national and grounded territorialities (cf. http://www.rockingthecity.com; http://www.nodel.org). The meanings of such practices are shaped in the context of illegality or alegality (like in the case of graffiti works), and in opposition to or rejection of the politics of the state. The cultural and social locations of such acts and the enactment of these practices by young, usually disenfranchised and minority youth, reflects—if not singularly, at least partly—processes of active opposition to a territoriality owned and controlled by a singular institution, that of the nation-state.

Conclusion

Mediated and interconnected space becomes a space of contestation, of complex and often conflicting articulations of identity. In having access to a variety of media and communication technologies diasporic people can connect to (or disconnect from) individuals and communities in their neighborhood or in distant places. By being more informed about the politics and the culture of the country of origin, other sections of the diaspora, the country of settlement, or their locality, diasporic individuals and groups can construct a world of *critical proximity*: they become aware that, either they like it or not, they are not just located in place, but they are positioned between places; they construct but they also feel the effects of interconnected and competing territorialities. Inevitably, diasporic identity is lived as multi-positioned in symbolic and geographical spaces. 'High mobility and increased mediation means that 'more and more people are living in a kind of *place-polygamy*. They are married to many places in different worlds and cultures. Transnational place-polygamy, belonging in different worlds: this is the gateway to globality in one's own life' (Beck, 2002, p. 24). In addition, and as contact with others they do not share the same past becomes a constant and daily reality, diasporas show that identity is not only about a location and a place but also about reinvention of the limits of presence, participation and representation in the complex spatial matrix where human subjects find themselves. The need to find a place in these spaces requires negotiation and development of performed identities that surpass the particularity of an inward looking identity. As a result, the relations that develop or are sustained across distance with family, kin, and communities, are more and more networked and mediated, bringing on major challenges for grounded and nation-bound territoriality, while opening opportunities for multiple, interconnected and mediated territorialities.

Note

1. The research leading to these results has received funding from the European Community's Seventh Framework Programme FP7/2007–2013 under grant agreement n° 217480. The research team is a consortium of five European universities (consortium leader: C. Slade). The author had led the team conducting research in London, Madrid, and Nicosia.

References

Aksoy, A. & Robins, K. (2000). Thinking Across Spaces: Transnational Television from Turkey. In *European Journal of Cultural Studies*, 3 (3), 343–365.

Anderson, B. (1983/(1991). *Imagined Communities: Reflections On the Origins and Spread of Nationalism*. London: Verso.

Appadurai, A. (1996). *Modernity at Large: Cultural Dimensions of Globalization*. Minneapolis, MN: University of Minnesota Press.

Bauman, Z. (1998). *Globalization: The Human Consequences*. Cambridge: Polity Press.

Beck, U. (2002). The Cosmopolitanism Society and Its Enemies. *Theory, Culture and Society*, 19 (1–2), 17–44

Beck, U. (2006). *Cosmopolitan Vision*. Cambridge: Polity.

Benhabib, S. (2008). *Another Cosmopolitanism*. Oxford: Oxford University Press.

Benjamin, W. (1997). *One Way Street*. London: Verso.

Brinkerhoff, J. M. (2009). *Digital Diasporas: Identity and Transnational Engagement*. Cambridge: Cambridge University Press.

Cohen, R. (1997). *Global Diasporas: An Introduction*. London: UCL Press.

De Certeau, M. (1984). *The Practice of Everyday Life*. Berkeley, Los Angeles and London: University of California Press.

Georgiou, M. (2006a). *Diaspora, Identity and the Media: Diasporic Transnationalism and Mediated Spatialities*. Cresskill, NJ: Hampton Press.

Georgiou, M. (2006b). Diasporic Communities Online: A Bottom Up Experience of Transnationalism. In K. Sarikakis & D. Thussu (eds.) *The Ideology of the Internet: Concepts, Policies, Uses*. Cresskill, NJ: Hampton Press.

Giddens, A. (1989). *The Consequences of Modernity*. Cambridge: Polity.

Gillespie, M. (1995). *Television, Ethnicity and Cultural Change*. London: Routledge.

Gilroy, P. (1993) *The Black Atlantic: Modernity and Double-Consciousness*. Cambridge, MA: Harvard University Press.

Hall, S. (1990). Cultural Identity and Diaspora in J. Rutherford (ed.) *Identity: Community, Culture, Difference*. London Lawrence and Wishart.

Harvey, D. (1990). *The Condition of Postmodernity: An Enquiry into the Origins of Cultural Change*. Malden, MA and Oxford: Blackwell.

Harvey, D. (2006). Cosmopolitanism and the Banality of Geographical Evils, *British Journal of Sociology*, 57(1), 529–564.

Holton, R. (1998). *Globalization and the Nation-State*. London, Macmillan.

Lefebvre, H. [with Elden, S. & Lebas, E.] (2003). *Henri Lefebvre: Key Writings*. London: Continuum.

Massey, D. (1993). A Global Sense of Place. In A. Gray & J. McGuigan (eds.) *Studying Culture: An*

Introductory Reader. London: Arnold.

Massey, D. (2005). For Space. London, Thousand Oaks, CA, New Delhi: Sage.

Moores, S. (2004) The Doubling of Place: Electronic Media, Time-Space Arrangements and Social Relations. In McCarthy, A. & Couldry, N. (eds.) *Mediaspace: Place, Scale and Culture in a Media Age.* London and New York: Routledge.

Morley, D. (2000) *Home Territories: Media, Mobility and Identity.* London and New York: Routledge.

Morley, D. & R. Silverstone (1995). Communication and Context: Ethnographic Perspective on the Media Audience.' In K. Bruhn Jensen & N. W. Jankowski (eds.) *A Handbook of Qualitative Methodologies for Mass Communication Research.* London: Routledge.

Portes, A. (1997). Immigration Theory in the New Century: Some Problems and Opportunities, *International Migration Review*, 31(4), 799–825.

Robins, K. (2001). Becoming Anybody: Thinking Against the Nation and Through the City, *City*, 5(1), 77–90.

Said, E. (1985). *Orientalism.* Harmondsworth, Middlesex: Penguin Books.

Sandercock, L. (2003). *Cosmopolis II: Mongrel Cities in the 21st Century.* London: Continuum.

Schiller, N., Basch, N. & Blanc, C. S. (1995). From Immigrant to Transimmigrant: Theorizing Transnational Migration. *Anthropological Quarterly*, 68(1), 48–63.

Silverstone, R. (1999) *Why Study the Media?* London, Thousand Oaks: CA, New Delhi: Sage.

Silverstone, R. (2004) Regulation, Media Literacy and Media Civics, in *Media, Culture and Society*, 26(3), 440–449.

Silverstone, R. & Hirsch, E. (1994). *Consuming Technologies: Media and Information in Domestic Spaces.* London: Routledge.

Urry, J. (2000) *Sociology Beyond Societies: Mobilities for the Twenty-First Century.* London and New York: Routledge.

Vertovec, S. (2009). *Transnationalism (Key Ideas).* London and New York: Routledge.

Werbner, P. (2008). *Anthropology and the New Cosmopolitanism: Rooted, Feminist and Vernacular Perspectives.* London: Berg.

CHAPTER TWELVE

Online Social Media, Communicative Practice and Complicit Surveillance in Transnational Contexts

MIYASE CHRISTENSEN

In contemporary media societies, everyday sociality and mobility are increasingly sustained through mediated forms of communicative practice. In the face of spatio-temporal constraints there is an even greater tendency in transnational communities to engage with various modes of both source-to-user and interactive technology in order to forge and maintain ties with "home," family and friends. In this chapter, I reflect on the complex process of media and information and communication use and their integration in the daily social and cultural lives of transnational migrants, by taking people of Turkish origin living in Sweden as a case in point. Secondly, I argue that there is a multi-layered interplay between the surveillant gaze—intricately embedded in communication technologies (mobiles, email lists, online fora, social network sites, etc.)—and everyday communication routines operating at the level of subjectivity, giving way to novel forms of communicative practice and reflexivity and *complicit surveillance.*

While younger members of migrant communities often abandon more traditional communicative domains—both online and offline—to avoid power hierarchies, forced allegiances (to home-nation and communal identity) and monitoring, they choose to enter other, newer, domains (e.g. social network sites such as Facebook or new online communal formations) for the very purposes of *seeing* and being *seen*. On the part of some individuals and groups, an additional goal is to form alternative power geometries: thus new forms of monitoring. The scope of surveillance, however, is never clear-cut, and surveillant practice increasingly assumes the form of permanent rather than temporary intervention. David Lyon (2001), for example, uses the metaphor "leaky containers" to describe the indiscreet data flow between and across various compartments of social life (leisure, work, state, etc.). Ultimately, newer forms of mediated interaction, and the complexities inherent in surveillance itself, restructure the nature of information-sharing and communicative practice in transnational sociality. Yet, the academic discourse of mediated communication and diasporic[1] communities, and the discourse of surveillance, media and power remain distinct and do not often come together. In this chapter, in light of a number of approaches to mediated communication and transnationalism on the one hand and to surveillance and privacy on the other, and based on empirical data on transnational communities of Turkish origin living in Sweden, I bring together the theoretical tropes of these two broad areas of inquiry and address questions related with representation, power, privacy and reflexivity, and decentralized, complicit surveillance, and how such notions and related norms embody themselves through the communication process in the context of social change. The empirical part of the study is based on qualitative research which was conducted in 2008 and 2009 and involves in-depth interviews with individuals of Turkish origin residing in Sweden.

Mapping the Field: Transnationalism and Mediated Communication in Everyday Contexts

In a post-9/11 global milieu where political polarization and cultural bias have increasingly become the norm rather than the exception, and multiculturalism is being replaced by integrationist/assimilationist tendencies throughout Europe, questions related with mobility, social control, identity and communicative action through mediation in particular contexts warrant fresh attention. While the general discourse of globalization produced within the last two decades has provided both a useful framework and an entry point to approach such contemporary socio-cultural questions, its totalizing logic has also been a hindrance in locating and scrutinizing the particularities that emerge out of global processes (Christensen, 2008). 'Transnationalism' both as an epistemological plane and contextual frame and the theories and research produced under its rubric accommodates refined definitions of a variety of issues related with groups and individuals with transborder and transcultural histories and experiences. In the experiential context, transnational ontology is marked by shared identities and the social, political and cultural visibility of transnational migrants in home and host countries (see Portes et al., 1999; Levitt, 2001).

As Ehrkamp (2005) argues, the conceptual literature on transnationalism is far from reconciling conceptualization of transnational migration and transnational communities with questions of immigrant incorporation or assimilation (see Kivisto, 2003; Nagel, 2002). Rather than becoming assimilated, identities are deterritorialized, at various levels, and reterritorialized in complex ways as a result of maintaining transnational ties (Smith, 2001) through various means, with new media and communication technologies providing one such avenue. Yet, the role and significance of global connectivity and instances of placeless sociality and deterritoriality notwithstanding, I share Portes et al.'s (1999) disposition toward taking the individual migrant and his/her networks as the unit of analysis and utilizing transnationalism as a departure point and an analytical tool to produce empirically situated research to address the various levels of transnationalism such as the political, economic and socio-cultural. A perspective that underscores the persistent significance of place and locality is additionally instrumental (see Smith, 2001) in scrutinizing the construction and negotiation of meaning in the context of the urban experiences of transnational individuals and communities given that most transnational migrants reside in urban areas as in the case of the migrants of Turkish origin in Sweden most of who live in the Stockholm area.

As global economic, cultural and human mobilities assume an ever increasing pace, both transnational/translocal relationships and the experiential realm of the everyday play central roles in identity construction in migrant contexts (see Glick-Schiller et al., 1995; Kivisto, 2003; Vertovec, 2001). The significance of ICTs and diasporic media in transnational life is well documented in research (see Georgiou, 2005, 2006; Bailey et al., 2007; Titley, 2008).

In the field of minority media research in Sweden, adequacies and inadequacies of media policies particularly targeting minorities to foster social inclusion are well-explored areas (see Christensen, 2001; Camauër, 2003). However, research on everyday mediations of diasporic media in general and of social networking media in particular amongst specific transnational communities in Sweden remains under-researched. In regards to Turkish immigrants and Turkish diasporic communities, the majority of the research comes from Germany (which hosts a large number of Turkish immigrants). Ehrkamp (2005), by way of an example, focuses on transnational practices and how Turkish immigrants create places of belonging in a German city. Although there are studies on the questions related with immigration in Sweden such as research focusing on sociocultural and educational adaptation of immigrant youth and ethnic identity (e.g., Vedder and Virta, 2005), research particularly on mediation, sociality and communicative practice within the Turkish diaspora in Sweden is virtually non-existent.[2]

As the results of this study indicate, there is an increasingly complex relationship between mediated practice, social space and power and surveillance, with online media assuming a variety of roles from an extension of offline entities (e.g. institutional websites) and a hybrid of online and offline connections to purely online territories which can simultaneously challenge and sustain offline 'power geometries' (Massey, 1993). Mediated communication (via ICTs and traditional media) assumes a greater role in the lives of certain transnational communities whose ontology is marked by geographic interruptedness and in-betweenness. The not-so-new shift from representation to presentation (or, from response to mediated practice) takes form in diverse ways. The space that opened due to the affordances granted by Web 2.0 is used for anything from everyday mundane tasks to establishing *and* maintaining social and cultural ties with the home and host countries; challenging established power hubs and their representative institutions within a given community; and, engaging with civic activities and citizenship practices (i.e. participation in public affairs, politics and social networking). Thus, in the social context, both the habitus and structural conditions of everyday life are simultaneously reproduced and challenged due to the dialectical relationship between mediation/pervasive mediatization and the contingencies of transnational materialities.

(Re)Mediations of Transnationalism: Online Social Media and the New Face of Surveillant Practice

In transnational contexts, mediated communicative practice and cultural citizenship assume further meaning by way of being a substitute for other forms of inclusion (such as political citizenship and communicative/cultural embodiments of national identity of the host country) to which transnational migrants often do not have access. Communicative action, however, is never singular and unchallenged but rather diverse and always contested within given transnational communities themselves. While the cultural and political space provided by the locale (i.e., city) itself offers little room—at least in central areas—for transnational transgressions of identificatory processes, the communicative space afforded by new communication tools such as online social media provides alternative avenues for representation and civic/cultural engagement.

Individuals and institutional representatives interviewed for this study pointed out that there exists considerable tension fields, inter- and intra-organizations, with certain groups and individuals perceived as power-grabbers and gatekeepers. Some new social formations (such as new associations or online portals, etc.) were named as challenges to such power matrices. Turkiska Riksförbundet (Federation of Turkish Workers), established in 1979, and its youth branch Turkiska Ungdomförbundet (Turkish Youth Federation), established in 1983, are two significant institutions that have taken on representative responsibility and leadership role within the community over the course of few decades. A representative from Turkiska Riksförbundet describes the mission of their organization, among other things, as helping Turkish people get in touch with their social environment in Sweden and learn about the country they live in. She further commented:

> The Turkish people in Sweden, even though not to the extent as those in Germany, constitute a closed group. They only interact with each other but not so much with the larger society. This goes mostly for those living in Stockholm, and most Turkish people live in Stockholm and other big cities. There are not many living in small towns. But those who live in smaller places might be interacting more with other people out of necessity as they don't have as large a Turkish group around them.... They also form groups based on where they come from in Turkey. For example, in Uppsala most people from Turkey are from Kahramanmaras. There are some from elsewhere but very few. Of course, what I say goes mostly for older people who only watch Turkish television and interact with others from their community. The youth are different. Those born or raised here are not very different from the Swedish youth (Female in her 30s).

Parallel to this, feelings of being rejected or discriminated against both by native Swedes and within the diasporic community itself[3] were often expressed during the interviews. Online social practice was pointed to as an alternative outlet, particularly amongst the younger female members of the community, where communication affords various modes of expressivity and where alternatives forms of belonging (or non-belonging through phantasmic online existence) is sought. Online social media sites such as Facebook are commonly used as a meeting space both for existing groups and institutions and new social constellation. At the time of writing, some examples include: 'Isvecli Turkler (Swedish Turks)'; 'Isvec Turkleri (Turks of Sweden)'; 'Turkar i Stockholm (Turks in Stockholm)'; 'Isvec'te Yasayan Turkler (Turks Living in Sweden)'; 'Isvec'teyiz (We are in Sweden)' moderated by the offline Sweden Idea and Culture Association; and, 'A Group for the Swedish Turkish'; constitute some of the popular Facebook groups. In addition, offline institutions such as Turkiska Ungdomförbundet and Turkiska Student och Akademikerföreningen (Turkish Students and Academics Association) have their own Facebook groups.

While the affordances of Web 2.0 and social networking sites allow for the emergence of such new formations, it would be a stretch to argue that power stays offline. Not only does offline power migrate to those online realms in the form of online mirrors of organizations such as Turkiska Ungdomförbundet but often times individuals (mostly younger people) active in offline circles take on visible roles online. Moreover, even in purely online spaces, new power geometries continuously emerge and merge, new tension fields take shape and representation and belonging is incessantly sought on the borderlands of the online and the offline *alongside* the apparent social continuities in the form of power hierarchies and social control.

One informant who took over an existing Facebook group as administrator to share information about/to promote Turkish cultural activities organized in Sweden commented

> When I took over I sent information about this new group to the other existing groups. But one of them[4] seems to delete all the messages I have been sending so I stopped sending them...I think there are those young people in that group who feel themselves as "Swedish" rather than Turkish and perhaps they don't want to be associated with Turkishness or Turkish cultural activities (Female in her 40s).

Strengthening existing ties or meeting new, like-minded others in their environment appear to be a major motivation for 'connecting' online and/or forging new groups. In that regard, online and offline sociality remain very much interlinked and mutually reproduced. The same informant continued

> The reason why I took over this Facebook group was to share information about artistic and cultural activities targeted to the Turkish diaspora. We have a real problem here. It is difficult to find people who are really engaged. Certain activities are organized and then nobody shows up and you can't even make back the money you spent for organizing that activity...I want to increase the number of the members in this Facebook group.

Another informant, a younger member of the diaspora, remarked that she only uses social media sites for personal communication, not for their group function as she does not want to take on an ethnic identity by way of becoming a member.

> After all, you don't need to become a member of those groups to search for information posted within the group...I don't want to take on an identity like that. And also Facebook is not reliable. You can become a member one day and the next day they can post something within the group that you don't agree with. But let's say you haven't checked the pages for ten days...So your name still appears there but you don't agree with that posting. So, I don't join groups that I don't really know (Female in her 20s).

Generational differences and educational, economic and cultural backgrounds constitute factors that influence the choice of media and communicative behavior. As was suggested by a number of the informants interviewed, territorial and institutional belonging remain as significant markers characterizing the communicative habits of the older members of the diaspora. Social control and monitoring between the older and the younger factions of the diaspora and those who stand on the different sides of the cultural/political spectrum (as well as *within* certain sub-groups) also remain prevalent.

Here, I argue that the discourse of surveillance studies and power adds a remarkable diacritic to research on transnationalism and mediated communication to reveal the complexities and new communicative patterns that emerge on the brink of the offline and the online. While, in surveillance studies, a heavy emphasis on top-down monitoring, cameras and "dataveillance" (Clarke, 1994) prevails, in reality a shift toward personalized, horizontal surveillance at the subjective level is most discernible. At the latter level, however, surveillance is much more complex, and invokes a multitude of questions that remain understudied. In the case of transnational communities and in the face of an increase in mediated practices of visibility/connectivity, the inception of new subjectivities, the re/positioning of the self and group identities and the shifting dynamics in the social power geometry (Massey, 1993) constitute areas in need of theoretical and empirical scrutiny. Due to social and spatial constraints, technology adoption takes place rapidly among transnational migrants, and online media use is often entangled with discourses of both freedom and surveillance.

The ways in which surveillance manifests itself (both in terms of scale and form) in the late-modern era shows signs of both continuity and discontinuity with the bureaucratic surveillance practices of the past. A decrease in spatial demarcation due to spatial ambiguity brought by technology and increased mobility stands in contrast to the bureaucratic surveillance and information systems of the early-modern era, and surveillance assumes a social, everyday, character. Yet, as Sassen (2002) argues, and as illustrated in various ways in the above examples, the unprecedented scale of transborder (spatial, cultural, economic) mobility in the late-modern era and deterritorialized aspects of life brought about by global forces run parallel to various forms of territorial anchoredness (spatially, culturally and institutionally) and reterritorialization. Such complex dynamics give way to specific surveillance practices both at the networked and diffuse bureaucratic/corporate and at the individual and communal levels producing new moralities of proximity and new regimes of monitoring and control. The territorially coded character of communal proximity within certain factions of the Turkish diaspora (i.e. formation of migrant neighborhoods in urban areas in Sweden based on the Turkish town of origin) was often brought up by the younger people during the interviews as a potentially restricting factor in everyday sociality—due to close monitoring—and in intra-group dynamics—due to strong in-group out-group sensibilities.

Such corporeal forms of spatial anchoredness and surveillance and social control need to be contextualized in relation to the kind of social ontologies and both offline and mediated practices they lead into. The perception, on the part of both young and high-cultural capital individuals, of online practices as alternative avenues for socialization and group formation where offline power and surveillance can be circumvented (at least potentially) is one such example. Such contingencies lead to both continuity and moments of rupture in social power hierarchies in the most complex sense by way of inevitably bringing in (1) large-scale, corporate surveillance; and (2) peer-to-peer monitoring and control to the social equation. It goes without saying that applications that package peer-to-peer surveillance in its leisurely, social and professional capacity elevate bureaucratic/corporate monitoring to an unprecedented level. In transnational, migrant contexts, social surveillance as an increasingly more personalized and horizontal regime operating at the subjective level gives way to the emergence of new power geometries which take shape between fluidity/mobility and fixity/locality and which are more elusive.

Urry's (2000) "mobilities" adds a secondary trope that makes room for a more nuanced understanding of how surveillance, sociality and power can be conceptualized in relation to the multivalent forms of mobility that are part and parcel of everyday existence. Social negotiation, as a whole, takes place in the midst of corporeal and virtual mobilities[5] with new borders and new proximities continuously emerging. Due to their scale and form, such intertwined mobilities also harbor forms of territorial anchoredness (spatially, culturally and institutionally), resulting in specific surveillance and communicative practices at the individual and communal levels. In understanding how identity, sociality and communal formations take shape between fluidity and fixity in transnational migrant contexts, considering 'mobile dwellings' on the one hand and the materialities of the urban structures and the kind of 'social morphologies' (Vertovec, 1999) such spatialities yield on the otherhand remains key. In the Swedish context, the city has heavier bearing in terms of shaping transmigrant social ontologies than the national. As the experiential realm of everyday existence, the city—and enabling/disabling factors therein—feeds into and from (re)mediations and transnational imaginary. Most of the individuals I interviewed for this study expressed strong feelings about living in Stockholm despite the ethnically segregated structure of the city. In fact, some further pointed out that they would not consider living anywhere else in Sweden as they perceive smaller Swedish towns as embodying a crushing sense of provinciality with no or little room for individual and communal expressivity. One commented that she uses online social media not to connect with those elsewhere but for the very purpose of finding like-minded Turkish people in Stockholm and building offline ties:

> I moved here from Izmir and couldn't really find others I'd like to hang out with through those Turkish associations. Most are from smaller towns in Turkey and they already have their own groups. I felt a bit unaccepted (female in her 40s).

This indicates how, through mediated sociality, everyday corporealities (both situated and distance) and spatially specific forms of sociality (family relations, national/cultural identity) are reconstituted and re-imagined, bringing in yet another set of enabling/disabling factors such as networked surveillance and electronically mediated social control which may or may not be conceded to by the user.

Furthermore, it should be noted that transnational communities have always been more vulnerable to essentialist and essentializing groupism that situates them in clear-cut, reductionist frames in the national imaginary of the host country. Thus, social practices of community formation and cultural expressivity through networked sociality need to be seen in that context. Online mediations allow for the emergence of new groups, such as the Facebook groups, which challenge or reaffirm existing ones. Here, power—engendered through communicative and surveillant practice—manifests itself through stigmatization *or* affirmation of certain groups and rivalry through online visibility in addition to the various offline manifestations of power. Online groups also allow for more seamless interaction and phantasmic appropriations of social identity formations for individuals who do not necessarily wish to subscribe to such identificatory categories by way of becoming group members and 'insiders' as was revealed during the interviews.

Privacy, Networked Sociality and Complicit Surveillance in Transnational Contexts

On the flip side of the coin, social flexibility attained through technology and the specific communicative/expressive habits that seep into everyday routines over time imply significant reformations in the norms that govern overall media use. For one, parallel to the dynamics discussed above, and at the individual level, current practices associated with the self-management of privacy and private information particularly on online social domains point to the rise of a changing understanding of privacy where privacy takes the form of 'personal asset' and symbolic capital. And, changes in the subjective positionalities as regards the norms that underlie such crucial values as privacy have important bearings on how we can approach surveillance and mediation at a conceptual/theoretical level. The ways in which such technologies and practices are adopted at the everyday level (and in specific social contexts) beyond analyses of the macro dynamics that drive social trends, remain lesser explored. As I suggested earlier, there is a multi-layered interplay between the surveillant gaze—intricately embedded in communication technologies (mobiles, email lists, online fora, social network sites, etc.)—and everyday communication routines operating at the level of subjectivity, giving way to a form of *complicit surveillance.*

In terms of the attitudes toward and experiences of information availability/exchange online, what is also discernable is further commodification of public intimacy and expression of privacy by the suppliers and 'objectification/symbolization' of privacy by the users. Particularly amongst the young people interviewed, privacy and the exchange of private information emerge as something to be displayed/made available or restricted/modulated in establishing hierarchies of intimacy, power and symbolic capital (*à la* Bourdieu) within one's circles. In that, the domain of private information is not understood as something to be tightly protected and fought for, but something to be tactfully managed and used toward personal and social ends.

Against the backdrop of in-group/out-group dynamics and social control practices in transnational contexts, there appears to be commonly practiced use patterns toward engendering 'personal enclaves in public domains,' particularly on social media sites:

> I know there isn't much privacy online but if I think about it too much, I can't communicate. So I prefer not to care that much...On Facebook, I created different groups for different categories of friends. Certain friends in Sweden are in one and my cousins are in another and so on and so forth. Therefore, my cousins cannot see the pictures I tagged for my friends. And I block some people from seeing anything or certain things. (Male in his 20s).

Similar preferences were noted by the other informants who took part in the project, pointing to an understanding of privacy in which the issue is not *whether* one's information is accessible, but by *who, to what extent* and *toward what gain or loss.* This brings to mind John Tomlinson's (1999, pp. 160–3) discussion (in reference to Giddens) of the capacity of technology to provide new forms intimacy and the changing attitudes. Here, we are reminded of earlier discussions on social (re)negotiations of notions of intimacy, publicness and privacy as in the example of the use of phones for private conversations (see Tomlinson, 1999, pp. 160–3). Morley (2003, pp. 451–3) also talks about how mobile phones 'dislocate the idea of home' and how the mobile phone call

> ...radically disrupts the physical space of the public sphere in a variety of ways, annoying others with its insistent demand for attention or imposing 'private' conversation on those near its user...what the mobile phone does is to fill the space of the public sphere with the chatter of the hearth...(2003: 451–3).

On that continuum, we could argue that the increase in the scale and the level of social appropriation of mediation and mediatization gives rise to increased infusions of publicness into the intimate/private domain and vice versa. In approaching and conceptualizing current patterns concerning the management of personal privacy and private information, therefore, both affordances inherent in technological applications and socially conditioned use patterns in a given social context need to be accounted for. In the case of social media such as Facebook and the emergent patterns of privacy modulation, there is a dialogical (albeit indirect) relationship between the industry-pushed affordances that enable such modulation (e.g. the creation of personal enclaves and new community constellations and communal identities) and a simulated sense of privacy control, and social norms and practices that originate at the user end which open the door for social control and close monitoring by way of providing flexibility and visibility.

Yet, particularly in the case of social media, the extent of surveillance remains murky despite a sense of control and awareness felt by the users—especially, the younger ones. In that regard, while conceptualizations of surveillance (with regard to social media) as being "participatory" and "empowering," as suggested by Albrechtslund (2008), capture well the non-linear character of social monitoring, a theoretical, analytical treatment of surveillance at a broader level is needed to see it in its entirety *and* with regard to its spatio-temporal contingencies. While it is tempting to see radically transformative values in the use of online media, as Couldry and McCarthy (2004) most correctly note in referring to Appaduarai (1990, cited in Couldry and McCarthy, 2004, p. 15), the historical depth and spatial reach of mediascapes cannot simply be captured within a discourse of sudden transformation and imminent liberation.

What is clear, however, is that most forms of surveillance nowadays do involve the use of personalized technologies and applications and are driven by a variety of social practices (such as online social networking) that are quickly, widely and willingly adopted, thereby necessitating 'involvement' and 'commitment' on the part of the individuals. Yet, the fact that the users themselves classify and stratify their data into various enclaves adds ease and convenience to collation of information, categorization and monitoring practices in the realm of social surveillance. The end result is an evermore complex entanglement of daily social and personal practice, and technology use, which makes it very difficult to differentiate between the various levels of the 'surveillant assemblage' (Haggerty and Ericson, 2000) such as industry-pushed surveillance for commercial gain, state/military intrusions (direct or indirect surveillance), community monitoring of groups/individuals and individually motivated personal monitoring of peers/others. Surveillance remains ubiquitous and diffuse, and social practice increasingly more complicit in the complex forms social control, further complicating power relations and social hierarchies in specific social contexts as illustrated in the case of the Turkish diaspora in Sweden.

While the term "complicit" has a moral tinge and implicates involvement in an act that potentially has incriminating consequences, I use it here to highlight two interlinked aspects of mediated surveillance in its current phase: (1) the increasing primacy of industry-motivated partaking (consensual or semi-consensual) in mediated social practice for both top-down and horizontal surveillance; and, (2) the role of agency in initiating/regulating the level of involvement from the point of entry onwards. In Foucauldian terms, power is explicit in the ability to see. In complicit surveillance power is also derived through the ability to control *what* is seen and *in what form*. In that sense, the panoptic and ubiquitous character inherent in the architecture of the surveillant assemblage is, to a degree, countered *and* blurred by these affordances of choice and reciprocity in the realm of the new applications through which the levels and forms of visibility and connectivity can be seemingly modified by the individual.

Power, Mediation and Reflexivity: Towards a New Understanding?

Ultimately, the discussion here on issues such as surveillance, privacy, identity, power and complicity boil down to the broader questions of agency, subjectivity and the possibility of 'social change' vs. 'reproduction of dominant structures' due to technology adoption in transnational contexts. In addressing these, I would like to briefly turn to a discussion of *reflexivity* (Lash and Urry, 1994; Giddens, 1992) and Bourdieu's (1977) *habitus.* While discussions around 'extended reflexivity' have a great degree of relevance in a social context where mobilities of various sorts and media use in general incorporate generative power, as Adams (2006) notes, dominant conceptualizations of 'extended reflexivity' attribute too significant a role to agency and voluntarism (see Beck, 1986, cited in Lash, 1993 and Giddens, 1990, 1994; Beck and Beck-Gernsheim, 2002; Lash and Urry, 1994). Bourdieu's conceptual scheme of habitus, on the other hand, has been criticized for its overt determinism and for lack of room to conceive social change. For Bourdieu, social structures are maintained/reproduced as a result of the social actions (voluntary and rational) by social agents even if they do not necessarily intend to do so (Calhoun, 1993).

Habitus is "the source of these series of moves [actions] which are objectively organized as strategies without being the product of a genuine strategic intention—which would presuppose at least that they are perceived as one strategy among other possible strategies," (Bourdieu, 1977, p. 73). And, *field,* in which habitus is engendered, refers to the "always existing, obligatory boundaries of experiential context.... We move through different fields but the collection of fields we confront tends to be common for different social groupings such as social class" (Adams, 2006, p. 514). A 'feel for the game' develops as an unconscious competence, and classed identities are reproduced (ibid). Hence, there is room for agency (agency-structure) but fields limit what we can do, thus constraining agency. In this model, the alternatives to choose from are not infinite but quite constrained and even though the actions are committed by the individual, the habitus reflects a common framework internalized by the individual to act in certain ways. As Calhoun (1993) discerns, difference (of eras, cultures, societies) is not a theme addressed in Bourdieu, either. Rather, a tendency of trans-historicity that does not account for specificity governs his analytical approach. (Yet, as Calhoun argues, Bourdieu's analytical apparatus can be used to understand and conceptualize specificity.) Reflexivity, in Bourdieu, *is* a habitus, emerges only in crisis situations (when self-awareness is heightened: e.g. switching from one field to another) and is part of cultural capital. So, reflexivity, in this account, is not autonomous.

Beck's (1986, cited in Lash, 1993) and Giddens' (1990) treatments of reflexivity and reflexive modernity, on the other hand, offer accounts of subjectivity and agency that are, in various ways, at odds with Bourdieu's non-autonomous reflexivity. Beck and Beck-Gernsheim (2002, p. 51) note the diminishing role of power of tradition and social structure and underscore rise of individualized society. Likewise, Lash and Urry (1994) point to the significance of reflexive subjectivity, and Giddens (1994) is known for his theoretical accounts of the ability of the individual [through agency] to affect social structures.

In an effort to charter a course between unbound agency and overbearing structure in understanding the role of subjectivity and social/communicative practice in the context of this study, I would argue in favor of a framework within which to identify how networked sociality and mediated action engender forms of reflexivity in habitus and field, opening the door to the possibility of *social change* in transnational contexts—and not inevitable *reproduction* as in Bourdieu. While the reflexivity I take on board here is more than a mere feature of habitus—as Bourdieu would see it—I would also argue against a view that fetishizes reflexivity. Through the increased use of mediated communication and an increasingly more complex power interplay between the online and the offline, fields are transformed, new 'feels' for 'new games' are created and power is negotiated in new symbolic terms. Continuity and rupture are both possible but need to be looked for in the light of empirical data.

The role of social structures and resources remains crucial, and reflexivity, in many instances, takes the form of expressive capacity, or 'expressive reflexivity.' In the transnational contexts, on which this study is based, while certain forms of power are negotiated (by some and to a certain extent), new destitutions are also created with the partial migration of representation/'produsage' and power onto newer, technological domains—thereby excluding certain others who do not have access to such technologically mediated sociality. Desire (Deleuze, 1992) on the part of the individual to enter into new playing fields to counter various power dynamics offline or online leads to a social picture in which complicity in reproducing social structures and practices (e.g. surveillance) that constrain/segregate/manipulate is also apparent but often conceded and to a degree modulated. Going back to Urry's mobilities, what is discernable in the context of the current study is that (1) movement between identificatory categories and social domains is occurring incessantly and in complex ways; and, (2) some habitus (as in the case of the transnational community studied here) incorporate a heightened sense of self-awareness and a more organic form of reflexivity that is routinized (even if it's not necessarily privileging) rather than occasional.

Final Thoughts

Within the scope of media and communication studies, greater attention has been paid to transnational communities vis-à-vis the role of transnational media (i.e. transfrontier broadcasting) (Naficy, 1993; Miladi, 2006) within the last decade. And, in cyber-culture studies and Internet research, essentializing approaches to digitally mediated communication often regarded online communicative activity as unique, isolated, self-contained, and, more importantly, purely *on*line. Recent research on the social and cultural role of mediated communication within and across diasporic communities in the everyday context of transnational life is revealing new social and communicative patterns, new moralities of proximity and privacy, thus posing new theoretical and methodological challenges.

While newer platforms such as online social networking sites might challenge existing power dynamics and communicative patterns, continuity is also visible. As Husband (2005) points out, there are demographic and economic factors over cultural predictability in influencing minority media (which serve audiences in broad ethnic terms). In other words, while formal characteristics and the nature of content are often attributed to a set of presumed characteristics inherent in the group itself, the use of one form or content over another and lack of diversity observed in traditional media outlets and institutional platforms are often products of material restrictions rather than cultural ones. The emergence of newer domains such as online social media and the flexibility they afford and the accompanying rise of new subjectivities and communal identities undermine the significance of such material scarcities that limit content which Husband makes note of.

With transnational life largely taking place in cities marked by increased mobility and complex forms of mediation, we cannot ignore the persisting significance of place and the mutually inclusive characters of offline and online practice. To avoid reductionist and/or non-contextualized accounts of communicative practice and to give research a clear ontology, I argue that the increasing interplay between individual and social practice on the one hand and the latest shape (re)mediation takes on the other needs be construed in the scope of social change (or the possibility of it) of varying scale and pace—particularly on the subject of transnational communities and mediation—as I attempted to realize in this chapter by way of drawing from transnationalism and diaspora studies, surveillance studies and sociological accounts of reflexivity.

Notes

1. Akin to Axel (2004), I take diaspora to mean a globally mobile category of identification rather than a community of individuals dispersed from a homeland; and, as constituted through a complex web of everyday social practices rather than displacement.
2. To say a few words about the Turkish diaspora in Sweden: there are about 65,000 residents of Turkish origin in Sweden, making them the 10th largest ethnic minority group in Sweden. Turkish influx to Sweden, mostly from rural Turkey, started in the mid-1960s with people coming for employment purposes, and continued until the first half of the 1970s. Turkish immigration to Sweden continues to this date with most people arriving for family reunion. Slightly more than half of the members of this community were born in Turkey, and the remaining are second-and third-generation members born in Sweden.
3. For example, people from a certain region in Turkey having prejudice against those from other towns or regions.
4. Name of the group is not disclosed here for the purposes of confidentiality.
5. Urry identifies five interdependent mobilities: corporeal travel (both daily and permanent); physical movement of objects (sending/receiving); imaginative travel via the images of people/places (the media); virtual travel; and, communicative travel (through messages, texts, phones, mobile, etc.).

References

Adams, M. (2006) "Hybridizing Habitus and Reflexivity: Towards an Understanding of Contemporary Identity?" *Sociology*, Vol. 40, No. 3, 511–528.

Albrechtslund, A. (2008) "Online Social Networking as Participatory Surveillance," *First Monday*, Vol 3, 3–3.

Axel, B. K. (2004) "The Context of Diaspora." *Cultural Anthropology*, 19, 1, 26–60

Bailey, O., Georgiou, M. & R. Harindranath (eds.) (2007) *Transnational Lives and the Media. Re-Imagining Diaspora* London: Palgrave.

Beck, U. & E. Beck-Gernsheim (2002) *Individualization: Institutionalized Individualism and its Social and Political Consequences.* London: Sage

Bourdieu, P. (1977) *Outline of a Theory of Practice.* Cambridge: Cambridge University Press.

Calhoun, C. (1993) "Habitus, Field and Capital: The Question of Historical Specificity," in C. Calhoun, LiPuma, E. & Postone, M. (eds.) *Bourdieu: Critical Perspectives*, pp. 61–88. Cambridge: Polity.

Camauër, L. (2003) "Ethnic Minorities and Their Media in Sweden. An Overview of the Media Landscape and State Minority Media Policy." *Nordicom Review* 2/2003, 69–88.

Christensen, C. (2001) "Minorities, Multiculturalism and Theories of Public Service." in U. Kivikuru (ed.) *Contesting the Frontiers*, pp. 81–102. Göteborg: Nordicom.

Christensen, M. (2008) "Rethinking European Media Landscapes: *D'où venonsnous? Que sommes-nous? Où allons-nous?*" in M. Christensen & Erdogan, N. (eds.) *Shifting Landscapes: Film and Media in European Context*, pp. 109–129. Cambridge: Cambridge Scholars Publishing.

Clarke, R. (1994) "Dataveillance: Delivering 1984'" in L. Green & Guinery, R. (eds.) *Framing Technology: Society, Choice and Change*, pp. 117–130. Sydney: Allen and Unwin.

Couldry, N. & A. McCarthy (eds.) (2004) *Place, Scale and Culture in a Media Age.* London: Routledge.

Deleuze, G. (1992) "Postscript on the Societies of Control," *October*, 59, 3–7.
Ehrkamp, P. (2005) "Placing Identities: Transnational Practices and Local Attachments of Turkish Immigrants in Germany." *Journal of Ethnic and Migration Studies* 31, 2, 345–364.
Georgiou, M. (2005) "Mapping Diasporic Media Cultures: A Transnational Cultural Approach to Exclusion." in R. Silverstone (ed.) *Media, Technology and Everyday Life in Europe. From Information to Communication*, pp. 33–53. Hants: Ashgate.
Georgiou, M. (2006) *Diaspora, Identity and the Media. Diasporic Transnationalism and Mediated Spatialities*. New York: Hampton Press Inc.
Glick-Schiller, N., Basch, L. & Blanc, C. S. (1995) From Immigrant to Transmigrant: Theorizing Transnational Migration, *Anthropological Quarterly*, 68, 1, Jan, 48–63.
Giddens, A. (1990) *The Consequences of Modernity*. Cambridge: Polity Press.
Giddens, A. (1992) *The Transformation of Intimacy*. Cambridge: Polity Press.
Giddens, A. (1994) "Living in a Post Traditional Society," in U. Beck, Giddens, A. & Lash, S., *Reflexive Modernization*, pp. 56–109. Cambridge: Polity Press.
Haggerty, K. D. & Ericson, R. V. (2000) "The Surveillant Assemblage," *British Journal of Sociology*, 51: 605–622.
Haggerty, K.D. & Ericson, R. V. (2007) "The New Politics of Surveillance and Visibility," in K. D. Haggerty & Ericson, R. V. (eds.) *The New Politics of Surveillance and Visibility*, pp. 3–34. Toronto: University of Toronto Press.
Husband, C. (2005) "Minority Ethnic Media as Communities of Practice: Professionalism and Identity Politics in Interaction." *Journal of Ethnic and Migration Studies* Vol. 31, No. 3, May, 461–480.
Kivisto, P. (2003) "Social Spaces, Transnational Immigrant Communities, and the Politics of Incorporation." *Ethnicities* 3:(1) , 5–28.
Lash, S. (1993) "Pierre Bourdieu: Cultural Economy and Social Change," in C. Calhoun, LiPuma, E. & Postone, M. (eds.) *Bourdieu: Critical Perspectives*, pp. 193–211. Cambridge: Polity.
Lash, S. & J. Urry (1994) *Economies of Signs and Space*. London: Sage.
Levitt, P. (2001) *The Transnational Villagers*. Berkeley: University of California Press.
Lyon, D. (2001) *Surveillance Society: Monitoring Everyday Life*, Buckingham: Open University Press.
Massey, D. (1993) "Power-Geometry and a Progressive Sense of Place," in J. Bird, Curtis, B., Putnam, T., Robertson, G., & Tickner, L. (eds.) *Mapping the Futures: Local Cultures, Global Change*, pp. 59–69. London: Routledge.
Miladi, N. (2006) "Satellite TV News and the Arab Diaspora in Britain: Comparing Al-Jazeera, the BBC and CNN." *Journal of Ethnic and Migration Studies*, Vol. 32, No. 6, August, 947–960.
Morley, D. (2003) "What's 'Home' Got to Do with It? Contradictory Dynamics in the Domestication of Technology and the Dislocation of Domesticity," *European Journal of Cultural Studies* 6, 435–458.
Naficy, H. (1993) *The Making of Exile Cultures: Iranian Television in Los Angeles*. Minneapolis: University of Minnesota Press.
Nagel, C. (2002) "Migration, Diasporas, and Transnationalism," *Political Geography* 20:(1), 247–256.
Portes, A., Guarnizo, L. E. & Landolt, P. (1999) "Transnational Communities: Pitfalls and Promise of an Emergent Research Field." *Ethnic and Racial Studies* 22, 217–37.
Sassen, S. (2002) "Introduction: Locating Cities on Global Circuits," in S. Sassen (ed.) *Global Networks, Linked Cities*, pp. 1–38. New York: Routledge.
Smith, M. P. (2001) *Transnational Urbanism: Locating Globalization*. OxfordBlackwell.
Titley, G. (2008) "Media Transnationalism in Ireland: an Examination of Polish Media Practices." *Translocations: Migration and Social Change*, 3, 1, 29–49.

Tomlinson, J. (1999) *Globalization and Culture,* Cambridge: Polity Press.

Urry, J. (2000) *Sociology Beyond Societies: Mobilities for the Twenty-first Century.* London: Routledge.

Vedder, P. & Virta, E. (2005) "Language, Ethnic Identity, and the Adaptation of Turkish Immigrant Youth in the Netherlands and Sweden." *International Journal of InterculturalRelations, 29,* 317–337.

Vertovec, S. (1999) "Conceiving and Researching Transnationalism," *Ethnic and Racial. Studies* 22(2), 447–62.

Vertovec, S. (2001) "Transnationalism and Identity." *Journal of Ethnic and Migration Studies,* 27(4), 573–582.

CHAPTER THIRTEEN

Cosmopolitan Capsules

Mediated Networking and Social Control in Expatriate Spaces

André Jansson

According to social philosopher Lieven De Cauter (2004), networked media technologies operate as a form of capsules—technologies that absorb and isolate people from one another. Expanding upon Michel Foucault's (1967/1998) theory of *heterotopia*, as well as Guy Debord's (1967/1995) critique of the *society of the spectacle*, the media are seen as a machinery of control that captures people's attention and desires, while simultaneously monitoring them through various forms of surveillance. In conjunction with other segregating forces of capitalist society, such as (sub)urban regeneration, the media thus reproduce a dualistic order where encapsulated 'other spaces' are becoming the normal state. This is what De Cauter calls the *capsular civilization*.

De Cauter's theory seems to be at odds with much thinking around cultural 'hybridization,' 'glocalization' and 'cosmopolitanization.' One might say that he stresses the dark side of network society, where networks are not primarily the means for connectivity and exchange, but for separation and dominance. As capsular technologies (e.g., cars, airplanes, computers, gated communities) are the preconditions for making use of networks, a gap opens between those in society who have and those who have not; those who are networked and those who are not. Furthermore, a gap opens between those for whom encapsulation operates as a protective cocoon, a means for distinctive connectivity and mobility, and those who are merely imprisoned and marginalized by the logic of encapsulation. These dualities clearly contradict most prophecies of network society as being the soil for a cosmopolitan society.

But does this mean that capsular civilization is the direct opposite of cosmopolitan society? Are we dealing with an either/or relationship? My point of departure here (see also Jansson, 2009a) is that the logic of encapsulation (or 'capsularization,' as De Cauter more often calls it) can be understood as an integral component of what Ulrich Beck (2002, 2004/2006) terms *cosmopolitanization*. This means that social encounters with 'other spaces' and 'other cultures' typically occur under conditions and within spaces *that are themselves encapsulated*. While many popular forms of media and mobility may open up new horizons of understanding, such 'banal' or 'vernacular' forms of cosmopolitan experience most often evolve as structural 'side-effects' of globalization processes (migration, media flows, consumer culture, etc.) rather than through value change (Beck, 2004/2006). People's life-worlds are pluralized and partly deterritorialized, but the processes are typically *made safe and secure*, and do not necessarily involve a more ethical turn to cosmopolitan*ism*. Rather, they reproduce the gap between groups who desire and enjoy deterritorialized lifestyles, and those who do not (see also Jansson, 2009a).

In this chapter I will discuss the socially embedded role of *networked social media* when it comes to reproducing this distinction. My discussion is theory-driven, exploring a theoretical terrain where cultural sociology, mobile sociology, surveillance studies and media studies intersect. But it also incorporates findings from an ethnographic case study conducted in 2008 within a Scandinavian expatriate community in Managua, Nicaragua. The community consists of professional people working within the development sector—a significant branch of global civil society (Kaldor, 2003), as well as an archetypical cosmopolitan context. The interview data gathered through this study provide valuable support for the theoretical assumption of a stratified glocal logic of cosmopolitanization, in which social media together with other means of communication attain a pivotal role for negotiating the logic of encapsulation.

My study involves two main conclusions, resonating with the introduction of two key concepts. First, I will argue that networked/social media use interacts with the accumulation of *cosmopolitan capital*, understood as a particular type of cultural capital (Bourdieu), shaped under the influence of network capital (Urry), working *simultaneously* as a means of global socio-cultural distinction, and as a counterforce to the logic of encapsulation. Secondly, I will argue that networked/social media use as such contributes to the negotiation of the ambiguous balance between *encapsulation* and *decapsulation* in the glocal setting, involving simultaneously the reproduction of safe, secure and sanitized *cosmopolitan capsules*, and a resource for temporary exit strategies. Before turning to these arguments, however, I will present the empirical context.

Expatriate Spaces: The Case of Development Workers

Studying expatriate groups can bring unique information about how people handle social renewal in general, and what particular significance the media have for maintaining and re-establishing a sense of belonging under mobile life conditions (see Moores, 2006; Moores and Metykova, 2009). Furthermore, among expatriates (as among many other globally mobile groups) the intersection of cosmopolitan and capsular social forces takes on a particularly distinct shape. The concrete expressions diverge, however, depending on positional and situational circumstances. It is worth noting that the community I have analyzed for this chapter belongs to the more exclusive category of 'expatriate professionals'—people who possess a repertoire of cultural, economic and other resources that sets them apart from most other migrants and trans-migrants.

Firstly, these people are engaged in *a voluntary form of mobility*, normally related to a specific organization or company, which means that there are good conditions for solving issues of dwelling and social services, as well as for becoming socially integrated through one's work-place. Secondly, expatriates live abroad for a *measured amount of time*, normally knowing that they have a home and a job to return to in one's country of origin. Thirdly, as opposed to people engaged in longer periods of global traveling, who take up temporary jobs in order to finance further traveling and leisure activities, the mobility of expatriate professionals is less experimental, foremost tied to *career opportunities or other kinds of competence-related incitements*. This, in turn, and fourthly, means that the expatriates' time abroad is normally tied to one particular location in which more or less *permanent living conditions* have to be established.

These conditions together position the 'expatriate professional lifestyle' somewhere in-between global tourism and permanent migration, suggesting that temporary foreign residence is to be regarded as an *exceptional* and *hybrid* condition. While the expatriate period marks a substantial time-space of a person's life biography it is also put 'within brackets' as a sacralized period during which professional tasks and ordinary daily routines blend with more or less extraordinary experiences, notably cultural encounters and events. Hence, the expatriate professional lifestyle is marked by a particular dynamic between the encapsulated security and predictability granted by the professional realm, and the influence of more unpredictable cultural experiences tied to the renewal and negotiation of social practice. The mobilities of these class fractions are also among those encouraged by the economic and political structures of society, and therefore ensured a high level of smoothness by means of efficient transportation, communication and surveillance technologies (see e. g., Graham, 2004; Lyon, 2007; Urry, 2007: Chs. 7, 9; Parks, 2007). Accordingly, many professional expatriate lifestyles, and typically those related to the global business élite, prove to be thoroughly encapsulated affairs, involving few problematic confrontations with the 'other,' and always high potential for *exit mobility* if conditions would get problematic at a particular location (see Urry, 2007, p. 201).

What makes the case of development workers particularly interesting is that the logic of encapsulation is paired with an explicit professional ambition to understand and interact with local populations in general, and socially disadvantaged people in particular. In other words, cosmopolitanism, understood as a *state of mind* (Hannerz, 1990), is an institutionalized precondition for the professional task, while also an expected outcome of the expatriate lifestyle—an outcome that, in turn, might be assigned a direct exchange value when applying for similar positions in the future (see also Eriksson Baaz, 2005, Ch. 3; Nowicka and Kaweh, 2009). The Scandinavian expatriate community referred to in this chapter corresponds in most respects with these characteristics. My six interviewees work for non-governmental organizations (NGOs) within the development sector, which means that they are in close contact with, on the one hand, Nicaraguan everyday society, spending significant amounts of time in 'the field' meeting local interest groups and inhabitants, and, on the other hand, the networks of the global civil society. They are also well-educated and speak Spanish fluently, which sustain their interaction with local citizens. All interviewees have previous experiences of living, working, and studying abroad. Some of them may even be called trans-migrants, working on a contract that is just one in a longer sequence. Together, their work experiences cover most parts of the so-called developing world.

Altogether, these characteristics point to the accumulation of what I further on will define as *cosmopolitan capital*, understood as a particular type of cultural capital (cf. Bourdieu, 1979/1984). It is much less related to economic capital. The latter condition, however, is highly relative, and has to do with the nature of international development cooperation: My interviewees earn little money compared to many other professional Westerners in transitional societies, for instance those within the diplomatic sector. But they are among the more affluent by Nicaraguan standards, affording a lifestyle and material standard that correspond to life in Scandinavia. This also implies that their dwelling conditions, as we will see, almost by definition obeys the segregating capsular logic of most Latin American cities, where the borders between gated middle-class communities and the *barrios* of the poor are sharp, even militarized. In other words, as subsequent discussions will show, as a professional Westerner in Managua it is difficult *not* to become encapsulated in one way or another.

I spent altogether four months in Managua, during a period similar to a sabbatical, while my wife worked for a Swedish development organization operating in Central America. This means that I lived within the expatriate community myself, sharing much of everyday life with people in similar situations, and could get an inside view of the issues I wanted to study. These experiences have been extremely valuable for both the design and the interpretation of the interviews. The selection of interviewees contains three men and three women in various family situations in the age-span of 25–60 years. Most interviews were conducted in domestic settings in Managua and lasted for 90–120 minutes. All names that figure in this text are fictive.

Cosmopolitanism and the Glocal Logic of Encapsulation

During the last decade there has been much theoretical discussion as to the status of cosmopolitanism. Who are the cosmopolitans, and what is the role of the (new) media for sustaining cosmopolitan values? In many of these discussions Ulf Hannerz's (1990) work on 'cosmopolitans' and 'locals' have been used as an important point of departure. In his seminal article 'Cosmopolitans and Locals in a World Culture,' Hannerz (1990) asserts that cosmopolitanism is to be regarded as a particular perspective of the world, or *a state of mind*, dynamically open to new cultural meanings, and willing and able to adapt to diverse contexts. As opposed to 'locals,' who are firmly anchored in pre-established cultural patterns and the social security of a territorialized home-place, 'cosmopolitans' are footloose people, who want to 'immerse themselves in other cultures' (ibid, p. 241).

To become cosmopolitan, then, takes a lot of cultural capital, understood as a broad interpretative repertoire, which can be gained only through education and cultural 'training,' that is, through mobile, real-life experiences of cultural diversity. Few groups live up to these requirements of open-mindedness, global mobility and cultural capital. Hannerz dismisses exiles and labor migrants, as well as tourists and transnational employees, arguing that these groups (and others) actually do not search for much else than an improved home-place, an international career, or temporary experiences of difference and adventure. Hannerz's cosmopolitans are thus exceptional individuals, who combine a universalistic intellectual orientation with a concern for concrete local matters as they travel the world both on and off the beaten track.

In the subsequent debates several commentators have criticized Hannerz's perspective for being elitist, practically setting aside the status of 'cosmopolitan' for only the happy few (who are most often white men from affluent parts of the world, equipped with sufficient amounts of economic and cultural capital) and for invoking a false one-dimensional continuum between 'cosmopolitans' and 'locals.' In response to Hannerz's view there have thus emerged alternative versions of cosmopolitanism, among which one can mention John Urry's (1995) tourism-oriented *aesthetic cosmopolitanism*; Mica Nava's (2002) consumer-oriented notion of *vernacular cosmopolitanism*, and Terhi Rantanen's (2005) discussion of *mediated cosmopolitanism*. The common denominator of these concepts, and others, is that they include a broader scope of glocal experiences as indicators of cosmopolitanism, implying that cosmopolitan individuals to a greater extent may be found also among tourists, urban consumers, and the everyday media user.

While these perspectives provide important insights as to the variety of cosmopolitan processes shaping modern society, they do not, however, speak of cosmopolitanism in terms of the above mentioned *state of mind*. One crucial advantage of Hannerz's original perspective is that his relatively exclusive conceptualization allows for a more critical interrogation into the power geometries through which contemporary cosmopolitanism operates (c. Massey, 1991). The obvious link between cosmopolitanism and cultural capital, which is inherent to Hannerz's view, implies that the expression of cosmopolitanism is also an expression of power of some kind—what we may tentatively term "the power to understand and evaluate transnational, or glocal, processes." Hannerz's sharper definition of cosmopolitanism is thus fruitful if we want to assess how basic cosmopolitan value components such as *cultural openness* and *global engagement* align with the dynamics of social space. As several empirical studies have shown, based on ethnography as well as survey techniques, if cosmopolitanism is defined as a particular state of mind, it proves to be an *ethos*, and indeed an *asset*, that is more likely to be acquired by well-educated people with a more privileged habitus, and only through a process of cultural learning—typically involving experiences from global travel and foreign settlement (e. g., Thompson and Tambyah, 1999; Nowicka, 2007; Olofsson and Öhman, 2007; Kennedy, 2009). It is important to stress here, thus, that positional factors related to cultural, social and economic capital do not *per se* constitute the cosmopolitan persona. Cosmopolitanism, I argue, must be understood as *a particular type* of embodied cultural capital (following Bourdieu, 1986), conditioned by a range of migrant mobilities, commodity flows and transnational mediations, which means that even though habitus is attributed a key role, one must not dismiss the significance of many different, or *discrepant cosmopolitanisms* (see Robbins, 1998).

What is also important to note here, as found in Olofsson and Öhman's (2007) study, is that cultural capital (education) tends to be a more significant factor than economic capital (money), stressing that cosmopolitanism shall be understood primarily as an active state of mind, or, more significantly, *a reflexive state of open-mindedness*, rather than a set of global experiences. This approach resonates with Ulrich Beck's (2002, 2004/2006) distinction between banal, or passive, forms of cosmopolitanization, and the more active enactment of cosmopolitan values in the shape of *dialogic imagination*. The latter mind-set integrates a willingness and socialized aptitude to see 'the other' within oneself, to rediscover the national as the *internalized global* (Beck, 2002, pp. 23, 35–6), which means that cosmopolitanism ultimately is a matter of combining cultural (self-) reflexivity with social recognition.

In my study among Scandinavian expatriates, the relevance of Pierre Bourdieu's analytical framework, as introduced in *Distinction* (1979/1984), proved to be very useful for understanding the structured and structuring logic of cosmopolitanism, pointing especially to the double nature of cultural capital, operating as a resource for both socio-cultural *dis*-embedding and *re*-embedding processes (see also Jansson, 2009b). One of my informants, Erik, a man in his 50s, is a typical representative of the footloose, open-minded people that Hannerz (1990) describes, having lived most of his life under more or less mobile conditions. During Erik's childhood and youth his family moved between several different locations in Sweden, due to the nature of his father's profession, and during his own professional career this pattern has taken on a global nature, predominantly related to work within the cultural and development sectors in Latin America. In the interview Erik told me that during his three decades within the development business he has not substantially altered the material conditions of his lifestyle, but rather reproduced and elaborated those cultural skills that make it easier for him to relate to new environments. This refers to an ability to, on the one hand, learn quickly how to 'find his way' in foreign places, and transcending potential cultural barriers of interaction, while also, on the other hand, identify those people and places who are similar to him, and with whom he may establish a deeper sense of community.

> *Erik:* I remember when I first moved to Peru and when I came back to Sweden and people said "oh gosh, that's so exciting and so different," and then I thought about it and said "no, it's not that different after all." I actually led pretty much the same life as I did in Sweden. It's some type of middle class kind of life with a certain intellectual orientation, kind of. The people you meet on a regular basis are very much people like yourself.

Erik also used the expression of "social cement" in order to explain how cultural habits tie together like-minded people regardless of their national or ethnic backgrounds—a statement that resonates with Bourdieu's understanding of the relationship between taste and cultural capital as a "system of matching properties, which includes people" (ibid, p. 174). This is an important reminder that cosmopolitan identity, as Craig Calhoun (2003) argues, must be understood as a *special sort* of belonging, emerging out of particular social circumstances, rather than as the *freedom from* social belonging. Viewing cosmopolitanism as a reflexive state of open-mindedness, as argued here, does not imply that it is to be regarded as a perspective 'from nowhere' (the universalist standpoint), but that it involves a critical understanding of precisely one's own perspective, and how that perspective relates to the perspectives of other groups and individuals. Erik's quote is an example of how this kind of self-reflexivity operates when the individual has come to terms with the significance of his or her own habitus, as well as the self-transforming role of cosmopolitanism as such (see also Nowicka and Rovisco, 2009, pp. 6–7). In other words, this is a cosmopolitan life trajectory—a trajectory into those realms of the cultural field that transcend and contest the boundaries of the nation-state and other territorial communities.

This is also where the logic of encapsulation becomes a critical issue for the cosmopolitan ethos. Obviously, the disembedding and re-embedding capacity of cultural capital may to a certain extent be described as a process of encapsulation, promoted and legitimized through institutional and administrative structures, ensuring that these cosmopolitan class fractions, when they move to a new destination, have access not only to matching socio-cultural contexts, but also to safe travel routes, beneficial meeting-places, and well-functioning media and communications infrastructures. The appropriation of such disembedding resources blends with the expression of cultural capital in an often seamless manner, implying that expatriate life conditions at a certain point take the shape of isolated enclaves—what we might call *cosmopolitan capsules*. For people working within the development sector the tension-field between these encapsulating forces—which are of a typically glocal nature, motivated as much by overarching professional needs for connectivity, as by very concrete risks related to medical and social problems—and the cosmopolitan ethos, is a subject of negotiation.

> *Michael:* There is a risk that after a while, when you have already been to a number of different places, you no longer have the energy to really appropriate a new setting. Then it's easier to stay within the bubble, especially in a country where you do not speak the local language. [...] You become almost like a lego soldier in the development business. Finally you hardly reflect upon where you are. [...] When I worked in Tegucigalpa I was rarely more than 100 meters away from armed security guards, regardless of where I went during the day. Outside the gym there were four armed guards. Four.

Michael has worked in many different parts of the world, and points to how the logic of encapsulation actually makes it easier to "stay within the bubble" than to engage in the realities of local life conditions. This testifies to Lieven De Cauter's (2004) understanding of encapsulation as a composite structural force: "Encapsulation is the inside shutting itself off from the outside through a convergence of factors:

urbanization and suburbanization, individualization, technologization and, especially under the pressure of the dualization of our society, a sort of *internal migration*: abandonment of the outside space and seclusion, *often out of necessity*, in protected enclaves" (ibid, p. 110, italics added). In a city like Managua, whose layout is markedly decentered, corresponding much to the "archipelago" metaphor used by De Cauter (ibid), with gated communities, megastores, hotels, and leisure centers, marking out the nodes in a vast suburban weave of nameless streets[1], the abstract encapsulation imperative attains a direct local articulation that everyone in the expatriate community must somehow relate to, and cope with, on a day-to-day basis.

> *André:* How is it to arrive here with lots of ambitions, but also having to stay within these enclaves?
>
> *Jens:* It's a big problem of course. And I continuously engage with doubts and feel like a swine. But at the same time there are things that I'm satisfied with. As a family we don't live in a gated community or behind those giant fences in an isolated house, but we live, well not in a barrio, but at least in a place where we can see the problems of the city. [...] Compared to many of the embassy people who are complaining themselves that they don't have any contacts with Nicaraguans. [...] So that is good although I am dissatisfied that we have to visit these enclaves after all.

What we are dealing with here, thus, is a multilayered structure of capsules. While Jens explicitly refers to the socio-material enclaves that structure the metropolitan area of Managua, the implicit production of cosmopolitan capsules, which I successively unravel in this chapter, involves the negotiation and appropriation of a variety of 'means of encapsulation.' The particular combination of such appropriation processes distinguishes the social and spatial trajectories of different cosmopolitan class fractions from one another, as well as from other groups. As we will see in the following discussion, networked media in general, and social media in particular, feed into this glocal logic of structuration, proving that cosmopolitan capsules evolve through a distinctive interplay between online and offline social practices.

Cosmopolitan Capital

As I suggested above, cosmopolitanism can only evolve through a process of socio-cultural learning. It is a view of the world, the dialogical imagination, that operates as a certain kind of cultural capital, and thus corresponds to a certain habitus. But cultural capital, if we follow Bourdieu, is not a unitary asset. The possession of cultural capital *per se* does not sustain glocal awareness and an engagement with 'the other,' but may also correspond to highly provincial value orientations. Thus, based on the above discussions, I here want to develop a notion of cosmopolitanism as a particular form of cultural capital, *cosmopolitan capital*, valid within the cultural (sub-)field of cosmopolitan class fractions. It is thus a resource that works both as a certain kind of interpretative, disembedding and re-embedding *competence*, and as a socially recognized and distinctive *possession* that may be exchanged into social capital at a global scale. Indicators of cosmopolitan capital are, typically, long time experience of international work and travel, involving transitional societies; professional engagement in organizations related to the political and diplomatic sphere or to global civil society; a global network of friends and associates; language skills, and knowledge related to particular cultural contexts and their history (see also Kennedy, 2009).[2]

These characteristics of cosmopolitan capital lead us back to Ulf Hannerz's (1990) cosmopolitan ideal type, confirming that in cosmopolitan society there are certain elite groups who attain greater possibilities than others to conceptualize, adapt to, and govern glocal socio-cultural transitions. At the same time, however, cosmopolitan capital is valid only within a certain, institutionally defined framework. As testified by my informants within the development sector, their competence is more or less useless as an exchange for economic capital in local or national settings. That is also why it is understood as important, as Michael points out, to make regular visits to one's home country, as a kind of "reality check." Furthermore, as Magnus Andersson (2008) found in a study of Bosnian immigrants in Sweden, bilingualism, civic engagement, and lived experiences of cultural globalization, are for many mobile groups not enough for reaching a more central position in the cosmopolitan power-geometry: none of his interviewees had "succeeded in transforming their cultural competence into a corresponding professional position" (ibid, p. 51). So while it is valid to understand cosmopolitanism as a sort of cultural capital, due to its foundation in symbolic competence rather than in material and economic possessions, the accumulation of cosmopolitan capital is obviously not independent from more economic-material types of assets.

Of particular relevance here is what has been called *motility capital* (Kaufmann, 2002), or *network capital* (Urry, 2007). Motility capital refers to people's potential to move, that is, the extent to which they possess the right resources for getting around spatial constraints (Kaufmann, 2002, p. 103). Urry's more recent conception is an explicit elaboration of Kaufmann's ideas, intending to map the ways in which various forms of mobility relate to socio-spatial inequalities. Network capital, then, is "the capacity to engender and sustain social relations with those people who are not necessarily proximate and which generates emotional, financial and practical benefit (although this will often entail various objects or technologies or the means of networking)" (Urry, 2007, p. 197). Urry asserts that the benefits in terms of social connectivity that network capital fosters is "non-reducible to the benefits derived from what Bourdieu terms economic and cultural capital" (ibid), and lists eight elements that in their combination define the particular stratification order of network capital. Among these eight elements are: appropriate documents and visas that enable safe movement; others-at-a-distance who offers invitations and meetings; location free information and contact points, such as email, mobile phones and answering machines; other communication devices, and appropriate, safe and secure meeting spaces (ibid, pp. 197–8).

Clearly, Urry's notion of network capital is acquainted with the concept of cosmopolitan capital. The accumulation of cosmopolitan capital benefits from the kind of cultural experiences that only world-wide mobility and connectivity can provide. It is even valid to say that cosmopolitan capital cannot exist without network capital. However, cosmopolitan capital is also something distinctly different from network capital, since mobility, or the capacity to move, does not automatically generate cultural reflexivity. Many class fractions that are rich in network capital tend to move within precisely the kind of socially disembedded capsules and corridors that De Cauter (2004) speaks about, implying that the 'means of networking' that Urry mentions in many respects operate as the 'means of encapsulation.' Cosmopolitan capital, on the contrary, regarded as a sub-form of cultural capital, involves an outspoken ambition and capacity to resist and counterbalance the forces of encapsulation.

Through this perspective we can grasp the significance of media use for the constitution and expression of cosmopolitanism. Media use relates to the circulation of both cultural capital, in its fostering and expression of cultural values and experiences, and network capital, being a concrete and distinctive enactment of the means of networking. This means, as several media scholars have pointed out (e. g. Tomlinson, 1999; Morley, 2000), that the media rarely hold an independent role in fostering cosmopolitan values, but interact with overarching socio-cultural (power) structures. In an Australian survey Phillips and Smith (2008) could identify "an interaction effect, wherein regular and intensive engagement with multiple sources of global consciousness (Internet, phone, media); regular tourism or business travel abroad and overseas social networks combine with a strongly cosmopolitan outlook" (p. 398). The precondition for this synergy to occur is thus the combined accumulation of network capital and cultural capital—implying that media are reflexively considered and appropriated in their dual capacity of both networking resource and symbolic resource.

Negotiated Encapsulation: Networking and Media Use in Glocal Context

Among the Scandinavian expatriates I interviewed in Managua the above mentioned 'interaction effect' proved to be the key reason to why these individuals often had lost, or were about to lose, contact with many of their friends in their countries of origin. Stina, who was spending her second period in Nicaragua, stressed that most friends back home did not show much interest or understanding as to why she lived in Nicaragua. The lack of understanding was most obvious among those who had neither traveled much, nor lived longer periods abroad themselves, and especially those who had not been "part of the communication" via MSN, email, or Skype. The same experience was described by Jens, who for a long time had been disappointed that nobody asked any questions or showed any interest in his trips, with the few exceptions of friends who were global travelers themselves. Successively he had basically stopped waiting for emails, and besides Skype contacts with close family members, Jens's most regular interaction with the Swedish cultural context occurred through political blogs, written by people with whom he could identify, and sometimes knew.

While this separation process occurs more or less involuntary, it also follows a logic of capital accumulation that echoes Bourdieu's general theory of distinction, as well as Hannerz's more particular distinction between 'locals' and 'cosmopolitans'—a process through which the sense of identification and cultural resonance between friends is successively dissolved. A crucial point here is that the abundant access to media and communication technologies among the expatriates I met did not compensate for the growing discrepancy in terms of global experiences and mobility—quite the opposite. Social media uses boosted the synergy of network capital and cultural capital. As all interviewees stated, global communication occurred mainly within the professional sphere and with other people on the move, rather than with people in their home countries. This shows how technological means of networking contribute to the encapsulation of online territories that converge with the global itineraries of cosmopolitan capital. While more or less ephemeral constructs, such territories attain a crucial significance for the establishment of communities of mobile professionals, feeding a cosmopolitan sense of belonging and connectivity that is distinct compared to most other expatriate or migrant groups (cf. Andersson, 2008; Hepp, 2009; Moores and Metykova, 2009).

What also follows from this discussion is that network capital must not be understood simply as 'media abundance.' While my informants regarded new technological means of networking as necessities for a mobile professional lifestyle and for the maintenance of social relations, the amount of media possessions were in general kept at a minimum, and handled in a manner that stressed *independence* and *control.* In particular, new social media platforms, such as Facebook, were handled with much reflexivity, even hesitation. Erik, for instance, described how he had indeed set up a Facebook account, but also realized that this form of interaction gave him a negative sense of losing control: "Facebook takes away some of the good things with digital media—that I control my filters." Similarly, another informant, Annika, explained that she "used Facebook for a while, just when I left for Nicaragua, when it was also very popular. But since it doesn't lead to any deeper contacts I got very bored with it." She also stressed that she did "not like the content that circulates," and found it problematic to lump together acquaintances from different realms and stages of her life biography within one single space. Sofia, who has lived and worked in a number of different countries before, described that she had almost stopped using Facebook, since many of her close friends and former colleagues in other parts of the world did not use it regularly either, and since the semi-public nature of Facebook created an unpleasant sense of dependence and social surveillance:

> *Sofia:* I found it very hard to publish anything on that Wall, and that I should be able to read what everyone was doing—while at the same time I found it quite fun of course to see what people at home [in Sweden] were doing and see what they were talking about, which they didn't post to me, but which I could check anyway. So I was in two minds for a start and started restricting my profile and included only those I knew well, and that not just anybody should be able to enter and see pictures, or that others should be able to publish pictures of me.

These reserved attitudes to Facebook highlight three important features of cosmopolitan capital. First, they show that cosmopolitan capital does not necessarily imply a particularly extensive or extravagant mode of media use. Even though many cosmopolitans are in charge of large amounts of network capital, this primarily denotes a *potential* for mobility and connectivity. My informants do indeed possess both the technological means and the cultural skills for becoming sophisticated Facebook users. But in this case there were also other factors that led them to a more restricted mode of appropriation, especially the perceived risk of getting dependent or exploited.

Secondly, we can once again see that a key component of the cosmopolitan ethos is the *resistance to encapsulation*, in the above example pointing to the 'digital enclosure' (Andrejevic, 2007) generated by Facebook—that is, the platform's dual significance as a means of networking and a means of encapsulation, or surveillance. Cosmopolitan capital, then, includes practical formulas for appropriating and using various means of networking in ways that do not reproduce the encapsulation processes of everyday life, and for making decisions that create opportunities for *decapsulation*. Typical avenues of decapsulation are de-mediated field trips or private excursions; participation in non-professional local events and communities, and the annual visit or weekly Skype call to friends and family in one's home country.

Thirdly, the attitudes of my informants suggest that cosmopolitan capital is typically linked to particular *modes of cultural distinction*. While the ethos of cosmopolitanism constitutes an embodiment of cultural openness and social concerns, it is also, and particularly within the private realm, objectified through tastes and symbolic markers that correspond to cultural status positions. The quiz shows, jokes, and film-clips that are circulated through Facebook are not surprisingly classified as a kind of standardized mass culture, even waste, that invades both time and space. In a similar manner, the distaste with the very mixture of social acquaintances, agglomerated in a standardized list of Facebook 'friends,' shows that there is a distinct cosmopolitan identity, foremost related to the professional realm, that must be continuously maintained.

These three points, in sum, reveal how cosmopolitan capital conditions a potential to resist encapsulation, while simultaneously generating socio-cultural processes that necessitate new forms of distinction and encapsulation. The cognitive and emotional ambiguities regarding Facebook, then, are somehow parallel to the ambiguities regarding the encapsulated enclaves of Managuá. While new means of networking (whether communication devices or meeting facilities) are considered a direct necessity for the appropriation of cosmopolitan capital, there is also a certain tipping point (depending on individual life conditions) where the accumulation of such resources is experienced as a coercive structure of encapsulation. Such experiences of dissonance and unease catalyze a particular form of cosmopolitan reflexivity; typically the desire to engage in decapsulating practices, such as going to places less affected by encapsulation and mediatization processes. One further consequence (and indicator) of this ambiguity is that one single communication device often occupies multiple, or ambivalent, functions within the everyday articulations of cosmopolitan capital:

> *André:* When did you get your mobile?
>
> *Stina:* During my first week here, maybe after four days. And it feels like a relief. I can manage without a mobile when I'm travelling, if I would go to Costa Rica then I'm not bringing my mobile, or when I travelled a couple of months in Argentina. But this is my stable point and I'm working and I think it brings security to both me and people around me. [...] Locally I'm texting quite a lot, like "what happens this week?," "going to the movies?"—as a coordination tool. [...] I'm also taking security measures. I call my landlady when I jump into a taxi late in the evening, or I make a fake call saying that "I'll be home in ten minutes" so the taxi driver can hear it.

Stina's story adds to a more localized, or *trans-local*, understanding of the negotiated encapsulation process (see Hepp, 2008, 2009). She is the youngest person in my selection, who travels alone and attains fewer professional benefits, including network capital. This also means that her everyday life is less safe, secure and frictionless; for instance, she has to use public transport, and visit Internet cafés for maintaining important local and global connectivities within the private realm. At the same time, this involves a freedom that she appreciates and entertains, both in the local neighborhood, which she is more integrated in than my other informants, and when traveling to other countries, sometimes off-the-beaten-track and off-line. Then, in Stina's case the structural pressure of encapsulation is relatively light, and, as told by the extract, a communication device such as the mobile telephone attains a more flexible role. It may operate *at the same time* as a *means of encapsulation* and as a *means of decapsulation*. At one level, the mobile may provide the sense of security that typically marks encapsulated lifestyles; particularly the fact that she is connected to other people within the trans-local (expatriate) community. At another level, it is precisely this sense of security that makes it easier for her to decapsulate, turning selected parts of the public realm into the significant 'somewhere' localities (Hepp, 2009), or 'interspaces' (Urry, 2007, pp. 176–7), that structure the mobilities and bind the nodes of her everyday life. In this way, the simulated phone call from inside a taxi in the streets of Managua synthesizes the ambiguous ethical composition of cosmopolitan capital, as well as the shifting socio-spatial role of communication technologies—be they cars or mobile telephones.

Conclusion

This study has generated a problematized and situated view of Lieven De Cauter's encapsulation thesis. Analyzing a particular cosmopolitan class fraction my study has shown, on the one hand, that encapsulation is indeed an overarching segregating force adhering to the glocal characteristics of network society, but also, on the other hand, that it is a force that meets social resistance on behalf of ordinary citizens. I have thus suggested that cosmopolitanism, understood as a reflexive state of open-mindedness, integrating cultural openness and global engagement, functions as an ethical counter-force to capsular civilization. While cosmopolitanism (as conceived here) celebrates cultural difference, it resists segregation and conservation: cultural encounters and re-amalgamations are considered as the normal state of society (in contrast to 'multiculturalism'). Nevertheless, as also demonstrated, the overarching trends of capsularization and cosmopolitanization are thoroughly intertwined, implying for instance that the institutionalized versions of cosmopolitanism found in the context of international development aid both *require* a certain level of exclusivity and protection in order to get exercised through professional praxis, and *produce* an exclusive sense of cultural community that in itself holds an encapsulating potential. Thus, I argue, potentially encapsulating forces such as mediated networking and surveillance, do not *per se* involve a threat to the spread of cosmopolitanism in society, but must be judged in relation to their contexts of implementation and appropriation.

In this connection my study has shown how a diverse ensemble of social media uses are entangled with the negotiation of encapsulation processes, globally as well as (trans-)locally, transcending the boundaries of online and offline spaces. The complexity of these patterns can be illuminated and understood only through a close, situated analysis of the particular socio-spatial ambiguities characterizing the actual type of cosmopolitan class fraction, as these ambiguities unfold in the glocal context. Having undertaken such an analysis among Scandinavian expatriates in Managua, my overarching point is that the most important key for reaching valid conclusions as to the multiple functions of social media is the *constitution and logic of cosmopolitan capital*. The concept of cosmopolitan capital brings us back to the fundamental question of socio-cultural reproduction, pointing to the structured unevenness in terms of cultural and technological resources, but also to the prospects for social change by way of cultural mobility and learning.

Notes

1. In Managua there are just a few main streets and roads that have names.
2. Paul Kennedy (2009) has recently discussed the role of cosmopolitan capital in a study of middle class transnational professionals in Manchester, pointing to patterns very similar to those outlined in this chapter. However, he does not provide a theoretical definition of the very concept of cosmopolitan capital.

References

Andersson, M. (2008). The Matter of Media in Transnational Everyday Life. In Rydin, I. & Sjöberg, U. (Eds.) *Mediated Crossroads: Theoretical and Methodological Challenges*. Göteborg: Nordicom.

Andrejevic, M. (2007). *iSpy: Surveillance and Power in the Interactive Era*. Lawrence: University Press of Kansas.

Beck, U. (2002). The Cosmopolitan Society and Its Enemies. *Theory, Culture and Society*, 19(1–2), 17–44.

Beck, U. (2004/2006). *The Cosmopolitan Vision*. Cambridge: Polity Press.

Bourdieu, P. (1979/1984). *Distinction: A Social Critique of the Judgement of Taste*. London: Routledge.

Bourdieu, P. (1986). Forms of Capital. In Richardson, J. (Ed.) *Handbook of Theory and Research for the Sociology of Education*. Westport, CT: Greenwood.

Calhoun, C. (2003). 'Belonging' in the Cosmopolitan Imaginary. *Ethnicities*, 3(4), 531–68.

Debord, G. (1967/1995). *The Society of the Spectacle*. Cambridge, MA: Zone Books/MIT Press.

De Cauter, L. (2004). *The Capsular Society: On the City in the Age of Fear*. Rotterdam: NAi Publishers.

Eriksson Baaz, M. (2005). *The Paternalism of Partnership: A Postcolonial Reading of Identity in Development Aid*. London: Zed Books.

Foucault, M. (1967/1998). Of Other Spaces. In Mirzoeff, N. (Ed.) *The Visual Culture Reader*. London: Routledge.

Graham, S. (2004). The Software-Sorted City: Rethinking the 'Digital Divide.' In Graham, S. (Ed.) *The Cybercities Reader*. London: Routledge.

Hannerz, U. (1990). Cosmopolitans and Locals in a World Culture. *Theory, Culture and Society*, 7(2), 237–51.

Hepp, A. (2008). Translocal Media Cultures: Networks of the Media and Globalisation. Paper presented at the *International Communication Association Conference*, Montreal, Canada, May 22–26.

Hepp, A. (2009). Localities of Diasporic Communicative Spaces: Material Aspects of Translocal Mediated Networking. *Communication Review*, 12(4), 327–48.

Jansson, A. (2009a). Cosmopolitanization or Capsularization? Making Sense of the Mediatization of Belonging. Paper presented at the 19th *Nordic Conference for Media and Communication Research*, Karlstad, Sweden, August 13–15.

Jansson, A. (2009b). Mobile Belongings: Texturation and Stratification in Mediatization Processes. In Lundby, K. (Ed.) *Mediatization: Concept, Changes, Consequences*. New York: Peter Lang.

Kaldor, M. (2003). *Global Civil Society: An Answer to War*. Cambridge, UK: Polity Press.

Kaufmann, V. (2002). *Re-thinking Mobility: Contemporary Sociology*. Aldershot: Ashgate.

Kennedy, P. (2009). The Middle Class Cosmopolitan Journey: The Life Trajectories and Transnational Affiliations of Skilled EU Migrants in Manchester. In Nowicka, M. & Rovisco, M. (Eds.) *Cosmopolitanism in Practice*. Farnham: Ashgate.

Lyon, D. (2007). *Surveillance Studies: An Overview*. Cambridge: Polity Press.

Massey, D. (1991). A Global Sense of Place. *Marxism Today*, June 1991, 24–29.

Moores, S. (2006). Media Uses and Everyday Environmental Experiences: A Positive Critique of Phenomenological Geography, *Particip@tions*, 3(2).

Moores, S. & Metykova, M. (2009). Knowing How to Get Around: Place, Migration and Communication. *Communication Review*, 12(4), 313–26.

Morley, D. (2000). *Home Territories: Media, Mobility and Identity*. London: Routledge.

Nava, M. (2002). Cosmopolitan Modernity: Everyday Imaginaries and the Register of Difference. *Theory, Culture and Society*, 19(1–2), 81–99.

Nowicka, M. (2007). Mobile Locations: Construction of Home in a Group of Mobile Transnational Professionals. *Global Networks*, 7(1), 69–86.

Nowicka, M. & Kaweh, R. (2009). Looking at the Practice of UN Professionals: Strategies for Managing Differences and the Emergence of a Cosmopolitan Identity. In Nowicka, M. & Rovisco, M. (Eds.) *Cosmopolitanism in Practice*. Farnham: Ashgate.

Nowicka, M. & Rovisco, M. (2009). Making Sense of Cosmopolitanism. In Nowicka, M. & Rovisco, M. (Eds.) *Cosmopolitanism in Practice*. Farnham: Ashgate.

Olofsson, A. & Öhman, S. (2007). Cosmopolitans and Locals: An Empirical Investigation of Transnationalism, *Current Sociology*, 55(6), 877–95.

Parks, L. (2007). Points of Departure: The Culture of US Airport Screening. *Journal of Visual Culture*, 6(2), 183–200.

Phillips, T. & Smith, P. (2008). Cosmopolitan Beliefs and Cosmopolitan Practices: An Empirical Investigation. *Journal of Sociology*, 44(4), 391–9.

Rantanen, T. (2005). *The Media and Globalization*. London: Sage.

Robbins, B. (1998). Comparative Cosmopolitanism. In Cheah, P. & Robbins, B. (Eds.) *Cosmopolitics*. Minneapolis: University of Minnesota Press.

Thompson, C. J. & Tambyah, S. K. (1999). Trying to Be Cosmopolitan. *Journal of Consumer Research*, 26, 214–41.

Tomlinson, J. (1999). *Globalization and Culture*. Cambridge, UK: Polity Press.

Urry, J. (1995). *Consuming Places*. London: Routledge.

Urry, J. (2007). *Mobilities*. Cambridge: Polity Press.

CHAPTER FOURTEEN

Reconfiguring Diasporic-Ethnic Identities

The Web as Technology of Representation[1] and Resistance

OLGA G. BAILEY

INTRODUCTION

For ethnic and diasporic groups the World Wide Web has the potential to create new forms of solidarity and, sometimes, forms of resistance[2] that otherwise might not be possible in the off-line world. Websites have become spaces of inclusion, participation, political activism, and create a sense of belonging[3] for many minority groups. Although we cannot forget that it is also a platform for racism and hate speech towards different minority groups. It seems that the online world has become a political space (Castells, 1997, p. 311) for innumerable social groups. Since many ethnic and diasporic groups are struggling for political and identity recognition, as well as for social justice, an analysis of how these groups represent themselves in online territories is an important aspect in understanding their participation in the off-line world. Obviously ethnic and diasporic groups do not necessarily use communication technology in a similar or more advanced way than anybody else. They appear to use it according to their needs and thus, in some ways, socially reshape it. This has resulted in new technological and cultural practices and politics, which are instrumental in new formations of ethnicity (Gillespie, 1995, p. 10). These new identity formations are possible because the online territory is not a single object but an unevenly distributed and complex set of interrelated technologies with multiple outcomes at differential spatial scales. In other words, a space which is produced, and which in turn 'produces subjectivity, folded in the off-line world (Crampton, 2003,p.6) suggesting a mutual process of production between physical space and abstract space, as a series of relations and as a process of becoming' (p. 12). Castells proposes that the Internet and, by extension online territories, are conceptualized as continuous with society, 'an extension of life as it is, in all its dimensions, and all its modalities' (2001, p. 118). For him, while the Internet has been appropriated by social practices in all its diversity, at the same time this appropriation does have a specific effect on social practices itself.

This chapter addresses the question of how ethnic and diasporic groups are representing their identities online. The aim is to uncover the various ways that migration, transnational connections and the web are redefining notions of home and contesting existing notions of belonging. By making cultural identifications and connections across borders they might be generating new forms of politics of resistance and solidarity. That is, by looking at diasporic identity representation on the web it might be helpful to examine how ethnicity has been used as a basis for organizing and challenging various forms of exclusion, and how ethnic and diasporic groups are crossing, contesting and reconfiguring local borders in terms of cultural belonging (Tastsoglou and Dobrowolsky, 2006). However, concomitantly to the politics of representation and solidarity, it might be possible to find that the representation of diasporic and ethnic groups online have been appropriated by online commercial enterprises in a conservative fashion where ethnic 'life style' becomes a mainstream commodity.

Diasporic Identities

In multicultural western societies, there has been a systematic exclusion of diasporas,[4] both ethnic and religious minorities, from the public sphere, including the mainstream media. Their misrepresentation, under-representation, and invisibility have generated an ongoing struggle over meaning between those at the periphery of society and the media. A great number of these groups have reacted by producing their own 'alternative media' including the use of the Internet. This is the result of processes of migration, of media misrepresentation of minority groups, and of changes in the media landscape mostly generated by new communication and information technologies.

Diasporic identities are mostly negotiated in the convergence of different cultural influences and constrained by different power structures. Their experiences are lived 'outside' and 'inside' a 'diaspora space' which is constructed by several axes of differentiation and inequality—nationality, class, gender, ethnicity (Brah, 1996). In many cases, they face discrimination, antagonism, celebration, as well as 'internal-group' pressures to resist or/and comply with a defined 'cultural identity.'

A useful starting point to situate diasporic subjects is suggested by Brah's definition of diasporas as networks of transnational identification encompassing 'imagined' and 'encountered' communities, or as 'multi-locationality' within and across territorial, cultural and psychic boundaries (1996, pp. 196–7); and Appadurai (1996) uses 'ethnoscapes' to speak of social formations connecting people living in different parts of the world despite the location of home. In these terms then, diasporic groups construct their ethnic communities locally as well as participating in various transnational encountered and imagined communities. They network with others in permanent or temporary alliances in a changeable and liminal zone that becomes 'home'; a space where cultural affinities and differences are constantly negotiated (Tastsoglou, 2006, p. 202).

'Diaspora' here is re-imagined within the context of the online world, the political nexus of building new identities and issues of participation and inclusion, both relating to senses of belonging. Online territories might provide experiential mediations that strengthen diasporic's understanding of their 'multi-locationality,' that is, their position in existing relations of power, a simultaneous situatedness within gendered, spaces of class, racism, ethnicity, sexuality, age' (Brah, 1996, p. 204). Marfleet (2006), while discussing migration, suggests that we should think of transnational communities of diaspora as 'networked communities.' For him the development of new technologies of communication has been fundamental in the advance of transnational communities. In the same vein, Appadurai (1996), suggests that everyday subjectivities are been transformed by the construction of 'public sphericules' arising from participation in the different spaces of online territories. These 'public sphericules' are constituted beyond the singular nation-state, 'as global narrowcasting of polity and culture which provide not only entertainment but, potentially, counter hegemonic views of current affairs and a proactive agenda of positive intervention in the "public sphere" (Cunningham, 2001, p. 133). Public sphericules as mediated spaces are defined by the identities of their audiences and might challenge essentialist notions of community. Online diasporic public sphericules are permeated by local and global forces and conditions thus creating one of the many 'heterogeneous dialogues' related to globalization (Appadurai, 1996), and becoming part of 'a complex form of resistance and accommodation to transnational flows' (Howley, 2005, p. 33). In this sense, ethnic and diasporic groups may be at the forefront of political innovation and social change.

It is argued that the 'articulation' and formation of cultural affinities or politics of belonging by the diasporic subject can perhaps be understood within the context of their online experiences: how the online space facilitates diasporic subjects to foster senses of belonging; how they intersubjectively organize and represent or perform identification (Anthias, 2002, p. 502) and develop a social agenda that leads to civic engagement and political participation. On the affective level, the Internet might alleviate the difficulties and challenges imposed by international migration, with all the internal and affective transformations on the self that such process entails, by offering a space where people can find others with common experiences.

Research elsewhere (Tastsoglou, 2006) indicates that multiple and overlapping spatial and symbolic attachments of various degrees of complexity are the rule among, for example, immigrant women living in Canada who demonstrate multi-dimensional geographies of belonging and citizenship, involving political and cultural practices, at the local, national and transnational levels. Such attachments and practices are 'eclectic, synthetic, syncretic, hybrid and multi-directional (as opposed to bi-directional ethnic transactions)' (p. 205). As such Tastsoglou (p. 213) suggests that they are 'cultural entrepreneurs,' they develop a 'double consciousness' (Werbner, 1997) of their diasporic state and the ability to build and experience a syncretic and hybrid culture thereby creating new cultural syncretism (Papastergiadis, 2000, pp. 125–7) or hybrid states of mind and living (Hall, 1992, p. 310). The specificities of migratory experiences cannot be generalized but the example above allows us to point out that migrants can be active agents of their lives even when living under difficult circumstances.

The Internet thus, becomes an important space for ethnic and diasporic groups to exercise new forms of agency. Following Mitra's (2008) argument that the Internet is not just a technological innovation but a discursive formation that has taken on a global scale, allows us to recognize that the 'space' of the Internet is constructed from discourses. The interconnection of discourses can produce a 'sense' of place that has no relation to the characteristics of traditional, material places where immigrants might feel threatened. The online territory as a 'discursive comfort zone' (Mitra, 2005, p. 383) becomes significant to diasporic groups as it provides opportunity to people articulate their voices and multiple identities which otherwise they probably would not be able to express elsewhere. Discursive zones then create a virtual place where people interact and have experiences of crossing cultural and imaginary borders, creating perhaps a sense of belonging; the psychological dimension of citizenship (Tastsoglou, 2006) or 'emotional' citizenship' (Bernal, 2006).

> By doing so, diasporic groups concomitantly might 'reaffirm traditions; challenge the hegemony of the dominant culture; reaffirm allegiance and belonging to multiple countries thus transgressing the boundaries of the nation-state; and engage in citizenship practices locally that derive however from transnational levels (Tastsoglou, 2006, p. 209).

Transnational Digital Spaces and Diasporas

In the light of the established literature about Internet and political participation, this chapter avoids a speculative celebration of the possible role of the Internet, and adopts a realistic approach. After all, when discussing diasporic and ethnic groups' participation in the online world, it could be argued that information and communication technologies are more likely to benefit those already advantaged with regard to income, education, technical skills and political links, which are important elements of the practice of participation. Considering the diversity of existing ethnic and diasporic groups, it can be argued that some of them are cosmopolitans[5] and well educated with the necessary social and cultural capital to participate in the online world, while many others are from poor areas of the developing world where these opportunities are likely to be unavailable to most of them. This resonates with Massey's argument that the geometries of communication and mobility reproduce power geometries, where the control of flows and connectivities are held by those possessing economic and cultural capital (Massey, 1991).

Nevertheless, the Internet as a new form of 'mass self-communication' (Castells, 2007) has created many possibilities for social actors (such as ethnic and diasporic groups) to counter, challenge and eventually change the power relations institutionalized in society. In many parts of the world, identity—gender, religious, national, ethnic—has become 'a political alternative project of social organization and institution building' (p. 249). Commentators such as Hepp (2009, p. 153) suggest that digital media are important in the process of articulating diasporic groups as these media facilitate connectivity across boundaries in an 'individualized networking.' In these networks diasporic and ethnic subjects forge a cultural identity while re-appropriating aspects of the home identity.

The issue of symbolic representation in the media/Internet is of paramount importance in the struggle of these groups to assert their voices in public spheres in order to alter the established values and interests in society. More precisely, because power relations are reproduced in the realm of socialized online communication, many ethnic and diasporic groups get involved in this space/ process of transnational communication to participate in 'the battle over minds,' that is, over meaning construction (Castells, 2007, p. 249) thus building their autonomy and potentially confronting the *status quo* of society. However, the engagement with the online space does not preclude the existence of minorities groups' struggles at the local level in face-to-face interactions. Their political practices, including other types of politics considered more ordinary and everyday, are embedded in the local, not in a virtual space, but in a combination of the 'spaces of flow' with the 'spaces of places.' In the digital age, the struggles of different social groups seem to have avoided isolation in 'the fragmented spaces of places and reached the global space of flows, building 'networks of meaning" (Castells, 2007, p. 250).

The use of technologies by minority groups is nothing new and does not seem to be any different with the Internet. Unfortunately there are far fewer studies detailing what minority groups do when online. In common with other people around the globe, diasporic and ethnic groups are appropriating Internet technologies selectively for specific reasons and aims. The available literature indicates that the Internet and social networks are generating new communication spaces and practices and creating new discursive communities as well as a variety of minority politics online—with distinct political stances—including commercial enterprises sites. The repertoire of these politics ranges from offering self-empowerment to identity construction processes, to information dissemination, that could be seen as a strategy to address the uneven politics of information of multicultural societies (Siapera, 2005, p. 16). In a similar line, while reflecting on the ways the Internet is used by British-born Chinese people, Parker and Song (2007) note that the increase in Chinese websites has facilitated greater connection between them and, most importantly, generated discussions about their experiences of hybrid identities, about issues of belonging and citizenship, thus helping in the construction of a collective identity. The sites analyzed by them—The British Chinese Online and DimSum—were found to 'encourage public intervention on the part of a group which has been severely under-represented in established political organizations' (p. 1057).

In this context, many ethnic and diasporic minorities can now voice their stories, problems, and achievements to a wider audience and get a response from others. In this sense it is a space for individuals and groups to exercise discursive power (Mitra, 2001) and connect many voices—local, national and transnational. In fact, Bernal contends that the Internet is the 'quintessential diasporic medium, ideally suited to allowing migrants in diverse locations to connect, share information and analyses, and coordinate their activities' (2006, p. 175). For her, we should rethink the argument that transnationalism and the Internet are creating a borderless world, and rather look at how they perhaps allow for a 'reconfiguration or remapping of boundaries, so that, for example, what might once have been outside the margins of the nation is now more effectively included with a larger framework of imagined community' (Bernal, 2006). Online spaces make problematic existing dichotomies such as center/periphery, inside/outside, majority/minority manifested in the off-line world. The online world then, has facilitated the cultural and political activities of many ethnic and diasporic groups, and their self-representation on the web might help to change prevailing constructions of ethnicity simply through its pervasiveness as a medium, suggesting that it contains many possibilities and diversity.

However, there are reservations to this point of view given that many ethnic minorities and diasporas have engaged with different technologies, but some are still under-represented and socially and politically excluded. The argument overlooks inequalities of access, uses, lack of technological skills and civic competencies.

Diasporic Representation Online

Online representation of ethnic and diasporic groups appears to be, in many cases, a political practice. Although the use of the Internet differs among ethnic and diasporic groups, it seems some groups use the Internet mainly for entertainment, to communicate with families and friends, while others use it for research or political purposes or a combination of them. In any case, the content of web pages and blogs, for example, is quite subjective—as opposed to the objectivity ethos of journalistic narratives—as it aims to express the concerns, practices and politics of a particular group. The issue of representation has two aspects: representation produced by mainstream media/online and self-representation using alternative media/online that potentially become texts and practices of resistance. Moreover, there is the commodification of ethnicity in some alternative commercial sites, which represent ethnic groups according to their marketing interests—particularly of collective identities. Self-representation online is a form of resistance to normalization because it is where diasporic groups work on themselves in a process of becoming a collective identity. It would seem unlikely that they could work on themselves as a cultural identity project if they were completely dominated by power (Weberman, 2000). Usually the object of diasporic resistance is to problematize something not perceived before as a problem, in order to allow new sets of knowledge and also new ways of being and new practices. However, these ways of being (diasporic) must already include the possibility of problematization. Thus problematization is an important part of resistance. In that way, resistance is an ongoing process (Crampton, 2003, p. 104).

The representation of diasporic groups online is produced by a combination of narratives that problematize issues that epitomize a particular time and history of the group or individual. They might be useful in understanding the dynamics of transnational communication as well as in charting the instabilities of identities and the elusive construction of home and community in the postmodern age (Bernal, 2006, p. 661).

In terms of access, in many ways, the web represents the existing inequalities between different ethnic groups in post-colonial societies. The Internet is perceived as both an instrument for potential empowerment as well as for established privileges (Leung, 2005, p. 33); on the one hand, many diasporic and ethnic minority groups are empowered to produce texts—self-representation—that can shape online spaces according to their needs and thus contribute to socially re-shaping the Internet, It also has potential for capacity building, and enhancing participation in alternative public spheres.[6] On the other hand, the invisibility and misrepresentation of ethnic groups—black/Latin/Asian—existing in the mainstream public sphere has its own manifestation in the online world. The online space hence should not be understood primarily as liberating for minorities groups or for its potential to end discrimination. It is crucial to recognize its potential to reproduce racism and prejudice that co-exist in the everyday of both online and off-line territories. For instance, when ethnic diversity is not made invisible, it is severely controlled through 'cybertypes,' racially stereotypical representations that serve to define white identity by what it is not. Indeed Zurawski (1998, quoted in Leung, 2005, p. 95) claims that inherent in the concept of the global village, which the Internet is said to epitomize, is the constraint of ethnic difference to the level of food, folklore, and traditional dance and music.

Diasporic Practices Online

The concern here is to illustrate the arguments about possible reconfigurations of diasporic and ethnic groups' identity online rather than to exhaust the online representational practices of diasporic and ethnic groups' which in itself might be an impossible task. I examine how self-representation is expressed and articulated through notions of identity, home, and belonging on diasporic online territories that might construct, de-construct, contest, revise and shift ethnic signifiers connected to the diasporic experience. The cases were chosen for their distinctive representation of home and belonging, the ways they might suggest forms of 'banal' resistance as well as comfortable 'discursive zones'; one is based on primary research on African diasporas online sites, and the other is a case discussed by Mallapragada (2006) on Indian-American web.

AfricanOz.com.au

This portal site was funded in 2004 by people of African and Australian origins. It is independent, although it offers advertising space to anyone with business interests in Africa. The content reflects the work of various contributors mostly from Australia. The aim is 'to improve community access to the web, showcase African-Australian arts and culture for everyone to enjoy, and raise the profile and understanding of Africa in Australia' (Africanoz.com.au). The site's policy is one of open access to, although the content is primarily of interest to the African diasporas in

Australia and the business public with whom they want to trade. The site declares to have a broad-based audience that includes Australians interested in Africa, African-Australian communities and people from media, business, government, community and cultural sectors. As a portal site, it offers diverse information on events, music, and opportunities for career development, study, and issues relevant to the community, a directory of websites and other links and resources. One of the sections 'community notices' presents jobs opportunities for the locals.

The news section brings news about Africa, published largely in Australia newspapers, but not limited to them, and has a quite visible link to the BBC's News Africa. The selection of news might contribute to shape a particular view of the world for the diasporas and perhaps helps to generate a space of communication where socio-cultural representation and collective identities are produced. More precisely, the site seems to find stories that have the potential to bring together the diasporic audience as both African as well as someone geographically distant from Africa, African-Australian. This process suggests a negotiation of identity narratives; presenting both a stable African identity, which is important to the migrant experience, and a redefined identity that is accepted by the new home. In offering stories that are close to Africans the site becomes an institutional voice that although not speaking on behalf of a 'community' it can speak for and to the African communities globally.

The list of events is updated showing what is happening in Australia: such as conferences, music festivals, political events, and so on. In this way the portal accomplishes a number of purposes: providing information that potentially connects people to the local and transnational African communities as well as with 'home,' and in the process becomes a third home. In addition, it also provides information for African-Australians on how to obtain expertise and knowledge for living with the uncertainties and challenges of everyday life in multicultural societies.

The portal as media has a hybrid identity (Bailey et al., 2007) in the sense that it is part commercial and part political: that is, it offers commercial products from business companies—travel agencies, rail companies, etc. In fact, they have their own 'African shop'; African food, music, books, and beauty products in line with the idea of bringing home to those who are away as well as perhaps 'branding' an ethnic culture thus structuring the relationship between individuals and collectives vis-à-vis diverse cultural patterns (Castells, 2009, p. 126) .

Nevertheless, it also has its own political identity, which can perhaps best be captured in the 'Facts and Features' section that works as a discursive space of representation of African-Australian people who are successful, such as fashion models, researchers, African charity founders, journalists, writers, singers, footballers, and politicians. They are presented through tales of life narratives, linking stories and experiences from back 'home' to the present, some of them quite poignant with memories of happiness, others of life-broken dreams due to political conflicts. These profiles seem to highlight how they have overcome the difficult circumstances of migration to succeed in a new life. Some of the youngsters, second-or third-generation African migrants, bring up the hardship of their parents as a force to move on; the importance of keeping the African values and cultures to bring the community together; and a sense of belonging. Some also reason about the importance of promoting African professionals and, in that way, assist families in the homeland. Others talk about the problem of racism, which exists in Australia and affects their diasporic lives with a sense of non-belonging. Most of the discourse in the 'facts and features' section epitomizes the general tension of migrants' experience; be part of a new culture and to maintain their cultural heritage which is sometimes expressed in their interest to connect to other migrants and use sites related to the homeland. The issue of misrepresentation of Africa was raised on the front page of the website in an interview with South African jazz musician Hugh Masekela, who emphasized that Africans should never forget where they come from—homeland—and their rich African heritage. For him, their culture 'has been compromised by conquest, religion and industrial exploitation. All these elements are in place to render and manipulate my heritage as being backward, uncivilized, pagan, heathen, savage and barbaric. I live to overturn this perspective' (Masekela, 2009). These feelings are somehow echoed in the interviews of other visible profiles of the African people in the site.

The portal also has a blog that presents articles and press releases related to African countries: for example, on the problem of piracy in Somalia and of tuberculosis and AIDS. Although the discourse on the site is not homogeneous, it could be suggested that the site appears slightly conservative as it does not seem to question the status quo. Nevertheless, it works as a discursive space to provide information and, most important, to facilitate a conversation among diasporic and non-diasporic people with similar interests. In that vein its political function is to represent the best of the African-Australian community while also empowering them with information and opportunities of self-development—school, courses, and jobs adverts. This might be seen as the 'banal' politics of daily life, not necessarily a strong resistance to mainstream culture and politics, but perhaps as a way of potentially participating and being part of the new home.

Slater and Miller (2000) in their seminal study of the Trinidadian migrants and their Internet use point out that Trinidadian immigrants rediscovered and re-imagined their homeland after using the Internet as it allowed them to interact with family at home. The AfricanOz.com territory suggests a dynamic interconnectivity between homeland (African continent), Australian home and the transnational home. In other words, the global online territory becomes 'local,' demarcated by a multiplicity of African diasporic and ethnic identities. The narratives presented in AfricanOz.com might offer some of the symbolic material in the construction of a new African identity that is outside of the discourse of the nation-state, as they are representing the African continent as the homeland and Africans in diasporic places. Moreover, parts of the African-Australian diaspora might identify with cultural narratives that perhaps go beyond the politics of identity to the affective realm where people might create emotional ties. However, cultural identification and affective ties are constructed not only on the Internet, but also in the off-line world, in a close process of interconnection between on/offline spaces. In that respect, AfricanOz.com seems aware of the importance of bridging the local community with the global cyber community. They support and advertise 'community' initiatives such as the 'African Families United Festival' and the Sydney's African Festival of Cultures presenting a uniform 'African family' which gets together to celebrate the different flavours of Africa as well as to expose the African community to the locals, Australians.

Indian-American web

In a provocative article, Mallapragada (2006) examines the representation of home on the Indian-American web territory. She explores advertisements of commercial web sites—Namaste.com, and Rediff.com featured in *Siliconindia* (No: 2000) a print magazine catering for professionals in India. Then she contrasts those with the DRUMnation.org and SAWNET[7] (non-profit site). One of her arguments relevant to this chapter is that the commercial sites participate in the construction as well as disruption of the 'Indian' identity by articulating diverse imaginations of home, such as household, homeland and homepage as well as the cultural, political and economic discourses of nation, family and community embedded in the national Indian culture. The commercially driven websites cater for people of Indian origin living in the United States with a well-educated and middle class profile. This web territory appears to reinforce the mainstream discourse of nation in the 'diaspora' as in the representation of traditional notions of Indian women's beauty, suggesting that 'immigrant women can maintain their 'Indian' traditions and cultural practices in their transnational locations' (Mallapragada, 2006, p. 214). In addition, women are associated with consumption, leisure, and the domestic, whereas middle class men are linked to issues of business, the public and transnational. The overall meaning seems to be that by consuming the products and services of these web pages one is investing in the home country, feeling at home and experiencing a sense of belonging. The thesis articulated is that in these sites' home page and the 'home' become the same i.e. web pages are bringing the home to one's home abroad and therefore the online territory also becomes 'home.'

By contrast, Mallapraga also presents the non-commercial site DRUMnatio.org which provides a diverse representation of home and belonging. It is produced by 'Desis Rising Up and Moving,' a non-profit organization dedicated to the struggles of the working-class and poor South Asian immigrants in New York. It represents the views of a marginalized group and thus presents the different social and economic locations of people in the Indian/South Asian diaspora. Moreover, due to its non-profit nature it is not visible to the wider society and/or to the American-Indian media circuit and relies on local networks to reach the migrant community, confirming Massey's (1991) argument that the geometries of communication and mobility reproduces existing power asymmetries, in this instance decoded on their existing political invisibility in the on-line/off-line worlds.

The representation of home and belonging differ from the commercial sites which cater for the transnational, middle-class Indian diaspora with a traditional nationalistic discourse. Here national representations of 'home' as the national, homeland place is resisted. That is, rather than a uniform idea of Indian, Bangladeshi or Pakistani identity, the site's identification is with South Asian working-class and poor immigrants of the region. Their position of home and belonging is 'regional' as it tells the story of exclusion and racism experienced by the working class South Asians immigrants working in the United States. Home and belonging refers to "the same neighbourhood—of the South-Asian region in the world and the working-class, migrant locations of global cities such as New York—that shapes their sense of being at home. For them collective sense of belonging is engendered by their 'out-side' status in the nation of America and in the South-Asia region" (Mallapragada, 2006: 219). Thus the online territory here has a very distinct border and boundaries with little of the borderless cosmopolitan identity or with the celebratory rhetoric of welcoming diversity found in multicultural, post-modern discourses.

Conclusion

The main point explored here is the way diasporic and ethnic groups are representing their identities online. The sites provide a good indicator of the complexity of the diasporic practices online but can not be generalized to the plethora of existing diasporic sites. These complexities might express the ontological tension of the fragmented and hybrid identities of the diasporic subject.

The representation of diasporas in the online territory seems to reinforce the uniqueness of each group as well as to highlight commonalities. Senses of belonging are generated through cultural specificities by means of the symbolic; language, collective memories, images and texts, although there are differences for each group. The Internet becomes a space for the articulation of different voices through narratives of identification, represent concurrent stories of success, positive and negative memories of the homeland, as well as experiences of racial discrimination and exclusion. The mediated space of the sites illustrate that diasporic groups are simultaneously engaging in discursive events with the homeland, the 'local' new home, and transnational home of the web. The online territory thus might become a 'comfort zone' because of its performative nature where multiple identity narratives can be performed while simultaneously engaging in discursive practices and emotional bond maintenance. This process might lead to potential practices of solidarity and resistance expressed in the ways home and belonging are contested, reconstructed and reframed in a constant play of disruption and/or acceptance of hegemonic notions of home and identities.

What seems to be common to these different online spaces is the need for diasporic and ethnic groups to represent themselves as a form of symbolic and political representation. In many ways, the diasporic and ethnic subjects can produce identities that contest existing mainstream stereotypes. Countless diasporic groups come from situations of political, religious or cultural repression or/and economic deprivation, and lack of free speech. Many are positioned on the fringes of society, invisible and isolated from their group, and these sites become not only a communicative space but an imagined community among the displaced, serving as an arena of civic engagement and dissent as well as entertainment. As suggested by Mitra (2005) the online world might be a 'speaking space' for those underprivileged groups in diaspora and without a voice in the public sphere. These spaces can potentially strengthen groups through their interconnected stories, and create an alternative public sphericule, a 'discursive comfort zone' where the 'ability to speak and a safe place to speak produce conditions that no other technologies have been able to create for immigrants' (p. 386).

This discussion indicates that some ethnic and diasporic groups bridge the local and the global as they cross virtual boundaries, inventing new forms of identification, informal citizenship, and political practices, creating new subjectivities, new social geographies and thus reconfiguring the online territory—a bottom-up space for ethnic minorities and diasporas—as well as a place which offers the possibility of changing their daily ordinary 'real' lives.

NOTES

1. Representation used as the production of symbolic meaning through language presented in different types of texts (Hall, 1997).
2. Resistance here follows Foucault's view on power, domination and resistance. Resistance does not

come from one single point and thus need not take the form of grand revolution to overthrow the existing powers (Foucault, 1980). Resistance 'can be translated into different practices, it can be socially organized in group or lived subjectively as personal commitment' (Lauretis, 1986: 3).

3. Belonging is define as the 'experience of cultural resonance between the social and the material realms of belonging and what governs the production of resonance is the embodied cultural predispositions of taste' (Jansson, 2009: 245).
4. The terms diaspora and ethnic group refer to people with different realities and intersected by historical and political differences. Identity here is understood as a process whereby the individual or group engage in struggles over meanings, constantly negotiated and contested (Siriphant, 1998).
5. Theoretically cosmopolitanism is understood as 'multiple rather than singular in origin, and something that emerges from the active engagement of social actors drawn from a range of particular socio-cultural contexts, rather than the gift of the West to the rest' (Holton, 2009: 42). Being cosmopolitan in the narrow sense used here relates to mobility—physical and imagined of elites and non-elites—and an open world-view that value cultural diversity and cross-cultural solidarity. For an in-depth discussion on cosmopolitanism see Holton, 2009.
6. The purpose is not to engage in a debate on the meaning and merits of the public sphere which has been discussed elsewhere (e.g. Calhoun, 1992; Dahlgren, 2006; Papacharissi, 2002). Rather the public sphere is used here as a metaphor for understanding the potential of the Internet in extending the political debate and enhancing the inclusion of diasporic and ethnic groups (for an analysis of the Internet and political participation please see Polat, 2005).
7. The SAWNET site is not relevant for this chapter's argument so it won't be explored.

REFERENCES

Africanoz.com.au (http://www.africanoz.com.au/, accessed May 2010)

Anthias, F. (2002). Where do I belong? Narrative Collective Identity and Transnational Positionality, *Ethnicities*, 2, 4, 481–514.

Appadurai, A. (1996) *Modernity at Large*. Minneapolis: University of Minnesota Press.

Bailey, O. Cammerts, B. & Carpentier, N. (2007). *Understanding Alternative Media*. England: Open University Press.

Bernal, V. (2006). Diaspora, Cyberspace and Political Imagination: the Eritrean Diaspora Online. *Global Networks*, 6, 2, 161–179.

Bhachu, P. (1996). The Multiple Landscapes of Transnational Asian Women in the Diaspora. In V. Amit-Talai & C. Knowles (eds), *Re-situating Identities: the Politics of Race, Ethnicity and Culture* (pp. 282–303). New York: Broadview Press.

Brah, A. (1996). *Cartographies of Diaspora: Contesting Identities*. London: Routledge.

Calhoun, C. (ed.) (1992). *Habermas and the Public Sphere*. Cambridge, MA: MIT.

Castells, M. (1997). *The Power of Identity*. Oxford: Blackwell.

Castells, M. (2001). *The Internet Galaxy*. Oxford: Oxford University Press.

Castells, M. (2007). Communication, Power and Counter-Power in the Network Society. *International Journal of Communication*, 1, 238–266.

Castells, M. (2009). *Communication Power*. Oxford: Oxford University Press.

Crampton, J. W. (2003). *The Political Mapping of Cyberspace*. Edinburgh: Edinburgh University Press.

Cunningham, S. (2001). Popular Media as Public 'Sphericules' for Diasporic Communities, *International*

Journal of Cultural Studies, 4, 2, 131–147.
Dahlgren, P. (2006). Doing Citizenship: The Cultural Origins of Civic Agency in the Public Sphere. *European Journal of Cultural Studies*, 9, 267–286.
DRUM Desis Rising Up and Moving,' URL http://www.drumnation.org/DRUM/Home.html (consulted May 2010).
Foucault, M. (1980). *Power/Knowledge: Selected interviews and Other Writings*. New York: Pantheon Books.
Gillespie, M. (1995). *Television, Ethnicity and Cultural Change*. London: Routledge.
Hall, S. (ed.) (1997). *Representation Cultural Representations and Signifying Practices*. London: Sage.
Hall, S. (1992). The Question of Cultural Identity, In S. Hall, D. Held & T. McGrew (eds.) *Modernity and Its Futures* (pp. 273–326). Cambridge: Polity Press.
Hepp, A. (2009). Differentiation: Mediatization and Cultural Change, In K. Lundby (ed.) *Mediatization, Concepts, Changes, Consequences*, (pp139–158). New York: Peter Lang.
Holton, R. J. (2009). *Cosmopolitanisms: New Thinking and New Directions*. Basingstoke, Hampshire, UK: Palgrave-Macmillan.
Howley, K. (2005). *CommunityMedia: People, Places, and Communication Technologies*. Cambridge, UK: Cambridge University Press.
Jansson, A. (2009). Mobile Belongings: Texturation and Stratification in Mediatization Processes, in K. Lundby (ed.) *Mediatization, Concepts, Changes, Consequences* (pp. 243–62). New York: Peter Lang.
Lauretis, T. (ed.) (1986). *Feminist Studies/Critical Studies*. Bloomington: Indiana University Press.
Leung, L (2005). *Virtual Ethnicity: Race, Resistance and the World Wide Web*. London: Ashgate.
Mallapragada, M. (2006). Home, Homeland, Homepage: Belonging and the Indian-American Web. *New Media & Society*, 8, 207–227.
Marfleet, P. (2006). *Refugees in a Global Era*. London: Palgrave Macmillan.
Masekela, H. (2009). Interview in AfricanOz.com (accessed July 2009).
Massey, D. (1991). A Global Sense of Place. *Marxism Today*, June, 24–29.
Mitra, A. (2001). Marginal Voices in Cyberspace. *New Media & Society* 3, 29–48.
Mitra, A. (2005). Creating Immigrant Identities in Cyberspace: Examples from a Non-resident Indian Website. *Media, Culture and Society*, 27, 371–390.
Mitra, A. (2008). Using Blogs to Create Cybernetic Space: Examples from People of Indian Origin, *Convergence; The International Journal of Research into New Media Technologies*, 14,4, 457–472.
Papacharissi, Z. (2002). The Virtual Public Sphere: The Internet as a Public Sphere. *New Media & Society*, 4, 9, 9–27.
Papastergiadis, N. (2000). *The Turbulence of Migration: Globalization, Deterritorialization and Hybridity*. Cambridge, UK: Polity Press.
Parker, D. & Song, M. (2007). Inclusion, Participation: and the Emergence of British Chinese Websites. *Journal of Ethnic and Migration Studies*, 33, 7, 1043–1061.
Polat, R. K. (2006). The Internet and Political Participation; Exploring the Explanatory Links. *European Journal of Communication*, 20, 4, 435–459.
Siapera, E. (2005). Minority Activism on the Web: Between Deliberation and Multiculturalism *Journal of Ethnic and Migration Studies*, 31, 3, 499–519.
Siriphant, T. (1998). Counter-Narratives: The Construction of Social Critique of Women Activists. *Gender Technology and Development* 2, March, 97–118.
Slater, D. & Miller, D. (2000). *The Internet: An Ethnographic Approach*. Oxford, UK: Berg Publishers.
Tastsoglou, E. (2006). Gender, Migration and Citizenship: Immigrant Women and the Politics of

Belonging in the Canadian Maritime. In E. Tastsoglou & A. Dobrowolsky (eds) *Women, Migration and Citizenship* (pp. 200–230). Hampshire, England: Ashgate.

Tastsoglou, E. & Dobrowolsky, A. (eds) 2006. *Women, Migration and Citizenship*. Hampshire, England: Ashgate.

Weberman, D. (2000). Are Freedom and Anti-Humanism Compatible? The Case of Foucault and Butler. *Constellations, an International of Critical and Democratic Theory,* 7(2):255-71.

Werbner, P. (1997). Introduction: The Dialectics of Cultural Hybridity, in P. Werbner & T. Modood (eds) *Debating Cultural Hybridity: Multi-Cultural Identities and the Politics of Anti-Racism* (pp. 1–25). Atlantic Heights, NJ: Zed Books.,

Zurawski, N. (1998). *Culture, Identity and the Internet*. (http://www.uni-muenster.de/PeaCon/zuraki /Identity.htm, Accessed April 2010).

Afterword

David Morley

Electronic Landscapes: Between the Virtual and the Actual[1]

Before commenting in detail on the chapters here, it seems best to offer some self-reflexive comments on my procedure in constructing this 'Afterword'. As someone whose own work has often been concerned with the varieties of interpretation of texts, I am well aware that the commentary which I offer below must be, in principle, contestable. Not only would it be possible to query my interpretations but, more fundamentally, my selection of only particular sections of these essays as comment-worthy. What follows is very much a personal response to only some of the themes articulated here: those which most closely resonate with my own contemporary research agenda, which has, of course, provided the framework within which I have made these selections and interpretations.

That agenda (see Morley 2009 and 2010) is partly premised on the argument that within communications and media studies, emphasis has, in recent years, fallen too exclusively on the virtual dimension of communications, to the neglect of the analysis of the material setting. Happily, from my point of view, this collection goes some way towards a more sophisticated investigation of the changing relations between the material and virtual, and addresses the question of how material geographies retain significance, even under changing technological conditions. In better addressing the articulation of the virtual and actual dimensions of communication, it also makes a valuable contribution to the project of avoiding a narrowly media-centric focus in our work. If, in 1995, Kevin Robins and I were concerned to address the new questions posed by postmodern geography (cf. Soja, 1989; Harvey, 1989), 15 years later, I am concerned that we do not mistake the emergence of virtual or 'electronic' landscapes for the death of material geography.

Theoretical Premises and Orientations

Judging from the premises on which these chapters are based, it is clear that most of the wilder technologically determinist fantasies concerning the role of online technologies in 'disembedding' us from the world of material geography have now been largely discredited. One key issue, as the 'Introduction' explains, is to 'think space and communication together' so that offline and online spaces can be understood simultaneously, as they are articulated in their material, symbolic and imagined dimensions. The further, underlying ethical question here is how to avoid the romanticisation of any 'grand narrative' of nomadology, fluidity and liquidity, while simultaneously disavowing a sedentarist metaphysics which, in overvaluing rootedness, can only then understand mobility as a morally retrograde form of inauthenticity.

While it would be foolhardy to ignore the significance of the affordances made available by contemporary technologies, nonetheless, the editors rightly warn us that we need to beware of the enchantments of 'ideologically fuelled metaphysics' of the rhetorics of techno-transformation (cf. Curran, forthcoming). Here one striking case concerns the much-advertised (and supposedly now imminent) 'death of television'. However, the problem is that, contrary to the rhetoric of the digerati, in the UK and elsewhere, the death of that medium has been somewhat exaggerated—as conventional forms of collective household TV viewing, far from decreasing as predicted, are actually increasing in some contexts (cf. London Business School, 2009).

Moreover, that which is 'new' is not necessarily the most significant. The editors are right to follow Vincent Mosco (2004) in recognising that it is only when new technologies lose the temporary bloom of the 'sublime' with which they are initially embellished, and become 'naturalised' that they have their most profound effects—when their relative invisibility reflects their 'taken-for-granted' place in the structures of everyday life (cf. also Edgerton, 2006). For an increasing number of people, the virtual is perhaps now best seen as a more or less banal overlay on their material lives, rather than some separate realm of wonder. Things like 'networked media practices' have moved, for some, from the category of the extraordinary to that of the mundane in the last decade. In that context, the question is how to understand the ways in which virtual and actual territories and practices are now intertwined in different settings or 'what kinds of multiplicities…will be co-constructed, with these new kinds of spatial configurations' (Massey quoted in editorial Introduction p3).

Here we potentially encounter all kinds of seeming anomalies and complexities: media can operate as means of re-territorialisation, as much as they can undermine existing territories; the ghosts of old material territories can reappear in virtual form—e.g., the provinces of the Hapsburg Empire, reborn as Mobile Phone Networks in the contemporary Balkans, as demonstrated in Lisa Parks' (2005) work. Similarly, de-territorialising technologies, designed to 'transcend' space, can turn out to principally function, in practice, not to extend cultural horizons but to produce reassuring 'discourses of the hearth' which provide virtual 'anchorage' amidst the anxieties generated by a world of physical hypermobility (Tomlinson, 2001).

From my point of view, one of the key contributions made by this collection derives from the editorial focus on the intersections of on-line and off-line activities as closely embedded within the material practices and settings of everyday life. The terminology here is of some consequence, and for my own part, rather than seeing these debates conducted within a terminology that counterposes the virtual with the real, I would argue that this distinction is better posed as one between the virtual and the actual (cf. Rowan Wilken, forthcoming). Once the matter is framed that way, we are better able to recognize the distinction between the immaterial and material worlds, without exclusively reserving the status of the real to the latter, and our attention can then profitably shift to understanding these different realms as different modalities of the real.

The further point here concerns the importance of resisting an over- generalised and abstract periodisation of technological development, which assumes that these matters work in the same way everywhere. As authors such as John Downing (1996) and more recently Brian Larkin (2008) have argued, most media theory to date has been very Euro-American centric, drawing its overall template from the particular experience of the techno-cultural conditions of the white, middle class Euro-American world—which we would be ill-advised to extrapolate to the rest of the globe. In this context it is very good to see that if, on occasion, the conditions of contemporary North American life are taken as an unquestioned norm, most of the chapters in this collection do achieve the editors' ambitions to avoid the trap of 'totalizing' logics, so as to scrutinise how local particularities emerge out of global processes. To this extent they offer nuanced analyses of the specific significance of new technologies in a variety of different cultural and sociopolitical contexts.

The problem with most discussions of 'new media' is their historical naivety, notwithstanding the fact that it is now some years since Carolyn Marvin (1988) definitively established that 'newness' itself is a historical constant. Happily, many of the essays here do benefit from a historical perspective, and are thus able to better link the 'moral panics' of our own day, in relation to the technologies which are new to us, to the earlier panics experienced in relation to technologies with which we now feel quite comfortable. To put matters like this is evidently not to think in terms of the immediate impacts of technologies, but rather, to invoke processes — and indeed, cycles—of their invention, innovation, dissemination, adoption, naturalisation and, as we shall see, 'domestication'. A number of these chapters are evidently informed by historical work— such as that of Lynn Spigel (1992) on the process of 'taming' television during the period of its entry to the American home in the 1950s, and by the work on the 'domestication' of the media in which I was involved in, along with Roger Silverstone and Eric Hirsch, in our research on the 'Household Uses of Information and Communication Technologies (HICT) ' in Britain during the 1990s (cf. Silverstone, Morley and Hirsch, 1992; cf. also Berker et al., eds., 2006). Certainly, none of the chapters here could be accused of operating with any simple notion of technological 'effects'.

Everyday Technology: Domesticating the New

In his opening essay, Christensen shows how, while the spectacular mechanics of violence have often found their way into representations of war—in the forms of spectacularised 'war porn' —the new technologies of blogging and online video also now allow the communication in real-time, often on a daily basis, of the banal routines of military life. These forms of more 'everyday' representation, as the American cartoonist Gary Trudeau has shown in his 'Doonesbury' strips on the experience of US soldiers serving in the current conflict in the Middle East, can be all the more shocking precisely because their very mundanity prevents us from consigning this world of military conflict to some entirely alien place in our cosmology—and thus connects us all the more closely to the troubling experiences within it.

While Christensen usefully alerts us to the place of blogs and video letters within a longer historical sequence of modes of communication, such as soldiers' letters home (of which the online video could simply be seen as a digital variant), he is also alert to the specificity represented by the capacity of these new technologies to communicate in 'real-time'. In doing so, they supply their recipients with an insistently synchronised experience of temporality (a parent at home, knowing from their son's blog, what time his patrol is due back to base, will be all the more disturbed if they fail to receive the customary reassuring message). In this respect we might also want to supplement Christensen's perspective by reference to the work of Johannes Fabian (1983) and Richard Wilk (2002) on the significance of media technologies in constructing the experience of 'coeval' temporality, as being every bit as important as their transmission of any particular content.

In the context of debates about the domestication of new media, Jonathan Lillie's chapter shows us how the 'cybersmut' panic of 1990s about growing access to online pornography needs to be grounded in the analysis of the context of reception. Indeed, he shows how it can be usefully situated within the context of how such encounters have been shaped by the everyday moral economy of the home and its regimes of social discipline and 'technologies of sexuality'. As he notes, 'far from being defenceless, the home is well protected by long established regimes of social discipline'. Thus his concern is with how the home shapes (and gradually domesticates) people's encounters with online pornography – in this case, partly via the technological forms of 'filtering' software (NetNanny; Surfwatch) designed to 'protect' the more vulnerable members of the household from harm. Lillie productively situates this not just in terms of earlier moral panics about pornography's entry to the home in other modalities (via print, film or video) but also within broader historical discussion of 'technopanics' in general.

Turning from these concerns with the domestication of problematic media technologies to the much-discussed issue of the transformation of active audiences into fan 'prosumers', Cornel Sandvoss' chapter usefully disaggregates the 'catch-all' category of fandom to focus on the varieties of fan, cultist and enthusiast engagements with texts. However, one key finding here concerns the fact that when fan activity is monitored closely, just as in most other areas of Internet life, the initial appearance of a widespread form of interactive communication is shown to be deceptive, in so far as a very small proportion of manically active enthusiasts (3.5% in this case) are responsible for a vast proportion (here 68%) of overall fan activity.

Throughout, Sandvoss displays a healthy scepticism in relation to the wilder claims of fan-theory, and his argument demonstrates well how the Internet is not the cause of fans' creative work, which precedes the emergence of the digital media; how the online part of most fans' activity would be incomprehensible outside of the broader context of their relation to their object of affection in material, offline form; and how the formation of fan communities was common well before the proliferation of the Internet. Indeed, at one point Sandvoss reveals the (partial) lineage of his perspective in classical 'uses and gratifications' theory, when he effectively updates Halloran's (1970) famous slogan about 'getting away from what the media do to people, to see what people do with the media' by arguing that we must get away from thinking about how the Internet has shaped fan culture and spend more time reflecting on how fan cultures shapes the Internet.

Moreover, he rightly notes that, although fan culture is generally associated with globalised forms of interpersonal communication, nonetheless, as the television market is still mainly structured in national form, the fan cultures which have grown up around television programmmes are themselves, on the whole, also still effectively national. Here, among other things, we see an interesting example of the intertwined relations of old and new media. Surrounded as we are by the rhetoric of globalisation, in this connection we would also do well to recall John Ellis' (1982) comments on television as the 'private life of the nation state'—a dimension of the issue which still remains pertinent today. Sandvoss is also alert to the fact that, in the context of what he calls the vast 'semiotic tundra' of the web, people can just avoid what they don't like—so fan communities often function as 'encapsulating': closed communities of the like-minded. In this respect Sandvoss' comments resonate very effectively with those of Zygmunt Bauman, quoted later by Thomas Tufte, when he observes that 'paradoxically, the widening of the range of opportunities, to promptly find ready-made 'like minds'... narrows and impoverishes, instead of augmenting our options'. Perhaps another way of looking at this phenomenon is simply to see it as the technological dimension of the ongoing process of the fragmentation of the public sphere into self-contained 'sphericules' noted earlier by Gitlin (1998).

Holly Kruse picks up the story of how one of the things which home Internet connectivity did was to bring a range of activities previously associated with the masculine public sphere into the conventionally feminised realm of domesticity. Alongside pornography, gambling and its associated, testosterone-driven rhetorics of masculine competition was clearly identified as another morally debased and profane activity at odds with 'family values' (cf. the recent TV adverts for online gambling in the UK, featuring the popular 'hard man' actor, Ray Winstone, which are couched exactly in these terms, as a 'matey/real men's' form of 'fan-on-fan' betting). As she notes, the dangers of masculinised public space invading the domestic haven in this way meant that allowing a child unsupervised access to the web came to be seen as being as irresponsible as leaving a child unsupervised in a public place. The problem was then how mothers (cf. Spigel, op. cit.) could regain control of a domestic space increasingly infiltrated by these masculine pursuits. In all these debates we see that, just as we found in the HICT research referred to earlier, the question is not only how to 'fit' the computer, as a new piece of technology into the home. Rather, the issue is how the perception of the technology itself is inevitably coloured by the problematic activities with which it comes to be associated (whether in the case of gambling here, or in the 1990s in the UK, as the 'wasting' of time by playing 'pointless computer games').

Spamming the Public Sphere

Whatever else the Net 'is', one thing we all know, from our daily experience at the keyboard, is that it is about spam. In his discussion of this issue, Kristoffer Gansing draws very effectively on Baudrillard's analysis of how technologies, in their moments of excess, can turn against themselves, so as to implode. He offers a compelling portrait of how spam—as, in his words, 'functional trash'—exposes the limits of online communication. If, as Baudrillard argues, we are obsessed with the perfect circulation of messages, and our success in that respect is conventionally measured by speed and capacity of transmission, then that 'success' is necessarily self-defeating: the more 'friends' you have on a social networking system, the less time you have to communicate with any particular one of them. The problem is that the logic of the system constantly drives towards greater speed—and communicative utopia is routinely figured as a state of nirvana where everyone is in constant interaction with everyone else. This is a model of communications whose theoretical deficiencies have been identified to devastating effect by John Durham Peters, in his magisterial *Speaking into the Air* (1999). To offer a more everyday analogy, the British teenager who had his Twitter feed 'adopted' by his favourite rap star in the summer of 2010 soon found that, far from this being the utopian moment he had dreamed of, its main consequence was that his own network simply went into melt-down, as multitudes of the star's other fans now began to follow his 'tweets'. Exactly the same logic is involved in denial-of-service (DoS) attacks, in so far as the force which is used to disable the target is precisely an 'overload' (in this case deliberate) of communications.

As Gansing shows, now that spam is variously recognised as accounting for 70–90% of all Internet traffic, Bill Gates' 2004 claim that spam would soon be a 'thing of the past' now looks not so much naïve as ridiculous. It is clear that, far from being an excrescence, or a marginal category, spam is central, both statistically, and in terms of functional principles, to what the Internet is about. Here Gansing's approach complements, very effectively, the analysis of the 'dark side' of Internet communication offered by Jussi Parikka and Tony Sampson in their recent collection of essays *The Spam Book : On Viruses, Porn and Other Anomalies* (Hampton Press, 2009)

In a context in which the Black Economy now represents a vast (and growing) proportion of world trade and where the profits of cybercrime are higher than those in any other sector, we have to recognise, as indicated in Misha Glenny's (2008) analysis of international crime, that the contemporary world is characterised as much by the circulation of 'Bads' as of Goods. This is a perspective which Gansing himself alludes to in his reference to Hawkins and Muecke's analysis of the 'cultural economics of waste'. Gansing is also careful to offer a historical perspective on these issues—in this case, rightly situating contemporary web-based phenomena, such as the '419' letter scams, in the longer history of earlier print based 'Advance Fee Frauds'. Of course, as he notes, the search for the precise historical origins of these scams (e.g., the rumoured role of underemployed Bulgarian computer programmers in the original development of viruses in the early 1990s) may ultimately prove impossible. However, Glenny (op. cit.) makes a good case for situating the origin of the '419' scams not simply in Nigeria, but more specifically among the disaffected Igbo of Eastern Nigeria (a people with a substantial history of involvement in earlier modes of long-distance trade) who felt, after the world had failed to help them during the Biafran tragedy, no moral qualms about taking their financial revenge on the *mugus* of the rich West. However, while this task is certainly now facilitated by the Internet, it is not specific to it: in their most developed modalities, these scams also depend on carefully orchestrated forms of physical theatre, such as meetings in prestigious hotels in the *mugu*'s own country with (fake) Ministers of State, whose physical presence is necessary to consolidate the 'trust' on which the scam depends (see the fictionalised portrayal of this process in Adaobi Trisia Nwaubani's 2009 novel *I Do Not Come to You by Chance.*

Technologies, Voices and Publics

Overall, in this section, the ambition, we are told, is to offer 'nuanced, empirically informed analyses of the ways in which online communication both affords and limits particular modes of social voice and presence in the public sphere …'. In this context Tufte's chapter is a particularly welcome contribution, in so far as, unlike so much work in the field, it is not based on presumptions which only apply in the affluent West. He rightly poses the issue of whether the theoretical concerns of Western-based media and technology studies have any universal value or are only relevant to the developed world of widespread Internet access (cf. Morley, forthcoming). If so, then their use in contexts such as East Africa would evidently constitute an ethnocentric imposition of themes, concepts and theoretical approaches of little relevance. More specifically, Tufte addresses the Internet's potential in Africa in relation to ideas about 'insurgent citizenship', and the potential of 'citizen media' in a context where there is a profound disjunction between the abstract idea of democracy and the actual experience of widespread insecurity

Like Sandvoss, Tufte is also sceptical of the often celebratory attitudes adopted in relation to the new social media. This is a welcome relief, given the current tendency in the field towards the over-emphasis on the role of communications technologies such as text messaging in political protest movements (such as those

which brought down the Marcos government in the Philippines) a tenc
has recently been subject to thoroughgoing critique (cf. Castells et al., 2007, pp. 186 et seq.). The same point is made in the editors' Introduction here, in relation to the dangers of romanticising the use of social networking media by young, cosmopolitan Iranian students in their protest against rigged elections in 2009. There is nothing in these networks (Twitter, Flickr, YouTube, etc.) which 'naturally' disposes them to progressive uses: the mob who misguidedly attacked the house of a hospital paediatrician at the height of a recent media-driven 'paedophile' scare in the UK, had organised themselves via their text message systems.

Moreover, rather than focussing on the supposed wonders of communication technologies themselves, Tufte demonstrates that simplistic communications strategies based on the use of new technologies to transmit 'health advice' are quite inadequate. As he shows, when problems like HIV and AIDS are so entangled with poverty, culture and gender roles, it is pointless to imagine that you can prevent the epidemic from spreading simply by only providing (via whatever technology) 'practical advice'. In this connection, we might think of the close parallel with Daniel Lerner's (now long-discredited) emphasis on using transistor radios to inculcate 'modern' methods of farming in the Middle East, as a potential 'stimulant' to agricultural development in the 1960s. This was an earlier example of an over-simplistic communications policy which, while central to US 'modernisation' strategies in the Middle East in that period, failed because it simply did not grasp the extent to which the 'problematic' behaviour was so deeply embedded in other discourses and structures that simply offering 'practical advice' of this kind was unlikely to change anything very much (cf. Lerner, 1958).

Turning from the potential of new technologies for Third-World development strategies to their potential for the politics of feminism, Liesbet van Zoonen offers a substantial critique of the early cyberfeminist investment of hopes in the transformative possibilities of these new technologies, in relation to conventional gender identities. As she shows, the enthusiasts imagined that the disembodiment and anonymity of the Internet would allow proactive experimentation with diverse gender identities. However, in effect, none of these hopes (as expressed variously by Sherry Turkle, Sadie Plant and Dale Spender) proved justified and nowadays, Internet usage and gender performances, by and large, still take place within the limits of dominant heterosexual gender discourses. Indeed the author's sobering conclusion is that, in itself, the Internet has not changed anything much, in this respect. Once again, the key factor seems to be the way in which this early cyberfeminist optimism failed to take into account the extent to which access to cyberspace takes place in a variety of material settings, which are themselves almost always heavily gendered. To this extent the early cyberfeminists attributed too much transformative power to the technologies themselves and failed to pay attention to how the socio-cultural context would determine their effectivity.

To return to the themes of my earlier discussion of the chapters concerned with the entry of pornography and gambling to the home, van Zoonen notes that at the point at which public attention began to recognize the downside of on-line anonymity (in relation to phenomena such as paedophilia and bullying) the earlier investment of feminist hopes in the Internet as a Utopian space came to look rather naïve. As the author argues in relation to the debates about 'Internet predators', the majority of such predators (or bullies) usually turn out to be from within a person's own social circle. The home, far from being a necessarily 'safe' place, is the most usual site of sexual abuse, and we must recognise the presence of dangers within the *Heimlich* sphere (Morley, 2000). Moreover, once we put these panics into historical perspective, we readily see that, despite the current focus on the 'dangers' of the computer screen, few of these issues have much to do with the technology itself. To this extent, van Zoonen is quite right to link current panics about children using social network media back to earlier fears for 'unsupervised' young women attending dance halls. These are, as they suggest, perhaps best seen as just a technologically updated version of the fundamentalist discourse of 'restraint, modesty and sexual discretion' as desirable traits in women.

Surveillance and Colonization

The vexed question of the 'nature' (if it has one) of what the Net 'is'— i.e, whether it is 'essentially' a democratizing force (as many self-interested digipreneurs have argued) or an oppressive one (as many techno-sceptics maintain) is fundamental. In relation to debates about the 'essential' nature of specific technologies, perhaps the most interesting finding in Laura Stein's chapter of new social movements' use of online technologies in the USA is that even among them, where one might expect a particularly high commitment to the democratic potential of Internet communication, only a small percentage actually use these technologies for interaction, dialogue or creative expression. The majority simply use them for the traditional purpose of transmitting information to their followers and members. Again, this is, to put it mildly, sobering evidence for anyone of a technologically determinist persuasion, who imagines that new technologies possess some kind of progressive or democratic essence (or 'bias') which will automatically assert itself over time, for the better.

It is now well established that many interactive technologies simultaneously function as modes of surveillance and, in this respect, David Phillips' chapter offers considerable insights into the specificities of the move to 'actuarial surveillance'. This involves, as he explains, the 'systematic, analytic, methodical creation of normativity' which, as a technique of knowledge production, renders us visible and functions as a form of population management which 'marks' (and raises queries about) all deviations from the norm. This, to put it in more mundane terms, is of course, increasingly how your credit card functions (identifying and querying any departures from what the surveillant technology has established as your normal pattern of expenditure). The associated risks (see below, re 'Advance Fee Frauds,' etc.) are why one has to deal with increasingly complex systems, originally designed to defend one's own interests, but necessarily involving a variety of passwords and proofs of identity, should one wish, for example, to vary one's shopping habits.

Phillips also recognises the increasing importance of geodemographic systems of consumer classification (e.g., ACORN/CACI in the UK) which classify places by reference to the types of persons who (predominantly) inhabit them ('You Are Where You Live'). Residential and other personal data are then cross-correlated, to produce predictive systems of consumer behaviour, based on these 'lifestyle clusters' in which data spaces and physical spaces are increasingly enmeshed. Importantly, Phillips also recognises the performativity of place—in the sense that place is also produced by the actions it mediates, and what constitutes appropriate behaviour within them is always, in principle, negotiable. In this context he also explores possibilities of resistant forms of creative response to (or 'hi-jacking') of surveillance systems (into forms of 'sousveillance') and of what he calls 'counter-normative identity play'. Perhaps more ambitiously, he argues that such strategies might even serve to denaturalize 'the ideology of unitary bodies in Newtonian space' in so far as they produce identities which may destabilise the assumption that database records 'refer unambiguously and unproblematically to pre-existing bodies'. These are certainly interesting speculations, but we would perhaps do well to exercise what Ulf Hannerz (1996) once called some 'unexciting caution' here, before deriving any generalised observations about the subversive potential of these (as yet) rather marginal practices of symbolic resistance.

One other area where netizens have attempted to subvert the authoritarian structure of social relations through innovative technological means is the development of P2P 'file-sharing' among music fans. In this connection, Patrick Burkart provides a stimulating update of Adorno and Horkheimer's critique of the Culture Industry by demonstrating the extent of the entertainment industry's attempts to commercialise the Internet. Thus he shows just how far, in the post-Napster period, the music industry has succeeded in the 'colonisation' of the net by techno-regulationist ideologies designed to achieve the 'cybernetic commodification' of fan behaviour .

As he says, if all this might be dystopian to a cyber-libertarian, it is 'Xanadu for the Internet regulationists'. Against the forces of commercial regulation, he poses the figure of the hacker and the culture-jammer who attempt to decolonise the existing copyright regime, evade online surveillance and playfully flout copyright laws, business confidentiality and state security. Evidently, the question here concerns the extent to which the power of surveillance systems can be effectively counterposed by these marginal modes of 'resistance'. While the full implications (for P2P activists) of the 'Pirate Bay' case in Sweden remain as yet, unclear (cf. Andersson, 2010) the vision of net-opposition offered here certainly has very substantial forces ranged against it.

The Varieties of Mobility: Tourists, Vagabonds and Diasporics

The discussion of issues concerning transnational and translocal patterns of culture and communications have long been bedevilled by a form of theoretical over-abstraction, which refers, in the singular, to The Postcolonial, The Diasporic or The Transnational condition. As I have argued elsewhere (Morley, 2007), the universalising and singularising tendencies of these modes of abstraction reduce all local differences to one template. They can be seen as what Michel Serres calls lazy modes of 'pass-key' analysis where, just as one key is used to open all locks, all questions are treated as having only one fundamental answer (cf. Serres and Latour, 1995). Conversely, what is encouraging here is the editors' concern with locating and scrutinising the particularities which emerge from these global processes, without adopting the totalising logic which would reduce them all to being seen as instances of the same phenomenon.

As someone troubled by the regrettably widespread tendency to over-emphasise the significance of processes of deterritorialisation, I welcomed Myria Georgiou's recognition that, while discourses of territoriality might seem an anachronistic in an era of online communications, nonetheless, territoriality remains deeply rooted in political conceptions of identity, especially when transposed into questions of passports, visas and citizenship rights. As she notes, if flows of culture and communication undermine national boundaries, nonetheless, nation states are still based on ideas of singular loyalty. It is in this context that she places the significance of diaspora where, as a result of their own (or their ancestors') mobility, migrants experience a kind of 'place polygamy' where they are effectively 'married into' different worlds and cultures simultaneously, and their participation in any one community is thus relativised. Here again, her arguments connect well with those of Aksoy and Robins (op. cit.) in relation to how Turkish migrants in London effectively exist in a space of 'in-betweenness' so that, if being Turkish makes a difference to how they consume British media, by the same token, being in Britain transforms their relation to diasporic Turkish media.

Georgiou may well be right to suggest that the dynamics of the processes involved here, are better captured by studying them at the level of the city, or even the locality, rather than in relation to the abstractions of national politics (cf. Kobena Mercer, 1994) on the ' problems of living with difference'). As Robins has noted (2001) the city is both an existential and experimental space and it is one which offers juxtapositions of different forms of cultural production (from pirate radio to the graffiti on the city walls) which can register the co-presence and proximity of a variety of forms of alterity. Here Georgiou's argument echoes, in part, Bailey's comments as will be discussed shortly, concerning 'regionally' defined communities in global cities where, rather than nostalgic dreams of diasporic unity, we potentially enter the realm of what Chantal Mouffe (2000) has called (necessarily) 'agonistic' dialogue with a variety of both corporeal, virtual and imagined others. Such dialogues are the lifeblood of any healthy form of democratic community, and will be crucial in enabling us to live out—both online and offline—the forms of 'critical proximity' which Georgiou rightly enjoins as the most appropriate way to inhabit the multi-dimensional and multi-cultural spaces of our contemporary world.

The editors are right to note the limitations of the widespread tendency to overemphasise the place-violating forces of technologically enhanced global mobility, to the neglect of the embeddedness of people's technological 'connectivity.' As Miyase Christensen rightly notes in her chapter, rather than accepting such simplistic accounts of de-territorialisation, we should note that mediated transnational activities usually take place at the juncture of the online and offline worlds .The issue is to understand both the connections and disjunctures whereby migrants inhabit fluid virtual networks of dispersed contacts while still being territorially anchored in the materiality of local spaces.

Taking the particular case of Turkish migrants in Sweden, Christensen's work demonstrates how loyalties based on physical co-proximity, originating in a rural village thousands of miles away, are often transposed to distant contexts. Given the simple mechanics of the process of chain migration, once a person is established in a given location, their friends and relatives are more likely to follow. Hence many migrants, even in their new 'host' country, still live closely with people from their place of origin where (much to the chagrin of some of the younger members of the community, as Christensen notes) corporeal forms of co-surveillance are often practised every bit as much as they were in their original home. However, she is every bit as alert to questions of change and adaptation as she is to continuities—for instance, in the way in which the younger migrants adopt social media and carefully 'manage' their levels of mediated visibility in strategic ways (cf. Hargreaves, 1995, on generational differences in communicative practices among Arab migrants in France). She is also right to insist that ' the affordances inherent in technological applications' such as online social media, cannot be accounted for by any 'discourse of sudden transformation or imminent liberation' but must, rather, be studied in terms of use patterns in a given social context. In this respect, I commend her attempt to steer a middle course between the structural 'over-determination' of Bourdieu's *habitus* and the romanticism of the proponents of the thesis of 'individualisation' and destructuration.

In relation to the need to move beyond singularised/abstract versions of 'The' migrant experience, André Jansson's chapter addresses the very particular characteristics of the western, professional, expatriate experience of 'fixed term'/temporary migration. His focus on 'Professional Westerners' in Managua provides 'a situated analysis of the particular socio-spatial ambiguities characterising this actual type of cosmopolitan class fraction.' Among the key distinguishing features of this particular lifestyle are the considerations that their migration is voluntary; that it is for a fixed/limited period of time; and that they have ready access to 'exit mobility' in the case of trouble.

Jansson also offers a very interesting exploration of the functions and roles played for this particular subcategory of migrant by a variety of 'technologies of encapsulation.' Here he draws effectively on De Cauter's recognition that technologies such as a computers, gated communities, cars and aeroplanes are all 'capsular,' in so far as they provide a protective cocoon, which not only connect people to (some) others, but simultaneously, separate them from problematic forms of alterity (cf. Cwerner, 2006) on the helicopter as 'technology of secession' from urban life for the rich). In this connection he elegantly weaves together De Cauter's theories with various models of 'cosmopolitanization,' in exploring the contradictory nature of the particular uses of new technologies made by this category of professional migrants.

These migrants are 'cosmopolitan' in outlook (and thus resistant, in principle, to 'encapsulation' within any mono-culture) and yet inevitably concerned for their own physical security. The mundane, but nonetheless pressing exigencies of managing everyday life in a problematic third world city are well-exemplified by his example of a respondent who makes a mobile phone call from within a taxi in the disorienting context of the 'nameless streets of Managua,' convince the driver that his passenger is directly connected to a secure 'elsewhere,' lest he should harbour any evil intentions,.

For many expatriates, their experience is one of rarely integrating where they physically live, while gradually losing touch with those at home, because of the profound differences in their respective daily experiences—and thus they end up speaking mainly to each other, whether on- or offline (cf. Nowicka, 2006, on this). In the end, despite their decisions to live (if temporarily) among the 'others' of the third world, as one of Jansson's respondents says, 'it's easier to stay within the bubble.'

If we turn to the other end of the social spectrum to address the situation of those whose migrancy is perhaps better described as that of 'involuntary vagabondage,' rather than 'voluntary tourism' (cf. Bauman, 1998; Hannnerz, 1995) Olga Bailey provides an interesting account of ethnic groups' online representation of their identities. She recognises the 'uneven distribution' of online territories and her account is informed by the fact that, in many Western societies, there has been a systematic exclusion of migrants from the mainstream media (cf. Hargreaves, op. cit., on the exclusion of Arab migrants from the French media) as a result of which some have reacted by producing their own alternative—and these days, online—media. Certainly, the UK case would fit this portrait. As Marie Gillespie (1995) noted in an earlier period, it was precisely the fact that British Asian migrants felt so ill-served by the mainstream UK media that meant they were among the earliest adopters of both video and satellite technologies, as ways of accessing culturally sympathetic material more suited to their needs.

Bailey's own perspective—and her invocation of Brah's notions of 'multilocationality' and Tastsoglou's model of multiple and overlapping spatial and symbolic attachments—is well supported by the work of Asu Aksoy and Kevin Robins (2000) on the use of communications and media among Turkish migrants in Britain. They demonstrate that these migrants' complex pattern of usage of local and transnational, broadcast and interpersonal media, along with their insertion in a highly developed system of mobility for the transport of persons and goods between Turkey and Britain, means that they are in effect, participants simultaneously, in material and virtual communities in a variety of locations. While also referencing the interesting work of Andreas Hepp (2009) on the specific uses made by migrants (in this case again, mainly Turkish) in Germany, of mobile and online communications technologies, Bailey raises the question of how diasporic and ethnic groups appropriate online technologies selectively, for specific purposes and wisely disavows any 'speculative celebration of the possible role of the Internet.' However, she also cites Bernal's contention that the Internet is 'the *quintessential* diasporic medium, ideally suited to allowing migrants in diverse locations to connect, share information and analysis and coordinate their activities' (my emphasis).

However, one does need to be cautious with any attempt to construct a homology between the experience of the migrant and the capacity of mobile online media. It is but a short step from that kind of proposition to a model of the migrant as an epistemologically privileged figure—an intellectual position which would replicate all the problems which ensued from Lukacs' elevation of a particular economic class of persons as best placed to see the 'essential truths' of a previous age.

(in his case, the proletariat; later, of course, supplanted by the vanguard party) . That problem aside, drawing on the work of Mallapragada, Bailey also argues that not all online communities necessarily function to link migrants back to the lost 'Heimat' of their nation state of origin. As she notes, we should recognise that some online sites construct a (present-tense oriented) 'regional,' rather than (nostalgic) national identification for working class immigrants of different national origins—who often now live in the same, poor neighbourhoods of global cities. In this case, we see vividly how the dynamic relations of online and offline territories are capable of taking a variety of different political inflections.

That these spaces now have to be understood in their virtual as well as material dimensions is perhaps the most basic proposition which unites the chapters in this volume. The further point, which I stressed earlier, concerns the need to investigate the changing relations between these material and virtual forms of territory—and in doing so, to avoid any simplistic periodisation between the worlds of the old and 'new' media. As I have argued elsewhere (Morley, 2009), rather than assuming that we have proceeded abruptly from one 'era' of communication to another, we need to investigate the continuities, overlaps and modes of symbiosis between technologies of symbolic and material communications. To do otherwise—and to imagine that the new technologies of our day are so totally transformative as to require us to entirely begin again, from some theoretical 'Degree Zero'—would be to fall back into the worst kind of technological determinism. It would also be to risk making the fatal mistakes, identified long ago by Michel Serres (Serres and Latour, 1995), not only of believing too readily in technicist ideologies of 'progress,' but also of imagining that we are 'not only right, but righter than it was ever possible to be before' for the 'simple, banal and naïve reason that we are living in the present moment' (1995, 48–9). In matters of technology, in particular, both 'presentism' and neophilia are dangerous temptations—and overall, the essays here are to be commended, not least, for marking out ways forward which will help us to avoid these dangers.

Note

1. This essay revisits some of the themes first raised 15 years ago by myself and Kevin Robins in *Spaces of Identity: Global Media, Electronic Landscapes and Cultural Boundaries*. London: Routledge.

References

Aksoy, A. & Robins, K. (2000) 'Thinking Across Spaces Transnational Television from Turkey,' *European Journal of Cultural Studies*. Vol 3.3.

Andersson, J. (2010) *Peer-to-Peer-Based File-Sharing: Beyond the Dichotomies of 'Downloading Is Theft'*

vs 'Information Wants to Be Free,' PhD Department of Media and Communications, Goldsmiths College, London University.
Bauman, Z. (1998) *Globalisation,* Cambridge: Polity Press.
Berker, T. et al. eds. (2006) *The Domestication of Media and Technology,* Maidenhead: Open University Press.
Castells, M. et al. (2007) *Mobile Communication and Society,* Cambridge, Mass: MIT Press.
Curran, J. (forthcoming) Technology Foretold' to Appear in N. Fenton (ed). *New Media, Old News* London: Sage.
Cwerner, S. (2006)'Vertical Flight and Urban Mobilities: The Promise and Reality of Helicopter Travel,' *Mobilities,* Vol 1.2.
Downing, J. (1996) *Internationalising Media Theory,* London: Sage.
Edgerton, D. (2006) *The Shock of the Old,* London: Profile Books.
Ellis, J. (1982) *Visible Fictions,* London: Routledge.
Fabian, J. (1983) *Time and the Other*, New York: Columbia University Press.
Gillespie, M. (1995) Television, Ethnicity and Cultural Change, London: Comedia/Routledge.
Gitlin, T. (1998) 'Public Sphere or Public Sphericules?' in T. Leibes et al. (eds) *Media Ritual and Identity,* London: Routledge.
Glenny, M. (2008) *McMafia: Seriously Organised Crime,* London: Vintage.
Halloran, J. D. (1970) *The Effects of Television*, St Albans, Herts.: Panther.
Hannerz, U. (1996) *Transnational Connections,* London: Comedia/Routledge.
Hargreaves, A. (1995) *Immigration, Race and Ethnicity in France,* London: Routledge.
Harvey, D. (1989) *The Condition of Postmodernity,* Oxford: Basil Blackwell.
Hepp, A. (2009) 'Localities of Diasporic Communicative Spaces: Material Aspects of Transnational Migrants' Mediated Networking,' *The Communication Review,* Vol. 12.
Larkin, B. (2008) *Signal and Noise: Media Infrastructure and Urban Culture in Nigeria*, Raleigh- Durham: Duke University Press.
Lerner, D. (1958) *The Passing of Traditional Society: Modernizing the Middle East,* Glencoe, Ill: Free Press.
London Business School (2009) 'The Future of Converged and On-Demand TV: Actual Consumer Behaviour' Proceedings of Seminar July 2009, accessed at *www.acbuk.net/conference.php*
Marvin, C. (1988) *When Old Technologies Were New,* New York: Oxford University Press.
Mercer, K. (1994) *Welcome to the Jungle,* London: Routledge.
Morley, D. (2000) *Home Territories*, London: Comedia/Routledge.
Morley (2007) *Media, Modernity and Technology,* London: Comedia/Routledge.
Morley, D. (2009) 'For a Materialist Non-Media-Centric Media Studies,' *Television and New Media,* Vol. 10.
Morley, D. (2010) 'Television as a Mode of Transport: Digital Technologies and Transmodal Systems' in J. Gripsrud (ed) *Relocating Television,* London: Comedia/Routledge.
Morley, D. (forthcoming) 'The Geography of Theory and the Place of Knowledge; Pivots, Peripheries and Waiting Rooms' in G. Wang (ed) *Communication Research Beyond Eurocentrism*, London: Routledge.
Mosco, V. (2004) *The Digital Sublime,* Cambridge, Mass: MIT Press.
Mouffe, C. (2000) *The Democratic Paradox,* London: Verso.
Nowicka, M. (2006) 'Mobility, Space and Social Structuration in the Second Modernity and Beyond' *Mobilities,* Vol. 1.3.
Nwaubani, A. T. (2009) *I Do Not Come to You by Chance,* London: Phoenix.
Parikka, J. & Sampson, T. (2009) *The Spam Book: On Viruses, Porn and Other Anomalies,* N. J.: Hampton Press.

Parks, L. (2005) Postwar Footprints : Satellite and Wireless Stories in Slovenia and Croatia in A. Franke (ed) *B-Zone: Becoming Europe and Beyond* Berlin, Institute for Contemporary Art.

Peters, J. (1999) *Speaking into the Air : A History of the Idea of Communication,* Chicago: University of Chicago Press.

Robins, K. (2001) 'Thinking through the City' in D. Morley & K. Robins (eds)) *British Cultural Studies* Oxford: Oxford University Press.

Serres, M. & Latour B. (1995) *Conversations on Science Culture and Time*, Ann Arbor: University of Michigan Press.

Silverstone, R, Morley, D. & Hirsch, E. (1992) 'ICTs and the Moral Economy of the Household' in R. Silverstone & E. Hirsch (eds) *Consuming Technologies,* London: Routledge.

Soja, E. (1989) *Postmodern Geographies,* London:Verso.

Spigel, L. (1992) *Make Room for TV,* Chicago: University of Chicago Press.

Tomlinson, J. (2001) 'Instant Access: Some Cultural Implications of Globalising Technologies' *University of Copenhagen Global Media Cultures Working Paper,* No. 13.

Wilk, R, Television, 'Time and the National Imaginary in Belize' in F. Ginsburg et al. (eds), *Media Worlds,* Berkeley: University of California Press.

Wilken, R. (forthcoming) *Teletechnologies, Place and Community,* New York: Routledge.

Contributors

Olga Guedes Bailey is a senior lecturer at Nottingham Trent University, UK. She is the program leader of the MA 'Media and Globalization.' She is the chair of the section 'Migration, Diaspora and Media' of the European Communication Research and Education Association—ECREA. She is also a member of the editorial board of the international journal *Communication Theory,* USA, and former managing-editor of the international journal *Body and Society,* Sage Publications. She has published essays on global audiences, environmentalism, journalistic practice, alternative media, race and representation, the politics of communication of ethnic minorities and diasporas in western societies, and on online citizen journalism. Her latest books include; a co-authored book entitled *Understanding Alternative Media* (UK, Open University, 2007) and an edited collection *Transnational Lives and the Media: Re-imagining Diasporas* (UK, Palgrave, 2007).

Patrick Burkart is an associate professor of communication at Texas A&M University. He is the author of *Music and Cyberliberties* (Wesleyan University Press, 2010) and *Digital Music Wars: Ownership and Control of the Celestial Jukebox* (with Tom McCourt, 2006).

Christian Christensen is professor of media and communications studies in the Department of Informatics and Media at Uppsala University in Sweden. His primary area of research is in the use of social media during times of war and conflict, but he has also published on the representation of Islam, post-9/11 documentary film and international journalism. His work has been published in the *International Journal of Press/Politics, Global Media and Communication, Media, War & Conflict, Studies in Documentary Film, Popular Communication: International Journal of Media and Culture* and the *British Journalism Review*. Christensen is the co-editor of a number of books, including the forthcoming *Understanding Media and Culture in Turkey: Structures, Spaces, Voices* (2011; New York: Routledge).

Miyase Christensen is professor of media and communication studies at Karlstad University and research fellow in the Department of Philosophy and History of Technology at the Royal Institute of Technology (KTH). She is the author and co-editor of a number of international articles and books, including *Shifting Landscapes: Film and Media in European Context* (2008); *Connecting Europe: Politics of Information Society in the EU and Turkey* (2009); and, *Understanding Media and Culture in Turkey: Structures, Spaces, Voices* (2011; New York: Routledge). Her current research focuses on social theory and globalization/transnationalization processes; social surveillance and the media; and, politics of popular communication. Ongoing research includes a funded project entitled *Secure Spaces: Media, Consumption and Social Surveillance* (with A. Jansson), and a second project on the environment and the media funded by FORMAS.

Kristoffer Gansing is a cultural producer, artist and media researcher working at the intersection of film, new media and experimental art practices. He is the co-founder and director of the media art festival the Art of the Overhead (2005) as well as an editorial board member of the artist-run channel tv-tv in Copenhagen. Since 2001, he's been teaching the theory and practice of new media at the K3, School of Arts and Communication, Malmö University, Sweden. He's also been conducting his PhD there (2006–2011), with a dissertation on "Transversal Media Practices" dealing with the articulation of the old and the new across the shifting boundaries of art, activism and the everyday in the cultural production of networked media culture.

Myria Georgiou teaches at the Departmentt of Media and Communications, London School of Economics (LSE). Her research focuses on the study of diaspora, transnationalism and the media, as well as on the city as a space of contact, communication, and conflict. She is currently conducting cross-European research with Arab audiences of transnational television. The project investigates patterns of television use among Arab speakers and their sense of cultural and political belonging. She is also writing a book, titled *Media and the City* (forthcoming, Polity Press).

André Jansson is professor of media and communication studies at Karlstad University, Sweden. He currently leads two research projects; *Rural Networking/*

Networking the Rural (FORMAS) and *Secure Spaces: Media, Consumption and Social Surveillance* (National Bank of Sweden). He has published several books and articles in the field of media and cultural studies, with a special focus on communication geography. Among his publications in English are the co-edited books *Strange Spaces: Explorations into Mediated Obscurity* (with Amanda Lagerkvist, 2009), and *Geographies of Communication: The Spatial Turn in Media Studies* (with Jesper Falkheimer, 2006).

Holly Kruse holds a doctorate in media studies from the Institute of Communications Research at the University of Illinois. Currently, she is a professor in the Department of Communications at Rogers State University. She has taught courses in communication technology and society, the Internet and new media, and gender and society. Her research has been published in the journals *New Media and Society*, the *Journal of Sport and Social Issues*, *Television and New Media*, *Popular Music*, *First Monday*, and others. Kruse is author of the book, *Site and Sound: Understanding Independent Music Scenes*.

Jonathan Lillie is assistant professor of digital media and online journalism at Loyola University Maryland, in the Department of Communication. He was a Park Doctoral Fellow in the School of Journalism and Mass Communication at the University of North Carolina at Chapel Hill. His research interests include multimedia journalism, the mobile Internet, information society discourses, web pornography, and development communication. Dr. Lillie is the managing editor of NMEDIAC, the *Journal of New Media and Culture*. He is co-curator of "New Media, Sex, and Culture in the 21st Century" at the Museum of New Art in Detroit, which includes research articles on the same topic published by NMEDIAC.

David Morley is professor of communications in the Department of Media and Communications at Goldsmiths College, University of London. His latest book is *Media, Modernity and Technology: The Geography of the New* published by Routledge in 2007.

David Phillips is associate professor of information at the University of Toronto. His work brings together queer theory, technology studies, media studies, surveillance studies, and political economy to study issues of surveillance, identity, and identification. He is the author of *From Privacy to Visibility: Context, Identity, and Power in Ubiquitous Computing Environments* (in *Social Text*), *Texas 9–1-1: Emergency Telecommunications and the Genesis of Surveillance Infrastructure* (in *Telecommunication Policy*), *Queering Surveillance Research* (in *Queer Online: Media Technology and Sexuality*) and numerous other works exploring the relations among information, economics, ideology, policy, culture, identity, and technology.

Cornel Sandvoss is a senior lecturer in the Department of Sociology at the University of Surrey, UK. Sandvoss specializes in audience research with a particular focus on fan audiences in realms of spectator sports, television, popular music and film. Among his publications are the books *Fans: The Mirror of Consumption* (Polity, 2005) and *A Game of Two Halves: Football, Television and Globalization* (Routledge, 2003).

Laura Stein is an associate professor in the Radio-Television-Film Department at the University of Texas at Austin. She writes about alternative and activist media, political communication, and communication law and policy. Her books include two co-edited volumes, *Making Our Media: Global Initiatives Toward a Democratic Public Sphere* (examining grassroots attempts to transform the policy and practice of information and communication media around the world) and *Speech Rights in America: The First Amendment, Democracy and the Media* (exploring the failure of neoliberal understandings of speech rights to protect democratic communication in the media).

Thomas Tufte is professor in communication at Roskilde University, Denmark (2004-). He is the co-director of Ørecomm (http://orecomm.net), director of the national research network 'Glocal NOMAD' (Network on Media and Development—http://glocalnomad.net/), and in 2009–2013 coordinator of the research project 'Media, Empowerment and Democracy in East Africa' (MEDIeA, www.mediea.ruc.dk). Recent publications include: *Youth engaging with the World. Media, Communication and Social Change* (2009, with Florencia Enghel); *Communication for Social Change Anthology: Historical and Contemporary Readings* (2006, with Alfonso Gumucio Dagron) and *Media and Glocal Chang: Rethinking Communication for Development* (2005, with Oscar Hemer).

Liesbet van Zoonen holds the chair in media and communication at Loughborough University (UK), and is professor of popular culture at the Erasmus University in Rotterdam (NL). She has published widely about gender and (new) media, and is currently involved in research about popular culture, citizenship and religion.

Index

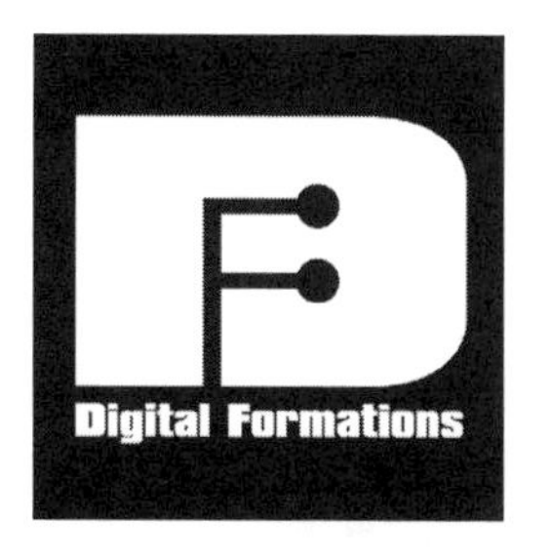

General Editor: Steve Jones

Digital Formations is an essential source for critical, high-quality books on digital technologies and modern life. Volumes in the series break new ground by emphasizing multiple methodological and theoretical approaches to deeply probe the formation and reformation of lived experience as it is refracted through digital interaction. **Digital Formations** pushes forward our understanding of the intersections—and corresponding implications—between the digital technologies and everyday life. The series emphasizes critical studies in the context of emergent and existing digital technologies.

Other recent titles include:

Felicia Wu Song
Virtual Communities: Bowling Alone, Online Together

Edited by Sharon Kleinman
The Culture of Efficiency: Technology in Everyday Life

Edward Lee Lamoureux, Steven L. Baron, & Claire Stewart
Intellectual Property Law and Interactive Media: Free for a Fee

Edited by Adrienne Russell & Nabil Echchaibi
International Blogging: Identity, Politics and Networked Publics

Edited by Don Heider
Living Virtually: Researching New Worlds

Edited by Judith Burnett, Peter Senker & Kathy Walker
The Myths of Technology: Innovation and Inequality

Edited by Knut Lundby
Digital Storytelling, Mediatized Stories: Self-representations in New Media

Theresa M. Senft
Camgirls: Celebrity and Community in the Age of Social Networks

Edited by Chris Paterson & David Domingo
Making Online News: The Ethnography of New Media Production

To order other books in this series please contact our Customer Service Department:
(800) 770-LANG (within the US)
(212) 647-7706 (outside the US)
(212) 647-7707 FAX

To find out more about the series or browse a full list of titles, please visit our website:
WWW.PETERLANG.COM